The Large-Scale Structure of the Universe

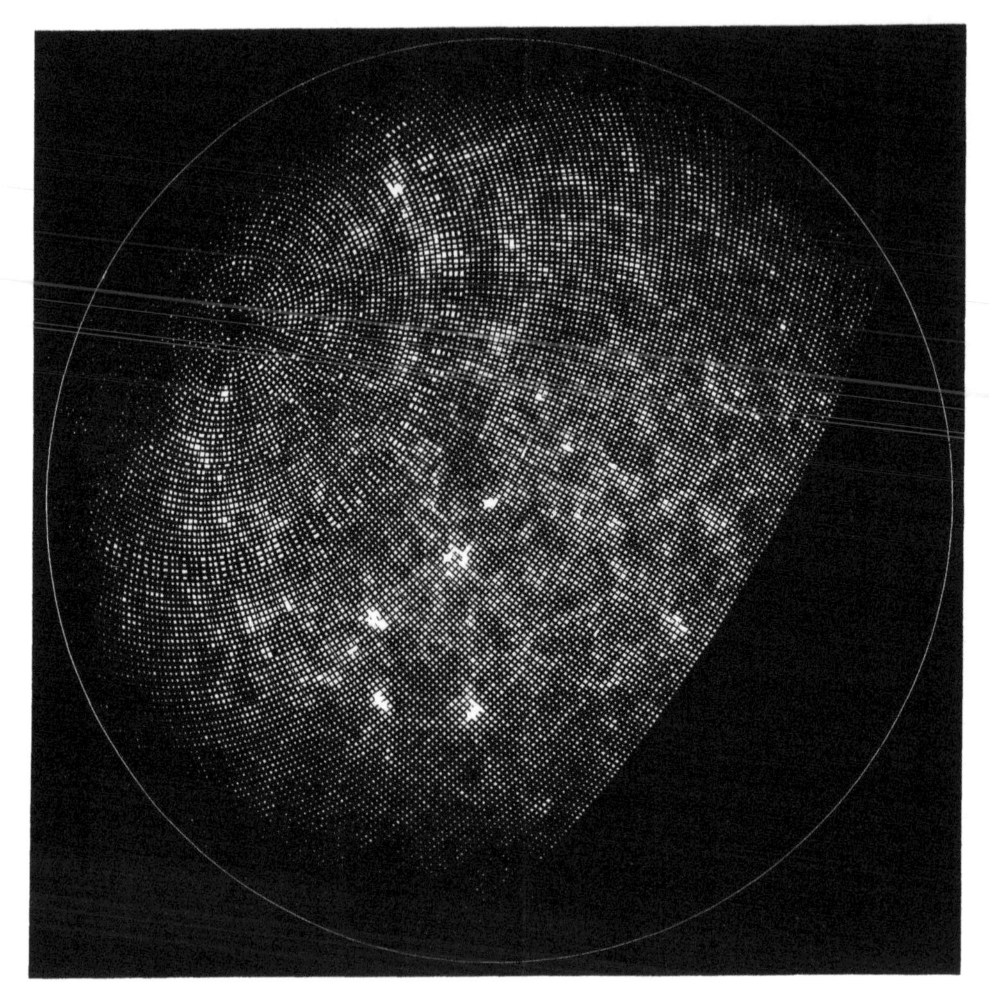

FRONTISPIECE: The large-scale pattern of the galaxy distribution. Each white square represents a sky cell about one degree by one degree in the Lick sample. The size of the white square is proportional to the number of galaxies brighter than 19th magnitude in the cell. The cells are arranged along lines of fixed right ascension and declination. The north pole of the galaxy is at the center of the map and the equator along the edge. (Map by J. A. Peebles and P.J.E. Peebles.)

The Large-Scale Structure of the Universe

by

P. J. E. Peebles

Princeton Series in Physics

Princeton University Press
Princeton, New Jersey

41 William Street, Princeton, NJ 08540
In the United Kingdom:
6 Oxford Street, Woodstock, Oxfordshire OX20 1TR

Cover image: Dark matter distribution in the universe, based on the Millennium Simulation, a very large cosmological N-body simulation (more than 10 billion particles). Courtesy of V. Springel, Max Planck Institute for Astrophysics, Germany

press.princeton.edu

First printed 1980
New paperback printing, 2020
Paperback ISBN 9780691209838

This book was originally published in the Princeton Series in Physics.
For more information on the series, please visit
https://press.princeton.edu/series/princeton-series-in-physics

The Library of Congress has cataloged the cloth edition as follows:

Peebles, Phillip James Edwin.
The large-scale structure of the universe.

(Princeton series in physics)
Bibliography: p.
Includes index.
1. Galaxies. 2. Cosmology. I. Title.
QB857.P43 523.1'12 79-84008
ISBN 0-691-08239-1
ISBN 0-691-08240-5 (pbk.)

Printed in the United States of America

To Alison

CONTENTS

PREFACE

From the first developments of modern cosmology people have recognized that an important part of cosmology is the large-scale clustering of matter in galaxies and clusters of galaxies. The point was largely eclipsed by the debate over homogeneous world models, but in recent years there has been a considerable revival of interest in the large-scale mass distribution and what it might tell us about the nature and evolution of the universe. The purpose of this book is to review our present understanding of these subjects.

Chapter I is a history of the development of ideas on the large-scale structure of the universe. As is usual in science the story is a mixture of inspired guesses and rational progress with excursions down paths that now seem uninteresting. What makes it somewhat unusual is the slow rate of development that has left ample time for the growth of traditions that are more than commonly misleading, and so it seems worthwhile to examine the evolution of the ideas in some detail. This is a history in the rather loose convention of scientists, that is, it is based on what I could glean from published books and journals. The few conversations I have had with participants have left me only too aware of how limited that is and how much more could be done. On the other hand, the published record is what was readily available to most people who might want to work on the subject and might want to learn what has already been done, though the actual use of the record was just as erratic in the 1930s as it is these days. I have tried to give a complete account of the important developments since about 1927 and have added enough more recent references to serve as a guide to the literature.

Chapter II deals with the behavior of a given mass distribution in the Newtonian approximation. This is only a limiting case of the full relativistic theory, but it is discussed first and in detail because it is a good approximation for most practical applications and is much simpler than the full relativistic theory. There is a considerable variety of methods and results in the analysis of the Newtonian limit. I have collected all those that seem to be useful and interesting.

The statistical pattern of the galaxy distribution is discussed in Chapter III. The descriptive statistics that have proved useful and are analyzed in this chapter are n-point correlation functions (analogs of the autocorrelation function and higher moments for a continuous function). The general approach has a long history but it is only in the last several years with the application of fast computers to the large amount of available data that the technique has been extensively developed and applied. This chapter surveys the main theoretical results and observational methods.

The n-point correlation functions have proved useful not only as descriptive statistics but also as dynamic variables in the Newtonian theory of the evolution of clustering. This is discussed in Chapter IV. The functions are generalized to mass correlation functions in position and momentum, and the BBGKY hierarchy of equations for their evolution is derived. This yields a new way to analyze the evolution of mass clustering in an expanding universe. Of course, the main interest in the approach comes from the thought that the observed galaxy correlation functions may yield useful approximations to the mass correlation functions, so the observations may provide boundary values for the dynamical theory of evolution of the mass correlation functions. The test will be whether we can find a consistent theory for the joint distributions in galaxy positions and velocities. The subject still is in a crude state because adequate redshift data do not yet exist. I present some preliminary considerations on how the analysis of the data might proceed.

The full relativistic analysis of the evolution of mass clustering is presented in Chapter V. The important application is to the behavior of the early stages of expansion of the universe when the high mean density would have made even modest density fluctuations strongly relativistic.

The last chapter describes some of the attempts to trace the links between theory and observation showing how the character of the matter distribution we observe developed out of reasonable conditions in the early universe. This is the main point of the subject, but it is not treated at length because I think there are too many options, all apparently viable but none particularly compelling. It seems likely that the game of inventing scenarios will go through several more generations before a secure picture emerges. Perhaps the best we can hope is that the final answer will draw on significant elements of the theory and observations as we now think we understand them.

I have limited the range of the discussion to length scales no smaller than the nominal size of a galaxy or else redshifts no smaller than the epoch at which mass concentrations comparable to present day galaxies appeared, thus excluding the structure and evolution of galaxies. I have excluded a few topics relevant to other areas of cosmology, such as the effect of mass clustering on the standard cosmological tests, and some obviously important subjects where I could find nothing very useful to report, such as the question of intergalactic gas clouds. I have omitted all discussion of the possibilities offered by nonstandard cosmologies not so much because I am sure the big bang picture is the most likely candidate as that I expect it is neither reasonable nor likely to expect that people will pay much attention to these alternatives until we have a much clearer picture of what the standard model has to offer and what it must deal with.

The choice of emphasis on topics within the boundary conditions, of course, reflects a personal judgment of what is promising. Perhaps the largest omission is the primeval turbulence picture. I have described its origins and some general and well-established results but have not discussed any specific scenarios. That seems reasonable because I doubt the merits of this picture, and there are others who can serve as better and more enthusiastic advocates.

I have provided a short guide to symbols and conventions in the appendix. It probably will prove best to look this over before reading much of the main text. I have given short summaries of concepts of cosmology as they appear in the text, but have left out details available in the standard books. References to my book, *Physical Cosmology,* are indicated by the letters *PC.*

ACKNOWLEDGMENTS

I list with special thanks the people who played the most direct roles in shaping this book: Charles Alcock, Marc Davis, Bob Dicke, Jim Fry, Margaret Geller, Ed Groth, Mike Hauser, Bernard Jones, Jerry Ostriker, Bill Press, Mike Seldner, Joe Silk, Ray Soneira, Juan Uson, Simon White, Dave Wilkinson, and Jer Yu. The process would have been considerably slower and the results less satisfactory without the skill and energy of Marion Fugill.

The first concrete steps toward this book were taken while I enjoyed the hospitality of the Physics Department at the University of California at Berkeley during the 1973–74 academic year. The first draft developed as course notes at Princeton University. I am grateful to John Bahcall for providing hospitality at the Institute for Advanced Study where the final draft was written. The work was supported in part by the National Science Foundation.

The Large-Scale
Structure of the
Universe

I. HOMOGENEITY AND CLUSTERING

1. Homogeneity and clustering

Modern discussions of the nature of the large-scale matter distribution can be traced back to three central ideas. In 1917 Einstein argued that a closed homogeneous world model fits very well into general relativity theory and the requirements of Mach's principle. In 1926 Hubble showed that the large-scale distribution of galaxies is close to uniform with no indication of an edge or boundary. In 1927 Lemaître showed that the uniform distribution of galaxies fits very well with the pattern of galaxy redshifts. The homogeneous model, when generalized to allow for evolution, yields a linear redshift-distance relation consistent with what Hubble was finding from his estimates of galaxy distances (as summarized by Hubble in 1929).

The evolving dynamic world model quickly won attention and in the following decades, before the idea became commonplace, it generated some lively discussions. The following sections trace the development of several questions. The first question is whether the universe really is homogeneous (after averaging over a suitable clustering length). Assuming it is, must we be content to say only that this happens to be a reasonable approximation to our neighborhood at the present epoch? Could the homogeneity of the universe have been deduced ahead of time from general principles? Or might it be a useful guide to new principles? The matter distribution in any case is strongly clumped on scales of stars, galaxies, and clusters of galaxies. This clustering is a fossil of some sort, a remnant of processes in the distant past as well as an on-going phenomenon. How does the clustering evolve in an expanding universe? What is its origin? What does it tell us about the nature of the universe?

2. Is the universe homogeneous?

In 1917 the phrase "the large-scale distribution of matter" was generally taken to mean the distribution of stars in the Milky Way galaxy. For example, the title of Eddington's (1914) book on the latter subject is *Stellar Movements and the Structure of the Universe*. It was considered well-established from star counts that the stars are concentrated in a flattened roughly spheroidal distribution, the Kapteyn system (after the astronomer mainly responsible for the laborious accumulation and analysis of the star count data). If the distribution had been homogeneous, the number of stars brighter than f would have varied as[1]

$$N(<m) \propto f^{-3/2} \propto 10^{0.6m}, \tag{2.1}$$

where m is the apparent magnitude. The star counts are different in different directions in the sky and increase with decreasing f distinctly less rapidly than would be expected from equation (2.1). The implication is that we are seeing the edge of the system.[2]

It is not clear how much Einstein in 1917 knew or was influenced by these ideas. He wrote of the distribution of stars as possibly being uniform on the average over large enough distances. He did not mention the arguments (marshaled at the time by Sanford 1917) that the spiral nebulae may well be other "island universes," other galaxies of stars, though it seems likely Einstein knew the general idea because he had discussed with de Sitter how matter might be distributed in the universe (de Sitter 1916). Einstein rejected the idea that the universe of stars might be a limited island in asymptotically flat space because a star escaping from the system would move arbitrarily far from all other matter yet preserve its inertial properties, contrary to Mach's Principle. At first he proposed that the line element might become singular outside the realm of the matter, but then hit on a much more elegant solution, a homogeneous closed world.

De Sitter was an astronomer (and a student of Kapteyn) and well aware that the stars are not uniformly distributed. He was willing to accept the island universe hypothesis and to speculate that these systems might be uniformly distributed through space (de Sitter 1917). However, he mentioned no tests of the uniformity idea.

It was known at the time that there are many more faint spiral nebulae, that is, nebulae of small angular size, than bright ones, and that there are hundreds of thousands of very faint objects that might be just like the bright ones but so far away that it is not possible to make out the spiral structure (Fath 1914, Sanford 1917, Curtis 1918). Hubble (1926) was the

[1]For stars of fixed intrinsic luminosity, those appearing brighter than f are at distances $< r \propto f^{-1/2}$, according to the inverse square law. For a homogeneous distribution the number counted would vary as the volume $\propto r^3 \propto f^{-3/2}$. The sum over stars of different intrinsic luminosities affects the constant of proportionality but not the power law behavior.

[2]It is now recognized that the counts in the direction of the Milky Way are strongly reduced by interstellar absorption, so the size of the star system was substantially underestimated. The counts in directions well away from the Milky Way are little affected by absorption, so the estimates of the thickness of the disc were quite reasonable.

first to ask whether the counts of these nebulae are consistent with the assumption that they are uniformly distributed through space. He used Seares' (1925) estimate of the limiting magnitude, $m \approx 16.7$, for Fath's counts of faint nebulae. He found that the number of these faint nebulae agrees well with what would be expected from the counts at $m < 12$ extrapolated according to equation (2.1).

The success of Hubble's test is impressive, for there are 600 times as many galaxies in the deep survey as at $m < 12$. And this stood in sharp contrast to the familiar behavior of the star counts; the indication is that the observations of stars reach the edge of the local star system while the observations of galaxies give no evidence of an edge to Sanford's "realm of the nebulae." Hubble put it this way in a later paper: "There are as yet no indications of a super-system of nebulae analogous to the system of stars. Hence for the first time, the region now observable with existing telescopes may possibly be a fair sample of the universe as a whole" (1934, p.8).

It is now known that the excellent numerical agreement Hubble found for these data is in part fortuitous because the galaxies at $m < 12$ are not a fair sample: there is a substantial excess of bright galaxies due to the local concentration in and around the Virgo cluster. Indeed Hubble clearly recognized that his result was only a preliminary indication, and over the next decade he undertook an extensive program of deep counts in selected areas (to be compared to the program of star counts except that far fewer astronomers were directly involved). A preliminary report was published in 1931, and in 1934 Hubble discussed in some detail the counts at limiting magnitudes ~ 19.1 and 19.6 The ratio of counts agrees well with the $10^{0.6m}$ law, as do the ratios of these counts to the number of Shapley-Ames (1932) galaxies at $m \leq 13$ (though again because of the local supercluster this latter result is in part fortuitous). In 1936 Hubble discussed counts to 5 limiting magnitudes in the range $m \sim 18.5$ to 21. The counts increase with m less rapidly than the $10^{0.6m}$ law, the discrepancy amounting to a factor 1.8 out of an observed ratio of counts of 19 over this range of magnitudes. Hubble tentatively concluded that the discrepancy is larger than would be expected in any reasonable relativistic world model and that this might indicate the relativistic theory is incorrect. The present tendency is to suppose that systematic errors in magnitude estimates and K-corrections (correction for the shift of the galaxy spectrum toward the red and out of the range of sensitivity of the photographic plate) could account for this relatively small discrepancy. Of enduring interest is Hubble's first point: to the depth of his survey there is no pronounced evidence of an edge to the realm of the nebulae.

It is surprising that the number-magnitude test (and the equivalent relation $N(>\theta) \propto \theta^{-3}$) was first applied to the counts of spiral nebulae and faint nebulae as late as 1926. It seems reasonable to suppose that Hubble was emboldened to try the test because he had just recently shown, by the identification of Cepheid variable stars of known absolute magnitudes, that the brightest spiral nebulae are galaxies of stars comparable to the Milky Way. He might also have been influenced by Einstein's and de Sitter's discussions of homogeneous world models, for he was at least familiar with de Sitter's attempt to guess at the mean mass density (and since Hubble had a much better distance calibration he found a much better estimate of ρ).

Hubble's results from 1926 to 1934 clearly were only preliminary though encouraging indications of homogeneity, but most theorists were quick to accept the evidence. Thus Einstein in 1933 wrote, "Hubble's research has, furthermore, shown that these objects [galaxies] are distributed in space in a statistically uniform fashion, by which the schematic assumption of the theory of a uniform mean density receives experimental confirmation" (1933, p. 107). Robertson, in his influential review of the Friedman-Lemaître cosmological models, said "we accept the data, due primarily to Hubble and Shapley, on the uniform distribution of matter in the large within the visible universe, and we extrapolate them to the universe as a whole" (1933, p. 82). In 1931 Eddington made the cautionary remark, " 'Lemaître's world' is also a model in that it represents the universe as a uniform spherical distribution of matter; there is no reason why the actual shape should not be highly irregular" (1931a, p. 415). But later in the same year he stated, "We no longer look for an end to the world in its space dimensions. We have reason to believe that so far as its space dimensions are concerned the world is of spherical type." (1931b, p. 447). It is perhaps not surprising that de Sitter was more cautious. He wrote in 1931, "It should not be forgotten that all this talk about the universe involves a tremendous extrapolation, which is a very dangerous operation" (1931, p. 708). And in 1932, he wrote "These wonderful observations [of galaxies from the Mount Wilson Observatory] have enabled us to make fairly reliable estimates of the distances of these objects and to say something about their distribution in space. It appears that they are distributed approximately evenly over 'our neighborhood' " (1932, p. 114).

In the 1930s there was a somewhat indirect running debate between Hubble and Shapley over the relative importance of departures from homogeneity. Both clearly emphasized that the galaxy distribution is strongly clumped on relatively small scales. For example, Hubble (1934) noted that the frequency distribution of nebular counts N found in different telescope fields is not Poisson, as would be expected if the galaxies were randomly distributed; the general clumping makes for a considerably broader distribution of counts. (He also made the interesting observation that the distribution of log N is remarkably close to Gaussian.) However, to Hubble the main effect clearly was the uniform distribution on large scales as revealed by very deep counts averaged over many sample fields. Shapley emphasized the great irregularities in the galaxy distribution: "the irregularities are obviously too pronounced to be attributed to chance; they are rather a demonstration of evolutionary tendencies in the metagalactic system" (Shapley 1933, p. 3). Having smaller telescopes at his disposal, Shapley and his colleagues studied the galaxy distribution at lesser depth but in greater detail across the sky. He noted that there is a considerable difference in the numbers of Shapley-Ames galaxies, $m \lesssim 13$, in the northern and southern galactic hemispheres, and he suggested that this north-south asymmetry might still amount to as much as 50 percent at $m = 17$ (Shapley 1934). The data suggested also that at $m \approx 18$ (in a magnitude system roughly consistent with that of Hubble) the galaxy density might vary across the sky by a factor ~ 2 on scales $\sim 30°$ (though there were problems with this because there were practical difficulties in transferring magnitude standards across the southern sky; Shapley 1938b). This led Shapley (1938a) to question whether the galaxy distribution really is close to uniform even when averaged over large scales and to suggest that the deviation from the $10^{0.6m}$ law in Hubble's data might be the result of large-scale density irregularities, not a failure of relativity theory. (Hubble in 1936 had mentioned but rejected the idea of large-scale irregularities.)

Shapley's remarks did not attract much attention. McCrea (1939) did point out that large-scale irregularities would raise problems for observational programs to measure the parameters in the standard cosmological models, then a subject much discussed particularly in connection with the possible role of the 200 inch telescope. Eddington (1939) and Tolman (1949), apparently independently, suggested that large-scale inhomogeneity may account for Hubble's results, and Omer (1949), at Tolman's suggestion, devised an inhomogeneous relativistic model (spherically symmetric about our position; § 87 below) that he could adjust to fit the galaxy counts. However, by the 1950s the possibility of large-scale inhomogeneity was largely displaced in the minds of cosmologists by the debate over homogeneous world models—evolving versus steady state versus Milne's kinematic cosmology—and, in the relativistic models, the possible values of parameters such as the cosmological constant, Hubble's constant and the time scale, the acceleration parameter and the open versus closed models. An example is Bondi's (1952) book on cosmology where the suggestion of Eddington and Tolman is noted but rejected as unprofitable. A second example where one finds a more cautious view is

McVittie's (1961) *Fact and Theory in Cosmology*. All the cosmological models discussed are homogeneous, but McVittie does stress the observational problems in establishing homogeneity.

Though the subject has not been very popular one can find occasional recent discussions of the question of large-scale inhomogeneity. De Vaucouleurs (1960, 1970, 1971) and van den Bergh (1961) have joined Shapley in observing that the traditional evidence from galaxy counts as functions of magnitude or position in the sky is at best slim. In 1965 Omer rediscussed his spherical model for large-scale inhomogeneity. Rees and Sciama (1968) used the spherical model in a discussion of the suggestion by Strittmatter, Faulkner, and Walmesley (1966) that clustering scales for quasi-stellar objects might be comparable to the distance to the horizon. Bonnor (1974) and Silk (1977a) also discussed this model, and Silk pointed to the indications of large-scale matter currents found by Rubin, Thonnard, Ford, and Roberts (1976) from a systematic survey of redshifts of galaxies. Kristian and Sachs (1966) explored another approach based on the assumption that all properties of the universe out to redshifts $Z \sim 0.3$, for example, can be usefully expanded in a power series about our position. Wertz (1971), Haggerty (1971), and Wesson (1976), stimulated by de Vaucouleurs, considered the possible dynamics of yet another picture, where the hierarchy of clustering continues to indefinitely large scales (§ 62 below).

De Vaucouleurs has made the interesting point that if the universe really is close to homogeneous on the scale of the horizon cH_0^{-1}, it is a remarkable break with the state of affairs on smaller scales: from subatomic particles on up we deal with objects—localized structures. De Vaucouleurs noted that this tendency to clump continues to scales at least as large as the local supercluster (the concentration of galaxies around the Virgo cluster, distance $\sim 10h^{-1}$ Mpc, of which we are an outlying part), and he could cite as indications of irregularities on still larger scales the angular gradients found in the Harvard survey (Shapley 1938a, b) and the large-scale correlation of rich clusters found by Kiang and Saslaw (1969), both effects pointing to strong clustering on scales $\sim 100\ h^{-1}$ Mpc. The indication is that if the clustering does terminate it does so perhaps suspiciously close to the largest depth of reliable observations and close to the largest possible scale consistent with the assumption that the universe is accurately uniform on the horizon ($cH_0^{-1} = 3000\ h^{-1}$ Mpc).

Direct observations of the large-scale galaxy distribution still are beset with the problem of controlling systematic errors when galaxy densities are compared over widely separated parts of the sky or at very different apparent magnitudes. Modern deep galaxy counts (Brown 1974, Kron 1978, Tyson and Jarvis 1979, Ellis 1980) are found to vary with magnitude roughly as $10^{0.45m}$ to depths comparable to cH_0^{-1}. The deviation from the $10^{0.6m}$ law is about what is expected from the K-correction. The best sample of the distribution across the sky is the Lick catalog (Shane and Wirtanen 1950; 1967). This gives counts to limiting magnitude $m = 19$ in $10'$ by $10'$ cells across two-thirds of the sphere. The effective depth of the sample is ~ 200 h^{-1} Mpc. There are large-scale density gradients, amounting to rms fluctuations $\delta\mathcal{N}/\mathcal{N} \approx 0.10$ in the surface density smoothed over 10°. One cause is purely local, the variation of absorption across the sky. It is difficult to decide how much might be true large-scale fluctuations in the space density of galaxies.

One convenient measure of the irregularities in the space distribution is the dimensionless autocorrelation function

$$\xi(r) = \langle \rho(\mathbf{r}_1)\, \rho\,(\mathbf{r}_1 + \mathbf{r})\rangle / \langle \rho \rangle^2 - 1, \tag{2.2}$$

where the angular brackets signify an average over the position $\mathbf{r}_1$ within the sample. An upper limit on large-scale clustering within the Lick sample is (Peebles and Hauser 1974)

$$\xi(50h^{-1}\mathrm{Mpc}) \lesssim 0.025, \tag{2.3}$$

and $\xi(50)$ is significantly less than this if variable absorption is important. One measure of the scale on which clustering is strong is the value of the lag r_0 at which the correlation function ξ is unity. In the Lick sample (Groth and Peebles 1977, § 57 below)

$$r_0 \approx 4h^{-1}\mathrm{Mpc}, \qquad \xi(r_0) \equiv 1. \tag{2.4}$$

Since this is small compared to the depth of the survey, ~ 200 h^{-1} Mpc, the indication is that within the Lick sample the progression of clustering observed on small scales does blend into a nearly uniform background.

Equation (2.4) describes a mean over the distribution, and one certainly can find spots in the Lick sample where the density stays higher than the mean over distances larger than r_0. Examples are provided by Abell's (1958) catalog of rich compact clusters. Galaxies, of course, tend to concentrate around Abell's cluster positions. This can be measured by averaging the galaxy space density over all shells, radius r to $r + \delta r$, centered on all Abell clusters. One finds that this mean density $n(r)$ is twice the overall average density in the Lick sample at distance (Seldner and Peebles 1977a).

$$r_a \approx 14h^{-1}\mathrm{Mpc}. \tag{2.5}$$

For the correlation among positions of Abell cluster centers, Hauser and Peebles (1973) estimate the clustering length

$$\xi_{cc}(r_s) \equiv 1, \qquad r_s \approx 30h^{-1}\text{Mpc}. \tag{2.6}$$

Kiang and Saslaw (1969) suggested r_s is closer to 100 h^{-1} Mpc from a reconstruction of the three-dimensional distribution using Abell's estimates of apparent magnitudes of brighter cluster members. This method has the advantage that it makes the apparent clustering larger (the angular correlation function, which must be unfolded to find $\xi(r)$, is smaller than ξ because of the overlapping of objects at very different distances), and it has the disadvantage that if the errors in Abell's magnitude scale vary systematically with distance, as seems possible, it will introduce spurious radial clustering.

The indication from equations (2.4) through (2.6) is that the clustering does blend into small fluctuations, $\delta\rho/\rho < 1$, well within the sizes of available samples. Of course, the samples are limited and deeper surveys are needed.

New methods of observation have provided some very deep glimpses into space and, indirectly, precise measures of homogeneity. Extragalactic radio sources, all or a fair fraction of which are galaxies, are distributed across the sky in a remarkably uniform way; the distribution of the 5000 4C sources (flux levels $S \geq 2$ Jy) is almost indistinguishable from random (Webster 1976b, Seldner and Peebles 1978). Because the number of 4C objects is much less than the number of Lick galaxies, the radio source data do not improve our upper limits on fluctuations in the density of objects across the sky at $\theta \lesssim 10°$. For example, the mean number of 4C sources found in a 3° by 3° cell is $\langle N \rangle \approx 2$, and the rms fluctuation in the number is close to Poisson, $\delta N/N \approx 0.7$. The rms fluctuation in the number of Lick galaxies is a factor ~ 3 smaller, $\delta N/N \approx 0.25$ (compared to the expected value $\delta N/N = 0.045$ if galaxies were randomly distributed). But since many of the sources are at distances $\sim cH_0^{-1} = 3000\ h^{-1}$ Mpc, we do have a strong new test of isotropy on large scales. The contrast with the distribution of bright galaxies is worth emphasizing; if the northern hemisphere were divided into two equal parts, the number of 4C sources in each would agree to $|N_1 - N_2|/(N_1 + N_2) \simeq 0.015$ (rms), while the number of Shapley-Ames galaxies would scatter by a factor ~ 2. A second important measure is the diffuse X-ray background (Wolfe 1970, Wolfe and Burbidge 1970; Fabian 1972). Since a substantial part of the flux comes from objects at modest redshifts—active galaxies and clusters of galaxies—this measures how the projected density of matter, integrated to the horizon, varies across the sky. The present limit on fluctuations in the projected density is $\delta f/f \lesssim 0.04$ at $\theta \sim 5°$ (Schwartz 1979, Schwartz, Murray, and Gursky 1976). Finally, the microwave background radiation is isotropic to $\delta T/T \lesssim 0.001$ on angular scales from 10′ to 180°. This does not measure the matter distribution directly because the radiation is thought to be very weakly coupled to matter in the present universe. But since the radiation temperature varies inversely as the expansion factor, or more generally as the redshift from source to observer, it does indicate the large-scale motion has been isotropic about us to an accuracy better than 1 part in 10^3.

These three sets of observations show that the matter distribution and motion are quite accurately isotropic on scales $\sim cH_0^{-1}$. This is a strong test of the standard homogeneous and isotropic world picture, but of course it is not complete because it leaves open the possibility that the universe is inhomogeneous but isotropic about a point near us. However, the galaxies at high redshift look much like the ones nearby, and in such a model an observer on any one of the enormous number of distant galaxies would find the universe is much less isotropic than we do. The more reasonable presumption is that the universe would appear isotropic on a distant galaxy, so the visible universe is accurately homogeneous.

We certainly do not have definitive evidence of homogeneity, and further developments in the tests will be followed with great interest. On the other hand, the observational situation has improved many times over since the 1920s, and the results must be counted as a spectacular success for the vision of Einstein and Hubble.

3. Physical principles

A. Prediction of homogeneity?

Might the homogeneity of the universe have been expected from general arguments and physical principles? In a sense the answer is yes, for Einstein did hit on a homogeneous world model as a way to satisfy some general considerations. He rejected the idea of an infinite material Newtonian universe on the grounds that the potential and hence star kinetic energies would be arbitrarily large. In the 1917 paper he gave two arguments against the idea that matter is concentrated like an island in otherwise empty asymptotically flat space. The first argument was that, given sufficient time, the system would relax, part contracting to high density (we would now say, to a black hole), part escaping with positive energy. Since Einstein supposed the global properties of the universe must be unchanging, this was unacceptable. It is not clear how seriously Einstein weighed this, for the universe could not be eternal in any case; for example, the solar system, given sufficient time (and if the sun does not explode), would relax in the way he envisioned for the island universe as a whole, and if energy is conserved, the stars must eventually stop shining. As mentioned in the last section, he did emphasize Mach's principle: a particle escaping this island universe would move into flat space, arbitrarily far from all other matter, but yet, according to relativity theory, its inertial properties would not change, contrary to the idea that inertia is generated by the matter in the universe. A discussion reported by de Sitter (1916) gives an interesting view of at least some aspects of Einstein's thoughts. Since flat space at infinity conflicts with Mach's principle, he considered the idea that the components of g_{ij} degenerate to singular values at the edge of the universe. Since observed objects give no evidence of strong space curvature, one would have to suppose, as de Sitter put it, that the g_{ij} become singular outside "hypothetical" masses that surround the known and ordinary realm of matter. The next year Einstein found a more elegant solution: replace the singular behavior of the g_{ij} at the boundary with the condition that the universe be closed—the three-dimensional analogy of the closed two-dimensional surface of a balloon. This universe is finite, with no flat exterior, no hypothetical masses, and indeed no edge. Einstein's brilliant argument from general principles thus led to a world picture that has stood the test of time and observation.

It is worth bearing in mind, despite this success, that such arguments tend to be matters of opinion. Many people have been attracted to another picture, an unlimited clustering heirarchy (§ 62 below). A review of such models and of the history of development of the idea is given by Mandelbrot (1977). In the scale-invariant clustering model that, according to Mandelbrot, can be mainly attributed to Fournier d'Albe (1907) the hierarchy scales so that the typical value of the mass within distance R of an observer varies as $M \propto R$. The size and mass of the universe are arbitrarily large, but the mean density M/R^3 converges to zero and the mean mass per unit area in the sky of an observer converges, so even if stars shine forever the surface brightness of the sky is not large and Olbers' paradox is avoided. The hierarchy is arranged so the virial velocity in clusters of size R is $v^2 \propto M/R$, independent of R. Thus although the mass is infinite, the peculiar velocities need not be high. This anticipated and countered one of Einstein's arguments against an infinite quasi-static universe. Also, it has been found that Fournier d'Albe's model gives a remarkably good approximation to the statistics of galaxy clustering on small scales. This is discussed in Chapter III below. Einstein (1922) felt that the hierarchical world picture (as rediscussed by Charlier 1908, 1922) was compatible with general relativity theory but not with his interpretation of Mach's principle. As discussed in the last section, the evidence from recent observations is that the hierarchical model in fact fails, and

Einstein's picture is a reasonable first approximation on scales larger than about 10 h^{-1}Mpc.

B. The cosmological principle

Milne (1933a) was the first to notice that although Hubble's law (recession velocity proportional to distance) was derived from the relativistic world model, it cannot be considered a very specific test of the model because it is the only functional form allowed by homogeneity and isotropy. (A particularly clear explanation is given by Milne 1934.) He proposed that this result might be extended and that one might be able to derive cosmology more or less complete by following such arguments from powerful general principles.

Milne referred to the homogeneity assumption by phrases such as "the extended principle of relativity" and "Einstein's cosmological principle," and soon fixed on the "Cosmological Principle." He clearly felt that this principle has a considerable *a priori* philosophical merit, perhaps even that it was logically necessary for what one means by the universe.[3] His program did not meet with much approval, though it was an important forerunner of the steady state model. His term, "the Cosmological Principle," was quickly taken up as an easy way to state and justify a central assumption. For example, it appears in the introductory comments in papers by Robertson (1935), Walker (1936), and, in a less positive way, de Sitter (1934). The cosmological principle is now firmly lodged in the lore of the subject.

The most interesting immediate reaction to Milne's ideas was that of Dingle (1933a,b). He and others objected to the idea that the cosmological principle is to be compared to a law of nature: "a principle coequal with the principle of relativity should be capable of universal application," at least, as he noted, within some substantial domain of phenomena (1933a, p. 173). Homogeneity could only apply in the average over many galaxies. Dingle pointed out that Einstein's field equations do admit strongly inhomogeneous solutions, and he took it to be "perfectly conceivable that an increase of telescopic power may reveal a variation of material density with distance." Dingle noted that the evidence of isotropy of the galaxy redshifts is far from complete because of the absence of observations in the Southern Hemisphere. Curiously, he did not mention the galaxy counts, though, since he had been in Pasadena, he should have been in a position to learn the status of Hubble's program.

[3]Milne 1933b, p. 185. In his book (Milne 1935), at least partly in reaction to Dingle's comments, Milne was careful to state the cosmological principle as an assumption or axiom, though he did argue it is necessary for an intelligible universe.

C. The instability of the universe

If the symmetry of the universe is not enforced by a principle then one might ask whether it always has been or will be as close to homogeneous as it is now. The history of ideas may be traced back to Einstein's 1917 paper.

Einstein had assumed as a matter of course that the universe is static. But the field equations of general relativity as originally formulated then indicate the pressure would have to be negative, $p = -\rho c^2/3$ (so the active gravitational mass density associated with pressure cancels that of ρ). To avoid this he modified the gravitational field equations, introducing a universal cosmic repulsion that varies in proportion to separation with strength determined by the cosmological constant Λ. With $|p| \ll \rho c^2$, the static model then requires (§ 97)

$$4\pi G\rho = \Lambda. \tag{3.1}$$

Nearly twenty years passed before it was clearly stated by Tolman that there is a serious problem with this—the model is unstable. The general point was sensed earlier by Weyl (1922) and Eddington (1924), who observed that a physical variable the density ρ is set equal to a constant of nature Λ. What happens if the matter is rearranged or if some of it is annihilated in stars thus changing $\langle p \rangle$?

In 1930 Eddington learned of Lemaître's (1927) work on evolving world models and recognized that it gives a partial answer. If an Einstein model were somehow perturbed so that the mean density is slightly less than $\Lambda/4\pi G$, the universe would expand, the density drop, and the expansion steadily accelerate. If the Einstein universe were perturbed so that ρ is slightly higher than $\Lambda/4\pi G$, the universe would collapse.

Lemaître and Eddington at first assumed the universe is expanding away from the Einstein model, and Eddington (1930) proposed that the balance of the initial Einstein world was broken, the expansion initiated, through the perturbation caused by the formation of galaxies "by ordinary gravitational instability." In the following several years there was rather extensive discussion of this, mainly by McCrea and McVittie (1931, and earlier references therein), who tried to decide whether this condensation into galaxies would inevitably produce general expansion rather than contraction. However, the topic soon went out of style as attention turned to models that do not trace back to the Einstein case.

McCrea and McVittie approximated a condensation as a distribution spherically symmetric about one point. This is a very convenient model because it permits a description of at least the rough outlines of a mass concentration like a galaxy while keeping the mathematics simple.

Lemaître (1931, 1933a,b) found the ultimate simplification: if the pressure can be neglected, the motion of each mass shell is the same as in some homogeneous world model. (Of course when mass shells cross, the motion of each follows an altered cosmological model.) Discussion of how a density irregularity might evolve thus is made simple: each mass shell goes its separate way.[4]

Lemaître pointed out that this result, which might at first sight seem remarkable, is in fact "obvious" at least for small-scale condensations because the general relativity description of a small region is equivalent to the weak-field limit, the Newtonian description.[5] The argument, in a fuller form than Lemaître gave, develops as follows (Callan, Dicke, and Peebles 1965). Suppose the mass $M(< r)$ within the shell of physical radius r centered on the condensation satisfies $GM(< r)/rc^2 \ll 1$, and suppose this mass inside the shell is temporarily removed. Then an earlier discussion by Lemaître (1931) describes the space inside the hollow: according to Birkhoff's (1923, p. 253) theorem, which generalizes Newton's iron sphere theorem, space must be flat, unaffected by the matter outside. The mass $M(<r)$ can be placed in this flat space and treated in the Newtonian approximation, so the iron sphere theorem applied once again indicates the acceleration of the surface of the sphere is the same as if $M(<r)$ were uniformly distributed within r. Thus the motion of the shell must agree with that of some zero pressure homogeneous world model. (A more general discussion of Newtonian gravity physics in relativistic cosmology is given in Sections 6 and 84 below.)

Tolman (1934a) discussed some interesting consequences of Lemaître's solution.[6] Einstein's static world model evidently suffers from an instability more general than that noted by Lemaître and Eddington, for if in the originally static case some matter were carried from one spot to another, the more dense spot would collapse, the less dense spot expand, and the universe would grow strongly irregular. For a generally expanding universe, since different mass shells can evolve independently, Tolman observed that there clearly is no "general kind of gravitational action which would necessarily lead to the disappearance of inhomogeneities in cosmological models" (1934a, p. 175). As Dingle also remarked, there

[4]Though the point is simple, it was by no means self-evident, as is illustrated by the lengthy computations by McVittie (1932), Dingle (1933*b*), and Sen (1934), all of whom assumed spherical symmetry but did not hit on Lemaître's trick.

[5]The agreement between Newtonian and relativistic descriptions of a spherically symmetric condensation was independently noted by McCrea and Milne (1934), but they offered no explanation of why it should be.

[6]Though Tolman may well have hit on Lemaître's result independently, he refers to Lemaître's prior discovery. Thus I find it curious that this often is called the Bondi (1947)-Tolman solution. Another standard reference is Einstein and Straus (1945).

appears to be nothing in Einstein's gravitational field equations that would guarantee that different parts of the universe must expand at the same rate or even that all parts of the universe, as observed by us, must be expanding, not contracting. Tolman noted that this conclusion might be modified by the effects of "more drastic kinds of inhomogeneities" than those spherically symmetric about one point and that nongravitational forces might promote homogeneity. He concluded that, pending the possible discovery of such effects, we should be cautious about extrapolating the observed behavior of "our neighborhood" to great distances in space or to the remote past or the distant future.[7] Similar cautionary remarks are expressed in his book and are contrasted with the views of Milne, "who would regard the homogeneity of the universe as a fundamental principle" (Tolman 1934b, p. 364).

The implications of Tolman's remark are worth emphasizing. For example, it appears to be conceivable that the part of the universe we see in the Northern Hemisphere could have been slightly too dense overall at the time of the big bang, so that it expands for 10^{10} years and then collapses, while the part we see in the Southern Hemisphere had slightly low density overall and so expands indefinitely. According to our present understanding of physical principles, this is a possible universe but one that would look markedly unlike what we observe.

How would a strong initial irregularity behave? A model that is easy to understand goes as follows. Suppose that at some very early time t_i the universe is everywhere homogeneous and isotropic, with uniform density and expansion rate, except within a spherical patch of radius $r_i > ct_i$. Toward the center of this patch the density is high and space is strongly curved, so unless conditions are specially adjusted space soon collapses to a singularity. How does space outside the patch behave? Birkhoff's theorem tells us there is no gravitational signal of what happens inside, and if $r_i > ct_i$, there is no pressure signal, so the exterior is unaffected, evolves as a homogeneous model (§ 87). If at the present epoch this patch came within the horizon, we would see a mass concentration, perhaps surrounded by an empty region (though the hole could be filled by interactions with neighboring irregularities). There is no observational problem with low mass black holes, but we can only account for the absence of black holes of extreme mass, like the absence of large density fluctuations on very large scales, by presuming that they were excluded by the initial conditions.

An important aspect of the puzzle is that in the model a light cone

[7]A tendency now is to distinguish extrapolations backward and forward in time (Peebles 1967*a*, 1972); in the spherical model it certainly can be arranged that the universe starts out highly irregular and grows homogeneous by adjusting the starting times for the expansion of each mass shell, but that requires very particular adjustment and so seems contrived.

traced back to the singularity encompasses only a limited part of the matter in the universe. The visible part increases with time, reaching zero as $t \rightarrow 0$.[8] A distant galaxy at high redshift is near our horizon and in the past would not have been "visible" from our position (unless there is some way to trace light rays back through the singularity). If causal connection is reckoned from the time of the big bang, galaxies at high redshift have not previously been in contact with us, and they have not been in contact with galaxies in other parts of the sky. How then do we account for the familiar appearance of the galaxies at high redshift? How do we account for the remarkable uniformity of the microwave radiation coming from parts of the universe that have not been in communication since the time of the big bang?

De Sitter (1917) noted the horizon effect in his cosmological solution. Tolman (1934b) derived the effect for Friedmann-Lemaître models but only briefly pointed to the conceptual problems it raises. Milne (1935, §§463–474) discussed it at length, mainly as an argument against relativistic cosmology. There was a revival of interest in connection with steady state cosmology and a new analysis by Rindler (1956). Misner (1967, 1968) and McCrea (1968) emphasized the importance of the causality puzzle, and Misner proposed an ingenious solution: the horizon would be broken if the early universe were not at all like the model, but chaotic. Perhaps, as Tolman had noted, "more drastic" irregularities could promote homogeneity. This idea, and the homogeneous anisotropic mixmaster model by which Misner illustrated it, has been an important stimulus, but one that must be applied with caution because the lesson of the spherical model certainly is that gravity promotes inhomogeneity, not homogeneity. If the early universe were chaotic how did it avoid becoming a tangle of black holes?

Tolman's 1934 discussion now seems reasonable, and the reaction to Milne's ideas stimulated some questions that seem interesting. Must the universe be homogeneous? Is it always that way? As it happened these questions attracted little notice; attention concentrated on the homogeneous models, their relative merits, and possible tests. The cosmological principle (and perfect cosmological principle) served a useful function in keeping the discussion focused on some well-defined and useful research problems. On the other hand, in elevating homogeneity to a principle

[8]The depth of the visible universe is $\sim ct \sim cH^{-1}$, and the number of baryons in the visible universe is $N \sim n(t)\,(ct)^3$. Since $n \propto a(t)^{-3}$ and $a \propto t^{2/3}$ in an Einstein-de Sitter model, $N \propto t$. The horizon, of course, also limits the propagation of a pressure wave. That is why in the preceeding discussion it was possible to ignore the radiation pressure gradient in the early universe.

people did tend to lose sight of the important observational and theoretical problems behind it.

The present situation is curious. Einstein did predict the large-scale homogeneity of the universe, and the observational developments mainly have agreed with the prediction. Einstein's idea was codified in the cosmological principle and has played a central role in cosmology. But it seems that with present dynamic theory we cannot account for this homogeneity; we cannot say whether Einstein's argument was a lucky guess or a deep insight into the way the universe must be. For now, it appears, we must accept it as an initial condition or, with Milne, as something to be added to the principles of physics.

4. How did galaxies and clusters of galaxies form?

A. The role of gravity

Lemaître (1933a,b, 1934) pointed out that if the evolving homogeneous and isotropic world model is a reasonable first approximation, then the next step is to account for the departures from homogeneity in structures like galaxies and clusters of galaxies. Like Jeans (1928) he supposed that in the remote past matter was uniformly spread through the universe and that gravitational instability caused the distribution to fragment into separate nebulae. Jeans had proposed that the size of each fragment would be comparable to the critical Jeans length (the minimum length at which the self gravitation of a developing irregularity exceeds the opposing pressure gradient). He had assumed as a matter of course that the density of the universe is independent of time, and he cited the opinion expressed by Newton on the point:

> It seems to me, that if the matter of our sun and planets, and all the matter of the universe, were evenly scattered throughout all the heavens, and every particle had an innate gravity towards all the rest, and the whole space throughout which this matter was scattered, was finite, the matter on the outside of this space would by its gravity tend towards all the matter on the inside, and by consequence fall down into the middle of the whole space, and there compose one great spherical mass. But if the matter were evenly disposed throughout an infinite space, it could never convene into one mass; but some of it would convene into one mass and some into another, so as to make an infinite number of great masses, scattered great distances from one to another throughout all that infinite space. And thus might the sun and fixed stars be formed, supposing the matter were of a lucid nature.

Einstein (1917) with many others felt that Jeans' static uniform Newtonian background model is not self-consistent. Lemaître had a more definite theoretical basis in the relativistic world models. He originally supposed that the expansion of the universe can be traced back to the static Einstein model in the distant past but soon turned to the Lemaître model (1933a) where the universe is assumed to expand from a dense initial state, decelerate until gravity and cosmic repulsion nearly balance, remain in this quasi-static phase for some length of time, and then resume expansion with Λ dominating ρ. Apparently one reason he liked the model is that it gives a preferred epoch to the formation of structures.[9] He supposed that in the dense early stage there were small irregularities in the matter distribution. In a patch where the density (evaluated when the local expansion rate has some chosen value) is slightly higher than average the matter may dwell in the quasi-static phase for a longer time and where the initial density is high enough the patch may collapse rather than resume the general expansion. Such a collapsing patch would end up as a galaxy. In larger volumes containing many protogalaxies, the initial density contrast must be smaller, and there are spots where the contrast is just such that the patch stays in the quasi-equilibrium phase for a very long time. He identified these patches with clusters of galaxies. The equilibrium between gravitational attraction and cosmic repulsion gives $\rho \sim (4\pi G)^{-1} \Lambda$ in such patches (eq. 3.1). This is the predicted density within clusters, the minimum density in a stable system. Lemaître (1934) argued that with current estimates of H and Λ the predicted density gave a reasonable fit to the typical density within a cluster.

Lemaître's approach was phenomenological; he asked whether small initial fluctuations could develop into irregularities that match in some detail what is observed, and he left for some deeper theory the origin of the initial fluctuations. The problem of accounting for the origins of galaxies and clusters of galaxies certainly is a worthy one, and the general approach Lemaître formulated now seems fairly useful: it is the subject of this book. But it is curious to note how little his ideas were discussed during the 1930s and how little they influenced the developments in the next several decades. One reason was his tendency to stick with the Lemaître universe while others were considering other models and many were arguing that Λ ought to be dropped. Another certainly was the excitement of the gathering storm over homogeneous models.

The next important development was Lifshitz's (1946) general analysis

[9]Another reason (Lemaître 1933b) was that the time since zero radius could be made larger than H^{-1}, thus relieving the time-scale problem resulting, as we now know, from an overestimate of Hubble's constant.

of linear perturbations in a Friedmann-Lemaître model. Unfortunately because he did not examine the details of joining the limiting behavior at high redshift where he assumed the relativistic equation of state $p = \rho c^2/3$ to the solution for $p \ll \rho c^2$ at low redshift, he decided that "we can apparently conclude that gravitational instability is not the source of condensation of matter into separate nebulae" (1946, p. 116). Novikov (1964a) was the first to point out that this is not quite right.

One can see the origin and resolution of the problem by the following heuristic argument. Consider an expanding model $\Lambda = 0$ cosmologically flat or close to it. The characteristic time for collapse or expansion is then

$$t \sim (G\rho)^{-1/2}. \tag{4.1}$$

If the velocity of sound in the matter (that we shall imagine behaves like a perfect fluid) is c_s, the critical Jeans length is (§ 16)

$$\lambda_J \sim c_s t. \tag{4.2}$$

If the density fluctuation occupies a patch smaller than λ_J, the acoustic response time r/c_s is shorter than the collapse time t, so the fluctuation oscillates like an acoustic wave. If $r > \lambda_J$, gravity dominates and the fluctuation can grow more prominent. Note in particular that if the universe is radiation dominated, $p = \rho c^2/3$, the velocity of sound is $c/3^{1/2}$, and the Jeans length is comparable to the horizon,

$$\lambda_J \sim ct, \qquad p \sim \rho c^2. \tag{4.3}$$

Consider now a patch with contrast $\delta\rho/\rho = \delta(t)$ and physical size $r(t)$. If $p = \rho c^2/3$ and $r \gg ct$, then in linear perturbation theory one finds that the contrast grows as $\delta \propto t$ (§ 86) and r closely follows the general expansion, $r(t \propto a(t) \propto t^{1/2}$. If $p = 0$, $\delta \propto t^{2/3}$ with $r \propto a(t) \propto t^{2/3}$ (§ 11). In either case the potential energy per unit mass associated with the fluctuation is

$$\phi c^2 \sim G\delta M/r \sim G\rho\, \delta\, r^2 \sim \delta(r/ct)^2 c^2. \tag{4.4}$$

Using the results quoted above, one sees ϕ is independent of time. The perturbation to the geometry due to the density fluctuation is on the order of the dimensionless potential ϕ so, if linear perturbation theory is to be valid, ϕ must be much less than unity. But then equation (4.4) indicates that, when $r = ct$, $\delta \ll 1$. That is, when the fluctuation appears on the horizon, the contrast δ must be small. After this epoch, if $p = \rho c^2/3$, $r < \lambda_J \sim ct$, and so δ is forced to oscillate like an acoustic wave: it cannot grow large. However, Novikov pointed out that if $p \longrightarrow 0$ while r still is larger than ct, then δ can continue to grow after it appears on the horizon and can finally develop into a stable system ($\delta \gtrsim 1$). When this happens, the object has energy $E \sim -\phi c^2$. Since we require $\phi \ll 1$, the object is nonrelativistic, which, of course, is what is wanted.

Lemaître's spherical solution gives another useful way to think of the behavior of the perturbation. Suppose pressure gradients may be neglected. Then Lemaître (1933a,b) showed that the perturbed patch behaves like a section of a homogeneous world model. If $\Lambda = 0$, the cosmological equation is

$$\left(\frac{\dot{a}}{a}\right)^2 = \frac{8}{3}\pi G\rho - \left(\frac{c}{aR}\right)^2, \tag{4.5}$$

where the curvature of space in proper units is

$$R_p = Ra(t), \tag{4.6}$$

with R a constant. Suppose R^{-2} is positive, so the patch has negative energy. The expansion parameter can be chosen to agree with the proper radius of the perturbed patch,

$$r(t) = a(t). \tag{4.7}$$

The ratio of the size of the patch to the space curvature in the patch is then

$$r(t)/R_p(t) = R^{-1}. \tag{4.8}$$

If this number is small, the curvature within the patch can be likened to a wrinkle in the background geometry; if $R^{-1} \sim 1$, it can be likened to a knob (fig. 87.1). One notices that R^{-1} is independent of time so the perturbation to the geometry does not change: a wrinkle stays a wrinkle (as long as this simple spherical model applies). This corresponds to the result ϕ = constant (eq. 4.4) in linear perturbation theory.

When the patch stops expanding, $da/dt = 0$, equations (4.5) and (4.7) indicate the radius is

$$r_m \sim c(G\rho)^{-1/2}R^{-1} \sim ct_m R^{-1}. \tag{4.9}$$

This is a relation between the time t_m when the protoobject breaks away from the general expansion (eq. 4.1), the radius r_m of the object when this happens, and the parameter R^{-1} that measures the perturbation to the geometry. For galaxies and clusters of galaxies $r_m \ll ct_m$, so $R^{-1} \ll 1$, and these objects would form out of wrinkles in the geometry (eq. 4.8). This corresponds to the condition $\phi \ll 1$.

The conclusion from this discussion is that, as Lemaître showed, one can think of small density fluctuations in the early universe growing into prominent irregularities like galaxies. However, the consequence that was only fully recognized later, is that in this picture one must accept the idea that the universe had primeval wrinkles (Novikov 1964a, Peebles 1967a).

Another aspect of the linear perturbation theory result was noted by Lifshitz (1946) and by Bonnor (1956, 1957, 1967). The density contrast $\delta\rho/\rho$ in an Einstein-de Sitter model grows as $t^{2/3}$, much less strongly than the exponential growth one usually associates with an instability. Bonnor noted as an example that if one starts the calculation at $t_i = 1$ sec, then one finds $\delta\rho/\rho$ grows by the factor $(t_0/t_i)^{2/3} \sim 10^{12}$. If at t_i the matter were hydrogen atoms distributed uniformly at random, the density fluctuations on the scale of a large galaxy (10^{11} $M_\odot \sim 10^{68}$ atoms) would be $\delta\rho/\rho_i \sim N^{-1/2} \sim 10^{-34}$. The growth factor thus is inadequate by many orders of magnitude.

Though all the steps in Bonnor's calculation are valid, one can revise the conclusion. The choice $t_i = 1$ sec is an impressively small value, but we must nevertheless suppose that the universe did not begin then, that it and the density irregularities had a still earlier history. If $\delta\rho/\rho$ at the chosen hypersurface $t_i = 1$ sec happened to agree with the thermal fluctuation value $N^{-1/2}$, then the fluctuations traced back to $t_i \ll 1$ sec would have to have been much smaller or much larger than this, depending on the initial velocities. Either case would be puzzling. The other side of this is that, if $(\delta\rho/\rho)_i$ were given, one could always choose t_i small enough to secure the wanted amplification to fit present fluctuations (Zel'dovich 1965a, Peebles 1968, Nariai and Tomita 1971). But finally there is no known reason to assume $(\delta\rho/\rho)_i \sim N^{-1/2}$ at any chosen t_i. Though the relaxation time may be very short, the maximum distance over which particles or energy can be shared is limited by the horizon, which at $t_i = 1$ sec contains only about the number of baryons in the sun. In sum, because we do not know how initial conditions were set up across the horizon at the time of the big bang, we do not know the growth factor in the gravitational instability picture; we cannot say whether in this picture galaxies could have formed.

In the 1950s and early 1960s cosmologists generally tended to accept the conclusions of Lifshitz and Bonnor. Perhaps most important was the effect on Gamow's thoughts. He had earlier adopted the instability picture and, with Teller (1939), had given a heuristic analysis of the effect. Gamow was very excited to learn of Lifshitz's work (according to the recollection of

J. A. Wheeler) and quickly accepted it. Apparently this was reenforced by the results of his own calculation with S. Ulam and N. Metropolis (reported by Gamow 1952 but not published). He then turned to primeval turbulence (§ 4D below).

The instability picture certainly was not abandoned during the 1950s. For example, Hoyle (1949b) used it as an argument against the big bang model: one might have thought we ought to have seen dense patches left over from very early epochs because, as he argued, the expanding universe is unstable. Raychaudhuri (1952) used the spherical model to argue that one can find a middle ground between the conclusions of Hoyle and Lifshitz.

A good illustration of the rather confused state of affairs is the discussion at the 1958 Solvay Conference on the Structure and Evolution of the Universe. In a report on the theoretical situation in cosmology, Adams, Mjolsness, and Wheeler (1958) accepted Lifshitz's conclusion and proposed that condensations like galaxies form "during the stage of contraction towards the end of the previous oscillation" of the universe. Lemaître (1958) described his ideas on cluster formation (which now included the thought that there is an ongoing exchange of galaxies entering and leaving clusters, a concept that since has not seemed promising). He mentioned that Bonnor had worked on this subject, but made no comments on the objections he and Lifshitz had raised. Hoyle (1958, p. 61) suggested the instability picture is not very promising:

> The formation of galaxies presents a curious problem, for the universe combines both expansion and condensation. This apparent contradiction is overcome in Lemaître's cosmology by arranging for the formation of galaxies to have occurred at an epoch when the universe was quasi-stationary. No such provision is made in other forms of relativistic cosmology, the origin of the galaxies being by-passed with the rather vague hypothesis that islands of higher density were present within the expanding cosmological material. At a certain stage these islands are supposed to have resisted the general expansion and to have condensed into stars. How and why this condensation took place is left in an equally vague condition.

He suggested that in the steady state model galaxies could form by thermal instability: where the density happens to be high the cooling time is low, so the pressure drops and the pressure gradient tends to push more material in to enhance the irregularity. Oort (1958) was largely unaware of all the debate on gravitational instability. In his report he considered reasonable processes for the formation of a system like a spiral or elliptical galaxy or a cluster of galaxies in an expanding universe. He did not use the jargon but he arrived at the conclusion that systems like the Virgo cluster might form by the gravitational instability process, while spiral galaxies require in addition something like primeval turbulence to account for their angular momenta.

Oort's remarks on cluster formation were taken up by van Albada (1960) who considered the evolution of the single-galaxy distribution function $F(\mathbf{r}, \mathbf{v}, t)$ in the self-consistent spherically symmetric potential well $\phi(r, t)$. He was able to find numerical solutions that commence as nearly uniform, expanding with the general expansion, and end up roughly reproducing the density run in a cluster as well as the observed tendency of the line of sight velocity dispersion to decrease with increasing projected radius. It is interesting to see, in the proceedings of the 1961 Conference on *Problems of Extragalactic Research,* the rather vigorous objections to van Albada's approach because of the slow rate of growth of irregularities in an expanding universe (van Albada 1962, p. 427). His response[10] was that in the solutions the galaxy distribution nevertheless does vary from nearly uniform at the initial time to strongly clustered at the final time and that the final state does bear some resemblance to a real cluster. This recalls Lemaître's original project to discover whether there is a self-consistent scenario of evolution that matches what is observed. As has been described here, people have objected in effect that if the gravitational instability picture were valid, it ought to be capable of giving an *ab initio* theory of galaxies, and that is not so within present fundamental theory. But we are left with the phenomenological approach.

B. Clustering without preferred quantities

Gravity physics with $\Lambda = 0$ does not involve any fundamental quantities of length, time, or mass, and the coupling constant G affords only variants of the one dimensionless relation as GM/rc^2 or $G\rho t^2$. The Einstein-de Sitter model ($\Lambda = p = R^{-2} = 0$) does not offer any fixed quantities either. It is not suprising therefore that in this model the dimensionless density contrast $\delta\rho/\rho$ varies as a power of time while preserving whatever initial spatial shape was given (in a pure mode: § 11), for this is the only possible functional form: the relation $\delta\rho/\rho \propto \exp t/\tau$, which often has been cited as what is wanted, is not possible because it requires the quantity τ that does not exist in the theory. It follows that in this model we cannot hope to predict the masses of systems that break away from the general expansion or when this happens. Though this point has not often been explicitly

[10] In his 1960 paper van Albada argued that the conclusion of Lifshitz and of Adams, Mjolsness, and Wheeler was mistaken. It is not clear, however, whether he considered the important role radiation pressure played in these earlier analyses.

discussed, it must have been apparent to many, to judge from the many schemes that have been proposed to introduce fixed characteristic quantities. And it must be counted as one of the reasons people have considered the instability picture unsatisfactory, as one sees, for example, in Hoyle's comments quoted above and in the detailed discussion by Harrison (1967a,b).

There are two ways to proceed. First one can consider how gravitation might be augmented by other effects, like fluid pressure, that in combination with gravity yield characteristic quantities like the Jeans length. That is reviewed in part (C) below. Second one can argue, as a virtue out of necessity, that the search for characteristic quantities is only a part of the problem and perhaps not even central to it. If we had derived the length ~ 10 kpc from the fundamental theory to account for the nominal sizes of large galaxies, we would still have to account for tight groups of galaxies at perhaps 100 h^{-1} kpc diameter, for the dense parts of rich clusters at $\sim 1\ h^{-1}$ Mpc, and for the pattern of clustering that extends beyond that to at least 40 h^{-1} Mpc. If the theory had predicted an exponential growth of $\delta\rho/\rho$ with time, then we would have had one characteristic time, but again the problem seems richer than that. Large galaxies generally are old: though there may be some young galaxies, the era of galaxy formation seems pretty well over. On the other hand, the density contrast in a supercluster of Abell clusters is not very large; so if the universe really is expanding and evolving, these systems could only have broken away from the general expansion quite recently.

Characteristic quantities certainly are important: galaxies appear as definite objects with definite properties to account for. However, it may be that continuity of phenomena is the more fundamental clue. Though a galaxy is very different from a supercluster of galaxies, the two can be considered extremes of a continuous range of objects. Just as one can trace a continuous progression from gas to liquid phase, one can find examples of galaxies with double or multiple nuclei, compact pairs of galaxies, compact groups, looser and richer galaxy associations, and so on through a more or less continuous spectrum.

Carpenter (1938) noted that in the scatter plot of radii and mean densities within clusters of galaxies there is a rather well-defined upper envelope representing the densest clusters found for each size of the form

$$n(r) \propto r^{-\gamma}. \tag{4.10}$$

This led Carpenter to speculate "that there is no basic and essential distinction between the large, rich clusters and the small, loose groups. Rather, the objects commonly recognized as physical clusterings are merely extremes of a nonuniform though not random distribution which is limited by density as well as by population. From this point of view, the term 'supergalaxy' is of questionable propriety, since it implies a distinctive and coherent organic structure inherently of a higher order than individual galaxies themselves" (1938, p. 355). De Vaucouleurs (1960, 1970, 1971) reexamined Carpenter's relation, adjusting the power law index to

$$\gamma = 1.7. \tag{4.11}$$

He remarked that the typical radii and densities of galaxies fit onto this relation, and he left as an open question whether there is a natural division or gap in the spectrum of clustering between a galaxy and a compact group or between a compact group and a rich cluster and so on. We are presented with a series of physically significant lengths if there is and with the continuity of the clustering phenomena if there is not.

Kiang (1967) arrived at the concept of continuous clustering from the attempts to model the distributions of galaxies and of rich compact Abell clusters of galaxies. Kiang estimated the autocorrelation among counts of Abell clusters counted in a mesh of cells across the sky, and he compared the results to a model of Neyman and Scott in which the clusters are in Gaussian-shaped superclusters, the superclusters being distributed uniformly at random. Kiang found that the best value of the supercluster radius (width of the Gaussian) varies with the lag angle θ of the correlation function at which the model is fitted to the data: the Gaussian supercluster model does not reproduce the shape of the cluster autocorrelation function. Earlier Neyman, Scott, and Shane (1956) had found the same problem in fitting this Gaussian model of galaxy clustering to the galaxy autocorrelation function in the Lick sample. They suggested that one may have to account for the clustering of clusters of galaxies, and they noted also that if clusters in the model overlap appreciably, the concept of an individual cluster may be only a convenient but oversimplified construct. Kiang was more direct: if there is no best value for the standard deviation σ in the Gaussian clustering model, then perhaps one should consider "the hypothesis, that clustering of galaxies occurs on all scales," with "no preferred sizes" (1967, p. 17).

The same point was made by Totsuji and Kihara (1969) who noted that the galaxy correlation function $\xi(r)$ (eq. 2.2) found by Neyman, Scott, and Shane for the Lick data approximates a power law at $10' \lesssim \theta \lesssim 3°$. They checked these results with their own estimates of the correlation function at 1° to 3.° They remarked that if the angular correlation function is close to a power law, then it is not very convenient to use Gaussian or exponential functions that have characteristic lengths to model the cluster shapes or to fit to the spatial autocorrelation function, as earlier workers had done (Neyman, Scott, and Shane 1956, Limber 1954, Rubin 1954). Totsuji and Kihara's fit to the power law model is

$$\xi \propto r^{-\gamma}, \qquad \gamma \approx 1.8. \tag{4.12}$$

This expression was independently discovered (Peebles 1974a,b) in the Zwicky catalog of galaxies (Zwicky et al., 1961–68). Like Carpenter's law (eq. 4.10), it certainly agrees with the idea that there is no preferred scale over a substantial range in the clustering.

The autocorrelation function is a useful measure of the nature of the galaxy distribution, but of course it contains only very limited information, so there is not a unique interpretation of a given $\xi(r)$. One systematic way to add more detailed information is to examine progressively higher order correlation functions. As will be described in Chapters III and IV, this approach proves convenient both for the reduction of the data and the theoretical analysis of clustering dynamics. The galaxy three-point function is known in some detail, and we have schematic estimates of the four-point function. The results (§ 61) are in good agreement with Fournier d'Albe's (1907) picture of a scale-invariant clustering hierarchy (§ 3a): when the distribution is viewed with resolution r, the mass appears in patches of size $\sim r$, typical density

$$n(r) \propto r^{-\gamma}, \qquad \gamma = 1.8. \tag{4.13}$$

This applies on scales as large as $\sim 10\ h^{-1}$ Mpc and down to and perhaps including that of an individual galaxy. At $r \gtrsim 10\ h^{-1}$ Mpc the indication is that the clustering pattern is starting to wash out into a uniform background (§ 2).

Carpenter's power law expression in equation (4.10) agrees with equation (4.13), but we must consider that this agreement is at least in part fortuitous because Carpenter had in mind separate and distinct clusters, not a clustering hierarchy. Carpenter's relation as adapted by de Vaucouleurs does describe a clustering hierarchy, and it is notable that the values of the index γ found by de Vaucouleurs (equation 4.11) and established from the correlation functions agree very well.

If the continuous clustering hierarchy picture is a valid first approximation, attempts to find theories of origin of specific objects may have been addressing the wrong question. Partly because of the continuity of the galaxy clustering, more importantly because of the scale invariance of the theory, there have been a number of discussions of possible theoretical aspects of continuous clustering hierarchies. Lemaître clearly saw that the growth of irregularities in an expanding universe would produce clusters within clusters (more specifically, galaxies within clusters of galaxies), but he did not pursue the general point. Layzer (1954) seems to be the first to have argued that this process would produce a clustering hierarchy. The scaling argument relating the shape of the power spectrum of assumed initial fluctuations in the early universe to the index γ in the clustering hierarchy was derived by Peebles (1965; § 26 below). Press and Schechter (1974) discussed how this scaling approach might be extended to estimate the frequency distribution of cluster density in clusters of fixed size $\sim r$. Saslaw (1968) observed that density fluctuations associated with gravity might be expected to vary as a power of r, and Kiang and Saslaw (1969) pointed out that this effect seems to be apparent in the distribution of Abell clusters. Totsuji and Kihara (1969) compared gravitational clustering in the expanding universe to fluctuations at the critical point of a fluid. In each case, since there are no fixed characteristic quantities, one looks for power law behavior, and that is what they found in the autocorrelation function estimates of Neyman, Scott, and Shane (1956). Peebles (1974a) arrived at the same conclusion from the autocorrelation function for the Zwicky galaxies. Chapter IV deals with some attempts to enlarge on these concepts.

It is tempting to compare the continuity observed here to the idea of the great chain of being where it was observed that there is a more or less linear and continuous progression from inanimate objects through the most primitive forms of life to the higher animals and on to angels. This fell out of favor roughly in the time of Darwin when it became apparent that a tree would be a better analogy, but it did play a useful role in making concrete the underlying unity among apparently diverse phenomena and the need for a unified theory. Of course it remains to be seen how useful continuity will prove here, and meanwhile there is no shortage of ideas on how to introduce characteristic lengths to supplement or replace it. Some of these ideas are reviewed next.

C. Cosmological parameters, fluid dynamics, and the primeval fireball

Jeans (1902, 1928) showed that pressure introduces a critical minimum length for gravitational instability. Gamow and Teller (1939) argued that at least roughly the same criterion applies in a uniformly expanding world model. Lifshitz's (1946) general treatment of linear perturbations of a Friedman-Lemaître model took account of matter pressure, and his results show that the Jeans criterion does apply though he did not explicitly state it. Bonnor (1957) showed how Lifshitz's results for the case $p \ll \rho c^2$, $r \ll ct$

could be derived from the Newtonian approximation, and he showed that the classical critical Jeans length applies in an expanding universe.

It is curious that in the 1930s Lemaître did not discuss how Jeans' argument would be modified when it was applied to an expanding world picture. As noted above, he concentrated on a characteristic number provided by the Lemaître model: clusters were supposed to form at the critical density $\Lambda/4\pi G$ during the quasi-static phase.

Gamow and Teller (1939) pointed out the possible role of another cosmological parameter. If $\Lambda = 0$, the time variation of the expansion parameter $a(t)$ is given by equation (4.5). If the universe is open, the constant R^{-2} is negative and so da/dt cannot vanish: the universe cannot stop expanding. In the early universe at small enough $a(t)$, the gravity term on the right hand side of equation (4.5) dominates (because the density varies as a^{-3}, faster than the curvature term $\propto a^{-2}$), so the model behaves like the Einstein-de Sitter case $R^{-2} = 0$. In the late stages of expansion of an open model, the R^{-2} term dominates and gravitational deceleration is unimportant so small density irregularities stop growing. Thus the redshift Z_c at the epoch of the transition between the early behavior where gravity dominates and the later free expansion roughly marks the end of the epoch of formation of bound clusters (§ 11).

Gamow and Teller also tried to find an argument to explain why objects should not form well before Z_c so that the epoch of formation would be fixed, but nothing convincing emerged. In 1948 Gamow suggested that in a hot cosmology blackbody radiation could inhibit the formation of objects at high redshift. He and Alpher (Gamow 1948a,b, Alpher, Bethe, and Gamow 1948) had remarked that in an expanding mixture of matter and radiation at a temperature ~ 1 MeV, neutron capture could build up a reasonable distribution of elements (PC, § V a). The mass density due to the radiation varies with time as a^{-4}, one power of a faster than the nonrelativistic matter density because the energy of each photon varies as $\nu \propto a^{-1}$, so there is a critical epoch Z_{eq} where the mass densities in radiation and nonrelativistic matter are equal. Gamow supposed that bound systems might not be able to form at $Z \gg Z_{eq}$ because the radiation would resist it and that the Jeans length for the matter at $Z \sim Z_{eq}$ fixed the size of galaxies.

It is now recognized that the role of radiation is somewhat more complicated than this. There is a second important redshift. $Z_{dec} \sim 1300$ where the radiation temperature is $T \sim 4000$ K and the free electrons and protons combine to form atomic hydrogen. Prior to Z_{dec} the mean path for Thomson scattering of the radiation is short, so matter and radiation behave like a single fluid. By an interesting (and perhaps ultimately very significant) coincidence the present measurements indicate Z_{dec} is roughly the same as Z_{eq} (eq. 92.42). If $Z_{dec} > Z_{eq}$, then in the interval between these two epochs the matter moves free of the radiation, but the gravitational instability is suppressed because the mass density in radiation speeds the expansion (§ 12 below). If $Z_{eq} > Z_{dec}$, then in the interval between the epochs matter cannot fragment into lumps because radiation drag prevents it (§ 92 below). Either way, as Gamow recognized, the primeval fireball radiation plays a very important role in fixing the maximum redshift at which nonrelativistic bound systems can form (Peebles 1965).

How the process of decoupling of matter and radiation at Z_{dec} affects the spectrum of irregularities can be seen by writing the assumed fluctuations in the matter and radiation distributions just prior to this epoch as a linear combination of two modes,

$$\begin{aligned} &\textit{isothermal:} \quad \delta\rho_r = 0, \\ &\textit{adiabatic:} \quad \frac{\delta\rho_r}{\rho_r} = \frac{4}{3}\frac{\delta\rho_m}{\rho_m}, \end{aligned} \tag{4.14}$$

where the subscripts refer to radiation and (nonrelativistic) matter. These names were introduced by Zel'dovich (1967). In an isothermal perturbation the radiation distribution is uniform, the matter distribution more or less irregular. In the second case the entropy per baryon is uniform, as in a sound wave. If photon diffusion through the matter can be ignored and if the wavelength of the perturbation is $\lambda \ll ct$ so gravity can be ignored, the two types of perturbation evolve independently.

An adiabatic perturbation oscillates, like an acoustic wave, and the oscillations tend to be damped by photon diffusion. Thus there emerges a new characteristic mass fixed by the smallest adiabatic irregularities that can escape strong dissipation through the decoupling of matter and radiation. (This was independently discussed by Silk 1967, 1968a, Michie 1967, and Peebles 1967b). Depending on the parameters in the cosmological model, this characteristic dissipation mass can be anywhere from that of a giant galaxy to that of a small cluster of galaxies (§ 92 below), the latter applying in a universe with low matter density. In this latter case one might suppose that the first objects to form were protoclusters with mass comparable to the dissipation mass and that galaxies formed through fragmentation of the protoclusters. This scheme has been discussed in some detail by Zel'dovich and his colleagues (Zel'dovich 1978 and earlier references therein, §§95,96 below).

An isothermal perturbation with $\lambda \ll ct$ is locked to the radiation through Thomson scattering until Z_{dec} when the matter becomes neutral and can move through the radiation. The matter temperature at Z_{dec} is known because the matter and radiation are very close to thermal equilibrium, so the critical Jeans mass for the matter is known (Peebles 1965, 1969b, § 94 below),

$$M_J \approx (3 - 10) \times 10^5 M_\odot. \tag{4.15}$$

Isothermal fluctuations on scales smaller than this are dissipated because they tend to oscillate, and photon diffusion during decoupling strongly damps the oscillation. Thus if the initial spectrum of isothermal irregularities were fairly flat down to scales $< M_J$, the damping would impose a strong shoulder at M_J and the smallest early objects to fragment from the general expansion would have this mass. Peebles and Dicke (1968) suggested these objects might be primeval globular star clusters. Doroshkevich, Zel'dovich, and Novikov (1967) suggested they might become supermassive stars, which could serve to raise the temperature of the remaining matter, perhaps thereby increasing the Jeans mass to something more reasonable for a galaxy.

Hoyle and Gold (Hoyle 1958, Gold and Hoyle 1959), who were considering the steady state model, discussed the idea that galaxy formation might be promoted by thermal instability. One could imagine that the newly produced matter is hot. (For example, Hoyle and Gold noted that if neutrons were produced, the plasma temperature would be $kT \sim$ decay energy ~ 1 MeV). Then two order-of-magnitude conditions are required. The cooling time must be comparable to the Hubble time H^{-1}, so that the denser spots can cool relative to the mean, creating pressure holes that tend to push matter toward the denser spots. The velocity of sound in the gas must be high enough to cross the pressure hole in a Hubble time, that is, the Jeans length must be larger than the hole. The process has been discussed in some detail by Field (1965). It was applied in the big bang model to the scenario of Doroshkevich, Zel'dovich, and Novikov as the method of producing protogalaxies out of the gas heated by the first generation of superstars.

A somewhat related process is thermodynamic instability. If the matter-radiation mixture in the early universe passed through something like a critical point, the thermal fluctuations in the density might be appreciable and, augmented by gravity, might force the development of structures. In the hot big bang model the most promising phase for this would seem to be the epoch of the decoupling of matter and radiation. Where the matter density is higher than average, the plasma combines to hydrogen faster reducing the molecular weight and hence the matter pressure, which tends then to increase the density irregularity. However, the effect is found to be negligible (Saslaw 1967, Peebles 1969b). One might imagine there were more violent thermodynamic instabilities at very high redshift, but here the horizon at high Z encompasses only a relatively small baryon number, so the instability would have a negligible effect on fluctuations on the scale of a galaxy.

Magnetic fields of galaxies present an interesting problem. The first question is whether the field existed before the galaxy (Hoyle 1958) or was produced in a dynamo in the galaxy (Parker 1975). If the former occurred, the magnetic stress could have played an important role in the formation of galaxies. Some aspects of this are discussed in Sections 17 and 95.

Sciama (1955, 1964) discussed a sort of regenerative process in the steady state model: a galaxy moving through the intergalactic medium would tend to leave a wake by gravitational perturbation that might then contract to new galaxies. The same process must play some role in an evolving universe, and indeed it has been speculated that small gravitationally bound clumps of matter could trigger the formation of progressively larger ones all the way to clusters of galaxies (Press and Schechter 1974, Carlitz, Frautschi, and Nahm 1973). Such a bootstrap hypothesis is attractive because one could then start at the very early stages of expansion with a universe as smooth as is allowed by the discrete nature of matter, which perhaps is a natural initial condition, and then imagine that the particles trigger the formation of progressively larger clumps through nonlinear interactions. However, it proves difficult to see how this process could work in the conventional cosmology (§ 28 below).

Among the processes mentioned here by far the greatest attention has been directed to those involving the primeval fireball radiation. Indeed, the discovery of this radiation has greatly stimulated interest in the gravitational instability picture, for, as has been described, we are given the epoch $Z \approx 1000$ at which nonrelativistic objects can commence forming, we are given two low mass cutoffs in the spectra of linear perturbations at decoupling, and we are given the properties of the matter at decoupling. Of course, a good deal will have to be added to this before we have a theory of galaxies, but the situation does seem interesting.

D. Primeval turbulence

The main argument for this picture has been the rotation of galaxies (and the fact that a spiral galaxy looks strikingly like a fossil turbulence eddy). A second argument has been that turbulence can promote formation of density fluctuations. The major problem is that this effect is too efficient: it tends to produce large density fluctuations and dissipate the turbulence too soon.

The primeval turbulence picture was first discussed in modern cosmology by von Weizsäcker (1949, 1951). (A review of earlier speculation is given by Jones and Peebles 1972). Von Weizsäcker discussed the possible role of turbulence in a broad range of astronomical settings. He suggested the rotation of galaxies is a residuum of primeval turbulence and that "in an expanding universe gravitational instability would not be sufficient to form sub-systems, while turbulence could do it if its velocity v_t were large enough compared with the velocity of expansion v_{ex}" (1951, p. 176). It is not clear how familiar he was with the work and controversy on gravitational instability in expanding world models; in any case he gave only a schematic treatment of cosmology.

As was described in Section 4A, by the late 1940s Gamow had become convinced that gravitational instability in fact is not sufficient to form galaxies, and so he turned to von Weizsäcker's turbulence picture (Gamow 1952). He noted that the highly irregular space distribution of galaxies as revealed in Shapley's surveys and in the preliminary results from the Lick survey (§ 2) seems evocative of irregular primeval motions as in turbulence (Gamow 1953, 1954), and this led him and Rubin to a discussion of the galaxy autocorrelation function as a measure of the density fluctuations (Rubin 1954). It is interesting to note Gamow's cautionary remark on the other side of the problem: "Although Reynold's number for the universe is always sufficiently large to expect the presence of turbulent motion, it is, however, difficult to see how such a motion could originate in a uniformly expanding homogeneous material. Thus it may be well to introduce the primordial turbulence on a postulatory basis along with the original density of matter and the rate of expansion" (1952, p. 251).

The main problem with the primeval turbulence picture is that it is difficult to arrange things so the turbulence does not dissipate prematurely forming objects at $Z \sim Z_{dec}$ that would be denser than galaxies (Peebles 1971a). To see why this is, let us suppose first that radiation pressure may be neglected. Interesting turbulence velocities are on the order of the rotation velocity in a large spiral, $v_t \sim 300$ km s^{-1}. The flow is close to incompressible if the velocity of sound exceeds v_t, which means the temperature exceeds 3×10^6 K for a plasma of electrons and protons. If the temperature is much less than this, there is nothing to deflect the eddies: they tend to pile up in shocks. That is wanted during the epoch of galaxy building but of course must be avoided at higher redshifts. However, at $Z \gtrsim 10$ the characteristic time for cooling by drag by the microwave background is shorter than the expansion time (*PC* fig. VII-1), so it is hard to see why the plasma should have stayed hot. What is more, in the absence of dissipation v_t varies as $a(t)^{-1} \propto (1 + Z)$ (§ 90): the higher the redshift the greater v_t and the harder it is to see why matter did not pile up into lumps well before galaxies could have existed. Ozernoi and Chernin

(1967, 1968) pointed out that the primeval fireball radiation offers a partial solution. At redshifts $Z > Z_{dec}$ matter and radiation act like a single fluid with velocity of sound comparable to the velocity of light, so nonrelativistic turbulence is subsonic and catastrophic shock formation is avoided. There is still the problem that in the conventional models matter and radiation decouple at a redshift $\sim$ 1000, too early for galaxy formation. Thus a nice balance of parameters is needed if primeval turbulence is to assist but not overwhelm galaxy formation.

Another less direct problem is that the incompressible turbulence assumed to exist at $Z \sim 1000$ can be traced back to diverging fluctuations in space curvature at high redshift, $t \rightarrow 0$ (§ 90). This is an unsatisfactory situation: the assumed primeval currents cause chaotic fluctuations in the geometry at the time of the big bang, but yet it has been arranged for the mass density to be accurately homogeneous as it comes into view on the horizon.

E. Alternative scenarios

In all the above discussion it has been assumed that the general expansion of the universe plays a central role in setting the framework within which structure develops. There will have to be considerable progress in the theory and observations before it will be clear whether that is so, and meanwhile it will be well to bear in mind that very different approaches are conceivable. The most actively discussed of the alternatives might be labeled the white hole scenarios. As examples, McCrea (1964) and Hoyle (1965) discussed the idea that old galaxies might produce new ones by ejecting "embryos," or by "calving," perhaps in the manner of an iceberg. Hoyle and Narlikar (1966) considered the idea that, in the steady state concept, matter is produced not at a steady rate but in cyclic bursts, which might be concentrated in spots where there already is matter, thereby producing giant elliptical galaxies. Novikov (1964b) and Ne'eman (1965) have discussed the idea of time-reversed black holes, perhaps spots where the big bang was substantially delayed.

On the observational side, Ambartsumian (1958, 1965) and Arp (1970) have discussed the evidence of ejection of objects from nuclei of galaxies. That ejection occurs is undoubted. Much more difficult to establish is whether this is new matter or debris that settled in the nucleus and then exploded.

There has been relatively little detailed exploration of the possibilities offered by such alternative scenarios, and probably it is only realistic to expect there will not be until it becomes fairly clear to most people that the conventional lines of thought are not productive.

5. Summary

The inhomogeneity of the universe has played a curious role in the history of the modern cosmology. In the earliest discussions the clustering of matter attracted considerable attention, first as the possible cause of expansion of the universe away from the Einstein state, then in Lemaître's study of the evolution of irregularities in an expanding universe. In the interval from approximately 1935 to 1965 a number of people more or less independently reinvented Jeans' and Lemaître's concept that the universe tends to grow more irregular, but there was no thorough discussion: the homogeneous models held center stage. The present general revival of interest is due to the greatly improved observations in cosmology in general and in the measurement of the clustering of matter. The options for the theorist have been reduced somewhat and the problem made considerably more interesting.

In the early 1930s the large-scale homogeneity of the visible universe was accepted by cosmologists with what now seems undue haste, despite the warnings of astronomers. As it has turned out, however, precise tests have emerged and the results do agree with the assumption: the indication is that the large-scale structure of the universe is very simple. Following Einstein's argument from Mach's principle and Milne, one can think of this remarkable concept as a physical principle, though it is a principle of an unusual sort since it applies only in the sense of a space average and it addresses just the one phenomenon. Milne's cosmological principle quickly became a standard catch phrase and it still is. When it was introduced, it served another function: in reaction people asked whether homogeneity really is enforced by general relativity theory. The tentative answer at the time was that it is not, and despite later independent approaches the present indication still is that the large-scale homogeneity of the universe could not have been predicted from established fundamental theory. This leaves us with the view that homogeneity must be accepted as a phenomenon to be explained by some future deeper theory.

It also means that we cannot hope to deduce the existence of galaxies from known physical principles because we do not know how to specify initial conditions. So we are left with Lemaître's program: try to find the character of density fluctuations in the early universe that would develop into the irregularities we observe. This is a rich problem and becoming richer as we learn more about the details of the phenomena to be reproduced. There has been some resistance to Lemaître's program, partly due to misunderstandings, partly because the fundamental theory seemed not to be helping since it does not offer any characteristic quantities (if one takes $R^{-2} = \Lambda = 0$). In recent years the situation has developed in two rather different ways. The primeval fireball radiation does fix some characteristic quantities, and one approach has been to invent scenarios that make important use of them. The second development is based on the observation that, by some measures, galaxy clustering does not offer any characteristic quantities. Perhaps the scale independence of the theory should be exploited, with quantities like a galaxy radius added as a detail.

One could imagine several possible results from Lemaître's program. Perhaps our universe is as smooth as is consistent with the discrete nature of matter and the attendant minimum perturbations in the early universe. Perhaps the strongly nonlinear behavior of the present clustering has erased most details of initial conditions: the universe ends up looking much the same under a fairly broad range of initial values. A more traditional idea is that we shall be presented with fluctuations of rather definite character in the early universe and that this will be an important datum for some future fundamental theory of the big bang. Or the detailed analysis of theory and observation may reveal that the general concepts we have adopted cannot account for the phenomena, that the history of irregularities must be quite different from what has been assumed. This last result certainly would have to be counted as progress and possibly as a necessary prerequisite to forcing us onto other paths.

II. BEHAVIOR OF IRREGULARITIES IN THE DISTRIBUTION OF MATTER: NEWTONIAN APPROXIMATION

6. Newtonian approximation

Discussion of how irregularities in the matter distribution behave in an expanding universe is greatly simplified by the fact that a limiting approximation of general relativity, Newtonian mechanics, applies in a region small compared to the Hubble length cH^{-1} (and large compared to the Schwarzschild radii of any collapsed objects). The rest of the universe can affect the region only through a tidal field. Though the point was clearly made by Lemaître (1931), it has not always been recognized that the Newtonian approximation is not a model but a limiting case valid no matter what is happening in the distant parts of the universe. Because of the importance of this result, it is discussed here at some length. A different approach based on linear perturbations to the Robertson-Walker line element is described in Chapter V (§ 84).

In general relativity events in space-time are labeled by the four coordinates x^i. The numerical values of the coordinates have no meaning: the proper distance or proper time interval between events separated by coordinate interval dx^i is determined by the ten elements of the metric tensor $g_{ij}(x)$ through the line element

$$ds^2 = g_{ij}dx^i dx^j. \tag{6.1}$$

The change of coordinates $y^i = y^i(x^j)$ changes the components of the metric tensor and the coordinate interval dx^i between two given events according to the transformation equations

$$g'_{ij}(y) = \frac{\partial x^k}{\partial y^i}\frac{\partial x^l}{\partial y^j} g_{kl}(x), \qquad dy^i = \frac{\partial y^i}{\partial x^j} dx^j. \tag{6.2}$$

One sees that ds is unchanged. This is the invariant interval measured by a physical rod or clock (§ 81).

If space-time is flat, coordinates can be chosen so g_{ij} everywhere is the Minkowski form,

$$g_{ij} = \eta_{ij} = c^2, -1, -1, -1, \tag{6.3}$$

along the diagonal, with zeros off the diagonal. In curved space this is not possible, but one can simplify the description of a small patch of space-time by choosing coordinates so g_{ij} and its first derivatives agree with the Minkowski form at one point,

$$g_{ij} = \eta_{ij}, \qquad g_{ij,k} = 0, \tag{6.4}$$

for these are $10 + 4 \times 10 = 50$ equations to be satisfied by the choice of the transformation coefficients

$$\partial x^i/\partial y^j, \qquad \partial^2 x^i/\partial y^j \partial y^k, \tag{6.5}$$

$4 \times 4 + 4 \times 10 = 56$ in all, 6 more than needed. To make $g_{ij,kl} = 0$, we would add 10×10 equations for a total of 150 to be satisfied by the choice of $56 + 80 = 136$ transformation coefficients plus first and second derivatives. Since there are fewer coefficients than equations to be satisfied, this is not generally possible.

To make the components of g_{ij} satisfy equations (6.4) along the path of some observer, we must satisfy the 10 conditions $g_{ij} = \eta_{ij}$ along the path and the 10×3 conditions that the derivatives normal to the path vanish, 40 in all. We can choose the 16 coefficients $\partial y^i/\partial x^j$ and the 4×6 second derivatives of y normal to the path, just enough to let us expect we can satisfy the conditions. The path of the observer is given, say $x^i(\lambda)$, so the path in the new coordinates is now determined,

$$\frac{dy^i}{d\lambda} = \frac{\partial y^i}{\partial x^j}\frac{dx^j}{d\lambda}. \tag{6.6}$$

If, as will be assumed, the observer is freely moving, the path follows the geodesic equations of motion,

$$\frac{d}{ds} g_{ij} \frac{dy^j}{ds} = \frac{1}{2} g_{jk,i} \frac{dy^j}{ds}\frac{dy^k}{ds}. \tag{6.7}$$

The coordinates at the starting point can be oriented so the observer is at rest at $y^\alpha = 0$,

$$\frac{dy^i}{ds} = \delta_0^{\,i}, \tag{6.8}$$

and equations (6.4) and (6.7) give

$$\frac{d^2y^i}{ds^2} = 0, \tag{6.9}$$

so the observer stays at $y^\alpha = 0$.

In these locally Minkowski coordinates the line element in the neighborhood of the path is

$$ds^2 = c^2dt^2 - (dx^2 + dy^2 + dz^2), \tag{6.10}$$

so the coordinates are locally orthogonal with coordinate space and time intervals agreeing with proper intervals, and of course the acceleration of a free particle at the origin vanishes so the gravitational acceleration has been transformed away. The measured acceleration of a particle a small distance r away from the world line is described by the field equations as follows.

The metric tensor can be written as

$$g_{ij} = \eta_{ij} + h_{ij}, \tag{6.11}$$

where h_{ij} is small, $h_{ij} \sim r^2$, in some region around the path. In this region Einstein's field equations are simple because the standard weak field linear approximation applies. (This is so no matter how large the second derivatives of g because in the field equations second derivatives appear only to the first power, and nonlinear terms like $g_{,i}g_{,j}$ are negligible compared to $g_{,ij}$ at small enough x.) On using the fact that in the locally Minkowski coordinates all time derivatives of g_{ij} and $g_{ij,\alpha}$ vanish along the path $x^\alpha \equiv 0$, one finds from the standard weak field approximation (e.g. Landau and Lifshitz 1979, eq. 105.9)

$$\begin{aligned} R_{00} &= -\frac{1}{2}\eta^{ij}(h_{ij,00} - h_{i0,j0} - h_{j0,i0} + h_{00,ij}) = \nabla_r{}^2\Phi, \\ g_{00} &\equiv c^2 + 2\Phi. \end{aligned} \tag{6.12}$$

Then the zero-zero component of the field equations for an ideal fluid with mass density ρ, pressure p, and velocity $v \ll c$ becomes

$$\nabla_r{}^2\Phi = 4\pi G(\rho + 3p/c^2) - \Lambda. \tag{6.13}$$

For completeness the cosmological constant Λ has been added; apart from some special sections below, Λ will be set equal to zero.

The geodesic equations (6.7) in the limit $v \ll c$, $h \ll 1$ are

$$\frac{d^2r^\alpha}{dt^2} = -\Phi_{,\alpha}. \tag{6.14}$$

Equations (6.13) and (6.14) are the standard equations of Newtonian mechanics, except that if there is an appreciable radiation background, one must take account of the active gravitational mass associated with the pressure, and of course if $\Lambda \neq 0$, there is the cosmic force $\Lambda\mathbf{r}/3$ between particles at separation $\mathbf{r}$.

Equations (6.13) and (6.14) apply to any observer outside a singularity, though depending on the situation, the region within which these equations apply need not contain much matter. The region can be extended by giving the observer an acceleration g_α to bring the observer to rest relative to distant matter, which adds the term $g_\alpha r^\alpha$ to Φ, and then by patching together the results from neighboring observers. This works (the acceleration and potentials can be added) as long as relative velocities of observers and observed matter are $\ll c$ and $\Phi \ll c^2$ (eq. 6.12). For a region of size R containing mass $M \sim \rho R^3$ with density ρ roughly uniform, this second condition is

$$G\rho R^2 \ll c^2. \tag{6.15}$$

In the Friedman-Lemaître models Hubble's constant is (§ 97)

$$H \sim (G\rho)^{1/2}, \tag{6.16}$$

(if one assumes Λ is negligible and the density parameter $\Omega \sim 1$), so equation (6.15) indicates

$$R \ll cH^{-1} \sim 3000 \text{ Mpc} \sim 10^{28}\, cm. \tag{6.17}$$

That is, the region must be small compared to the Hubble length. Since the expansion velocity is $v = Hr$, this condition also says $v \ll c$.

The Newtonian approximation can fail at much smaller R if the region includes a compact object like a neutron star or black hole, but one can deal with this by noting that at distances large compared to the Schwarzschild radius the object acts like an ordinary Newtonian point mass. It is speculated that in nuclei of galaxies there might be black holes as massive as 10^9 $M_\odot$, Schwarzschild radius $\sim 10^{14}$ cm. If this is an upper limit,

Newtonian mechanics is a good approximation over a substantial range of scales,

$$10^{14}\ \text{cm} \ll r \ll 10^{28}\ \text{cm}. \tag{6.18}$$

7. Particle dynamics in expanding coordinates

It is often convenient to describe the matter distribution and motion in terms of the departure from the mean homogeneous and isotropic world model. In this background model the proper separation of two particles varies with time as

$$\mathbf{r} = a(t)\mathbf{x}, \tag{7.1}$$

where $\mathbf{x}$ is constant for the pair and, because of the homogeneity and isotropy, the expansion parameter $a(t)$ is a universal function of proper world time. To simplify the discussion it will be assumed for the moment that $p \ll \rho c^2$, $\Lambda = 0$. Then equation (6.13) implies that an observer in the background model at $\mathbf{r} = 0$ measures potential

$$\Phi_b = \tfrac{2}{3}\,\pi G \rho_b(t) r^2,$$

where $\rho_b(t)$ is the mean mass density. On using equation (7.1) with the equations of motion (6.14), one finds the cosmological equation

$$\frac{d^2 a}{dt^2} = -\frac{4}{3}\pi G \rho_b(t) a. \tag{7.2}$$

Equation (7.1) is a change of variables from proper locally Minkowski coordinates $\mathbf{r}$ to expanding coordinates $\mathbf{x}$ comoving in the background model. In these latter coordinates the proper velocity of a particle, relative to the origin, is

$$\mathbf{u} = a\dot{\mathbf{x}} + \mathbf{x}\dot{a}, \tag{7.3}$$

so the Lagrangian for the particle motion is

$$\mathcal{L} = \tfrac{1}{2}\, m\, (a\dot{\mathbf{x}} + \dot{a}\mathbf{x})^2 - m\Phi(\mathbf{x}, t). \tag{7.4}$$

The canonical transformation

$$\mathcal{L} \rightarrow \mathcal{L} - d\psi/dt, \qquad \psi = \tfrac{1}{2}\, m a \dot{a} x^2, \tag{7.5}$$

reduces the Lagrangian to

$$\mathcal{L} = \tfrac{1}{2}\, m\, a^2\dot{x}^2 - m\phi, \tag{7.6}$$

with

$$\phi = \Phi + \tfrac{1}{2}\, a\ddot{a}x^2. \tag{7.7}$$

The field equation (6.13) for the new potential ϕ is

$$\nabla^2\phi = 4\pi G\rho a^2 + 3a\ddot{a}, \tag{7.8}$$

where the gradient is with respect to $\mathbf{x}$ (while in equation (6.14) the gradient is with respect to $\mathbf{r} = a\mathbf{x}$). With equation (7.2) this becomes

$$\nabla^2\phi = 4\pi G a^2[\rho(\mathbf{x}, t) - \rho_b(t)]. \tag{7.9}$$

The equations of motion from (7.6) are

$$\mathbf{p} = ma^2\dot{\mathbf{x}}, \qquad \frac{d\mathbf{p}}{dt} = -m\nabla\phi. \tag{7.10}$$

The proper peculiar velocity of the particle is

$$\mathbf{v} = a\dot{\mathbf{x}}. \tag{7.11}$$

This is the velocity measured by an observer at the particle position and at fixed $\mathbf{x}$: that is, $\mathbf{v}$ is the motion relative to the background model. According to equation (7.10),

$$\frac{d\mathbf{v}}{dt} + \mathbf{v}\frac{\dot{a}}{a} = -\frac{\nabla\phi}{a}. \tag{7.12}$$

If $\phi \equiv 0$, $\mathbf{p}$ is constant and $\mathbf{v}$ decays as a^{-1}. This should not be attributed to some cosmic force; it is simply a result of the coordinates change. The peculiar velocity $\mathbf{v}$ is measured relative to an observer comoving in the background model. A freely moving particle always is overtaking comoving observers that are moving away from it due to the general expansion, so $\mathbf{v}$ decreases.

The source of ϕ is the contrast $\rho - \rho_b$ (eq. 7.9). This is as it should be because if there are no irregularities, $\rho - \rho_b$ vanishes and each particle remains undisturbed at fixed coordinate position x.

There may be, in addition to the irregular distribution of nonrelativistic matter, a smooth background sea of radiation or zero mass neutrinos. If $\phi \ll c^2$, the perturbation to this relativistic background by ϕ is negligible, so the density can be taken to be strictly uniform. Taking account also of Λ, one sees from equations (6.13) and (6.14) that equation (7.2) becomes

$$\frac{1}{a}\frac{d^2a}{dt^2} = -\frac{4}{3}\pi G[\rho_b(t) + 3p_b(t)/c^2] + \Lambda/3, \tag{7.13}$$

and equation (7.8) is replaced with

$$\begin{aligned}\nabla^2\phi &= 4\pi G(\rho + 3p_b/c^2)a^2 - \Lambda a^2 + 3a\,\ddot{a} \\ &= 4\pi G[\rho - \rho_b(t)]a^2.\end{aligned} \tag{7.14}$$

As before, the source for ϕ is the fluctuating part of the nonrelativistic matter density; the homogeneous relativistic part and Λ cancel out.

8. The peculiar acceleration

The solution to equation (7.9) for the potential is

$$\phi(\mathbf{x}) = -Ga^2\int d^3x'\,\frac{\rho(\mathbf{x}') - \rho_b}{|\mathbf{x}' - \mathbf{x}|}. \tag{8.1}$$

Since $\rho(\mathbf{x}) - \rho_b$ is supposed to fluctuate around zero in a statistically uniform way through space (a spatially homogeneous and isotropic random process) with correlation length $\ll cH^{-1}$, the integral converges to a definite value before $|\mathbf{x}' - \mathbf{x}|$ reaches the horizon. Of course, one can add to equation (8.1) a source-free part. The term $\phi_\alpha(t)x^\alpha$ would represent a uniform acceleration, which is removed by choice of the velocity reference frame. The term $\phi_{\alpha\beta}(t)x^\alpha x^\beta$ with $\phi_{\alpha\alpha} = 0$ could represent the tidal field in a homogeneous anisotropic cosmological model; however, the universe is observed to be isotropic to high precision, so this will not be discussed. It could also represent a gravitational wave with long wavelength (and by using a fuller relativistic approximation one can describe gravitational waves with shorter wavelength; § 83).

If one assumes gravitational waves can be ignored, the peculiar acceleration is (eq. 7.12)

$$\mathbf{g}(\mathbf{x}) = -\frac{\nabla\phi}{a} = Ga\int d^3x'(\rho(\mathbf{x}') - \rho_b)\frac{\mathbf{x}' - \mathbf{x}}{|\mathbf{x}' - \mathbf{x}|^3}. \tag{8.2}$$

If we specify that the integral is first over angles at fixed $|\mathbf{x}' - \mathbf{x}|$, then over $|\mathbf{x}' - \mathbf{x}|$, we can rewrite it as

$$\mathbf{g} = Ga \int d^3x' \rho(\mathbf{x}') \frac{\mathbf{x}' - \mathbf{x}}{|\mathbf{x}' - \mathbf{x}|^3}, \tag{8.3}$$

and in the point particle picture, where

$$\rho = \Sigma m_j a^{-3} \delta(\mathbf{x} - \mathbf{x}_j), \tag{8.4}$$

we have

$$\mathbf{g}(\mathbf{x}) = \frac{G}{a^2} \sum m_j \frac{\mathbf{x}_j - \mathbf{x}}{|\mathbf{x}_j - \mathbf{x}|^3}. \tag{8.5}$$

The sum in general is not well-defined; the answer depends on how the terms are ordered. The prescription from equation (8.3) is that the sum is in order of increasing $|\mathbf{x}_j - \mathbf{x}|$. Under the assumption that the particle distribution is a spatially homogeneous and isotropic random process with correlation length $\ll cH^{-1}$, this sum converges to a definite value well within the relativistic horizon.

Because of the boundary condition on the sum the usual law of conservation of motion of the center of mass is modified (Clutton-Brock and Peebles 1980). Consider a model universe that is homogeneous everywhere except inside the surface Σ. The acceleration of particle i in Σ is given by equations (7.12) and (8.5), and this equation summed over all particles in Σ is

$$M \frac{d}{dt} a\mathbf{V} = \frac{d}{dt} \sum m_i a\mathbf{v}_i = \frac{G}{a} \sum_i m_i \sum_j m_j (\mathbf{x}_j - \mathbf{x}_i)/|\mathbf{x}_j - \mathbf{x}_i|^3. \tag{8.6}$$

Because of the conditions on the sums, we cannot exchange the order and conclude that the expression vanishes. We can translate the sums to integrals with density $\rho(x) = \rho_b + \delta\rho(x)$ to get

$$M \frac{d}{dt} a\mathbf{V} = Ga^5 \rho_b \int d^3x \int d^3x' \, \delta\rho(\mathbf{x}')(\mathbf{x}' - \mathbf{x})/|\mathbf{x}' - \mathbf{x}|^3. \tag{8.7}$$

The integrals are over the region of Σ. Here we can exchange the order of integration, and we see that the integral must vanish only in the limit $\Sigma \longrightarrow \infty$.

Finally, let us generalize equation (7.6) to the Lagrangian for all particles in the universe. Equations (8.1) and (8.4) give

$$\mathcal{L} = \Sigma \tfrac{1}{2} m_j a^2 \dot{x}_j^2 - U,$$
$$U = -\frac{Ga^5}{2} \int d^3x_1 d^3x_2 (\rho_1 - \rho_b)(\rho_2 - \rho_b) / |\mathbf{x}_1 - \mathbf{x}_2|, \tag{8.8}$$

where the integral excludes $\mathbf{x}_1 = \mathbf{x}_2$. The factor of $\frac{1}{2}$ appears in U because each particle appears twice, from the integral over $\mathbf{x}_1$ and the integral over $\mathbf{x}_2$.

9. Two Models: the Vlasov Equation and the Ideal Fluid

Two models for the matter are standard and convenient. One can imagine that the particle mean free path is short and matter can be described as an ideal fluid. In the other extreme one supposes that the particles interact only by gravity, so that the mean free path is very long; the particles might be stars or galaxies. This second case will be discussed under the further assumption that the particles can be considered moving in the potential ϕ of a smoothly varying particle density function. The more general case considering the individual interaction of each particle with every other particle is discussed in Chapter IV.

A. Vlasov equation

One imagines a sea of identical particles, each moving without collisions in the potential ϕ of the smooth space density function. The distribution of particles in position and momentum (eq. 7.10) is

$$dN = f(\mathbf{x}, \mathbf{p}, t) d^3x d^3p. \tag{9.1}$$

The proper mass density is

$$\begin{aligned} \rho(\mathbf{x}, t) &= ma^{-3} \int d^3p f(\mathbf{x}, \mathbf{p}, t) \\ &\equiv \rho_b(t)[1 + \delta(\mathbf{x}, t)], \end{aligned} \tag{9.2}$$
$$\rho_b \propto a(t)^{-3}.$$

Here m is the particle mass, the factor a^{-3} is the conversion to proper space density, ρ_b is the mean mass density, and δ is the dimensionless density contrast.

By Liouville's theorem f is constant along a particle trajectory in phase space (eqs. 7.10):

$$\frac{\partial f}{\partial t} + \frac{\mathbf{p}}{ma^2} \cdot \nabla f - m\nabla\phi \cdot \frac{\partial f}{\partial \mathbf{p}} = 0. \tag{9.3}$$

This with equations (9.2) and (7.9) completes the description.

A standard way to deal with equation (9.3) is to take velocity moments. The result of integrating over $\mathbf{p}$ and using equation (9.2) is

$$a^3\rho_b \frac{\partial \delta}{\partial t} + \frac{1}{a^2} \nabla \cdot \int \mathbf{p} f d^3p = 0. \tag{9.4}$$

The last term from equation (9.3) vanishes by integration by parts. The local mean or streaming velocity is

$$\mathbf{v} = \frac{\int (\mathbf{p}/ma) f d^3p}{\int f d^3p}, \tag{9.5}$$

which in equation (9.4) gives

$$\rho_b \frac{\partial \delta}{\partial t} + \frac{1}{a} \nabla \cdot \rho\mathbf{v} = 0. \tag{9.6}$$

The first moment of equation (9.3) is the result of multiplying the equation by $\mathbf{p}$ and integrating over momentum,

$$\frac{\partial}{\partial t} \int p_\alpha f d^3p + \frac{1}{ma^2} \partial_\beta \int p_\alpha p_\beta f d^3p + a^3\rho(\mathbf{x}, t)\phi_{,\alpha} = 0. \tag{9.7}$$

This expression in equation (9.4) yields

$$\frac{\partial^2 \delta}{\partial t^2} + 2\frac{\dot{a}}{a}\frac{\partial \delta}{\partial t} = \frac{1}{a^2} \nabla \cdot [(1 + \delta)\nabla\phi] + \frac{1}{\rho_b m a^7} \partial_\alpha \partial_\beta \int p_\alpha p_\beta f d^3p. \tag{9.8}$$

The mean value of the product $v^\alpha v^\beta$ for the particles found in a small patch around $\mathbf{x}$ is

$$\langle v^\alpha v^\beta \rangle = \frac{\int f p^\alpha p^\beta d^3p}{m^2 a^2 \int f d^3p}, \tag{9.9}$$

and this in equation (9.8) gives

$$\frac{\partial^2\delta}{\partial t^2} + 2\frac{\dot{a}}{a}\frac{\partial\delta}{\partial t} = \frac{1}{a^2}\nabla\cdot[(1+\delta)\nabla\phi] + \frac{1}{a^2}\partial_\alpha\partial_\beta[(1+\delta)\langle v^\alpha v^\beta\rangle]. \quad (9.10)$$

B. Ideal fluid

The standard equations for an ideal fluid are

$$\left(\frac{\partial\rho}{\partial t}\right)_{\mathbf{r}} + \nabla_{\mathbf{r}}\cdot\rho\mathbf{u} = 0,$$

$$\rho\left[\left(\frac{\partial\mathbf{u}}{\partial t}\right)_{\mathbf{r}} + (\mathbf{u}\cdot\nabla_{\mathbf{r}})\mathbf{u}\right] = -\nabla_{\mathbf{r}}p - \rho\nabla_{\mathbf{r}}\Phi. \quad (9.11)$$

The subscript $\mathbf{r}$ indicates the spatial variable is the proper distance $\mathbf{r}$ from some chosen origin, while $\mathbf{u}$ is the proper velocity relative to the origin. This is related to the peculiar velocity $\mathbf{v}$ by (eqs. 7.3, 7.11)

$$\mathbf{u} = \dot{a}\mathbf{x} + \mathbf{v}(\mathbf{x}, t) = (\dot{a}/a)\mathbf{r} + \mathbf{v}(\mathbf{r}/a, t). \quad (9.12)$$

On changing variables from $\mathbf{r}$ to $\mathbf{x} = \mathbf{r}/a$, one sees that the first term in the mass conservation equation (9.11) becomes

$$\left(\frac{\partial}{\partial t}\right)_{\mathbf{r}}\rho(\mathbf{r}/a(t), t) = \frac{\partial\rho}{\partial t} - \frac{\dot{a}}{a}\mathbf{x}\cdot\nabla\rho, \quad (9.13)$$

while equation (9.12) in the second term gives

$$\nabla_{\mathbf{r}}\cdot\rho\mathbf{u} = \frac{1}{a}\nabla\cdot\rho\mathbf{v} + \frac{3\dot{a}}{a}\rho + \frac{\dot{a}}{a}\mathbf{x}\cdot\nabla\rho. \quad (9.14)$$

The sum of equations (9.13) and (9.14) is

$$\frac{\partial\rho}{\partial t} + \frac{3\dot{a}}{a}\rho + \frac{1}{a}\nabla\cdot\rho\mathbf{v} = 0. \quad (9.15)$$

This is the mass conservation equation in the expanding coordinates. The result of applying the same change of variables to the second of equations (9.11) is

$$\ddot{a}\mathbf{x} + \frac{\partial \mathbf{v}}{\partial t} + \frac{1}{a}(\mathbf{v} \cdot \nabla)\mathbf{v} + \frac{\dot{a}}{a}\mathbf{v} = -\frac{1}{\rho a}\nabla p - \frac{1}{a}\nabla\left(\phi - \frac{1}{2}a\ddot{a}x^2\right). \quad (9.16)$$

The potential has been replaced with ϕ (eq. 7.7), and one sees that this cancels the leading term on the left side. The results are the fluid equations in expanding coordinates,

$$\frac{\partial \mathbf{v}}{\partial t} + \frac{1}{a}(\mathbf{v} \cdot \nabla)\mathbf{v} + \frac{\dot{a}}{a}\mathbf{v} = -\frac{1}{\rho a}\nabla p - \frac{1}{a}\nabla\phi,$$
$$\frac{\partial \delta}{\partial t} + \frac{1}{a}\nabla \cdot (1 + \delta)\mathbf{v} = 0. \quad (9.17)$$

The second equation follows from equation (9.15) with the definition of the dimensionless density contrast δ in equation (9.2).

Equations (9.17) can be compared to equations (9.6) and (9.7). As in equation (7.12), there is a term $(\dot{a}/a)\mathbf{v}$ that comes from the expanding coordinates, and the source for the peculiar acceleration $\nabla\phi$ is the density fluctuation $\rho_b\delta$. As was remarked in Section 7, these equations apply when $\Lambda \neq 0$ and there is a uniform background of relativistic matter, where $\rho_b(t)$ is the mean density due to the nonrelativistic particles or fluid.

Equations (9.17) can be combined into one expression that corresponds to equation (9.10). The result of multiplying the first equation by ρ and the second by $\mathbf{v}$ and adding is

$$\frac{\partial}{\partial t}(\rho v^\alpha) + \frac{1}{a}\frac{\partial}{\partial x^\beta}(\rho v^\alpha v^\beta) + 4\rho v^\alpha \dot{a}/a = -\nabla p/a - \rho\nabla\phi/a. \quad (9.18)$$

The divergence of this equation is

$$\frac{\partial^2 \delta}{\partial t^2} + 2\frac{\dot{a}}{a}\frac{\partial \delta}{\partial t} = \frac{\nabla^2 p}{\rho_b a^2} + \frac{1}{a^2}\nabla \cdot (1 + \delta)\nabla\phi$$
$$+ \frac{1}{a^2}\frac{\partial^2}{\partial x^\alpha \partial x^\beta}[(1 + \delta)v^\alpha v^\beta]. \quad (9.19)$$

To apply equations (9.10) or (9.19), one must find some way to deal with the last term in either expression. In the linear approximation discussed next the term is dropped. It can be approximated by using the assumption that the distribution in $\mathbf{p}$ has zero skewness about the mean (van Albada 1960, Davis and Peebles 1977). Other applications of these equations are discussed in Sections 18 and 27.

10. Linear perturbation approximation for δ

It is assumed here that the matter is only slightly perturbed from the background cosmological model. This may be true at some epoch in the early universe, and, as discussed in Section 28 below, the results may give a good description of the behavior of matter on large scales even when there is strongly nonlinear clustering on small scales.

If

$$\delta \ll 1, \qquad (vt/d)^2 \ll \delta, \tag{10.1}$$

where d is the coherence length for spatial variations of δ, v is the characteristic fluid velocity and t is the expansion time $\sim(G\rho_b)^{-1/2}$, equations (9.17) and (9.19) can be reduced to the linear perturbation equations

$$\frac{\partial^2\delta}{\partial t^2} + 2\frac{\dot{a}}{a}\frac{\partial\delta}{\partial t} = \frac{\nabla^2 p}{\rho_b a^2} + 4\pi G\rho_b\delta, \qquad \frac{\partial\delta}{\partial t} + \frac{1}{a}\nabla\cdot\mathbf{v} = 0. \tag{10.2}$$

Equations (9.6) and (9.10) that describe particles with long mean free path become in this approximation.

$$\frac{\partial^2\delta}{\partial t^2} + 2\frac{\dot{a}}{a}\frac{\partial\delta}{\partial t} = 4\pi G\rho_b\delta, \qquad \frac{\partial\delta}{\partial t} + \frac{1}{a}\nabla\cdot\mathbf{v} = 0. \tag{10.3}$$

Here there is, in addition to the streaming velocity $\mathbf{v}$, the rms particle velocity v_0 and from (9.10) the condition

$$v_0 t/d \ll 1. \tag{10.4}$$

In the ideal fluid case there is a microscopic rms particle velocity v_0, and when it exceeds this limit, δ oscillates like an acoustic wave (§ 16). Of course, this is not possible if the particle mean free path is longer than d because the particles move directly across the irregularity and tend to erase δ.

A simple way to derive solutions to equations (10.3) when there is negligible relativistic background mass density is to use Lemaître's observation that in a spherically symmetric perturbation with zero pressure, each mass shell moves like a separate homogeneous world model (§ 3C). Thus the fractional difference between the $\rho(t)$ for homogeneous models with slightly different parameters is a valid $\delta(t)$. Two solutions are generated this way, which is a complete set (Zel'dovich 1965a, Guyot and Zel'dovich 1970).

It is readily seen that this method reproduces equation (10.3). The neighboring world models are $a_b(t)$, $\rho_b(t)$ and $a(t)$, $\rho(t)$ with

$$a = a_b(1 - \epsilon(t)), \qquad \rho a^3 = \rho_b a_b^3, \qquad \delta = \rho/\rho_b - 1 = 3\epsilon. \tag{10.5}$$

The cosmological equation for $a(t)$ is (eq. 7.13)

$$\frac{d^2a}{dt^2} = -\frac{4}{3}\pi G\rho a + \frac{\Lambda}{3}a. \tag{10.6}$$

The result of substituting equations (10.5) in (10.6) and keeping only the terms linear in ϵ is

$$\frac{d^2\epsilon}{dt^2} + 2\frac{\dot{a}_b}{a_b}\frac{d\epsilon}{dt} = 4\pi G\rho_b\epsilon, \tag{10.7}$$

as before. It follows that if $a(t, \alpha)$ represents a family of solutions to equation (10.6), labeled by the parameter α, then a solution to equation (10.3) is

$$\delta \propto \frac{1}{a}\frac{\partial a}{\partial \alpha}. \tag{10.8}$$

Using $\rho_b \propto a^{-3}$, one sees that the first integral of equation (10.6) is

$$\frac{\dot{a}^2}{a^2} = \frac{8}{3}\pi G\rho_b + \frac{\Lambda}{3} - \frac{R^{-2}}{a^2}, \tag{10.9}$$

where the constant of integration is the curvature term R^{-2}. This gives

$$t = \int^a \frac{da}{X^{1/2}} - C, \qquad X = \frac{8}{3}\pi G\rho_b a^3 \frac{1}{a} + \frac{\Lambda a^2}{3} - R^{-2}, \tag{10.10}$$

where C is the second constant of integration. The results of differentiating this equation with respect to the constants of integration, at fixed time, are

$$0 = \frac{1}{X^{1/2}}\frac{\partial a}{\partial R^{-2}} + \frac{1}{2}\int^a \frac{da}{X^{3/2}}, \qquad 0 = \frac{1}{X^{1/2}}\frac{\partial a}{\partial C} - 1, \tag{10.11}$$

and so, according to equation (10.8), the two solutions are (Heath 1977)

$$\delta_1 = \frac{X^{1/2}}{a}\int^a \frac{da}{X^{3/2}}, \qquad \delta_2 = \frac{X^{1/2}}{a}. \tag{10.12}$$

It is a straightforward though tedious calculation to verify by direct substitution that δ_2 is a solution to equation (10.3) (if a satisfies equation (10.9) and $\rho_b \propto a^{-3}$). Then δ_1 is easily checked by writing $\delta_1 = f(a)\,\delta_2(a)$ and substituting in equation (10.3).

It might be noted that the behavior of δ in this linear approximation is local: the evolution at a given spot $\mathbf{x}$ depends only on the initial values of δ and $d\delta/dt$ at $\mathbf{x}$. That is so even though the gravitational field $\mathbf{g}$ depends on an integral over the mass distribution because what is relevant is the divergence of $\mathbf{g}$, which of course depends on the local density. It is evident, therefore, that the same local behavior will be found in the full relativistic theory (Chapter V below). The evolution of δ is nonlocal in second order (§ 18).

11. Solutions for $\delta(t)$: $p = \Lambda = 0$

If the pressure is negligible, equation (10.2) agrees with (10.3) for the density contrast:

$$\frac{\partial^2\delta}{\partial t^2} + 2\frac{\dot{a}}{a}\frac{\partial\delta}{\partial t} = 4\pi G\rho_b\delta. \tag{11.1}$$

The solutions discussed here are for models with $\Lambda = p = 0$ with no relativistic background. Then the cosmological equation (10.9) is

$$\frac{\dot{a}^2}{a^2} = \frac{8}{3}\pi G\rho_b - \frac{R^{-2}}{a^2} = \frac{8}{3}\pi G\rho_b\left[1 + (\Omega_0^{-1} - 1)\frac{a(t)}{a_0}\right]. \tag{11.2}$$

Here a_0 is the present value of the expansion parameter, and the density parameter Ω_0, which is fixed by R^{-2}, is the ratio of the present mean mass density to the mass density in an Einstein-de Sitter universe with the same Hubble constant $H_0 = \dot{a}_0/a_0$ (Appendix, eqs. 97.3, 97.12).

A. Einstein-de Sitter model

At small enough a, where

$$1 + Z \equiv \frac{a_0}{a} \gg |\Omega_0^{-1} - 1|, \tag{11.3}$$

equation (11.2) reduces to the Einstein-de Sitter case

$$\dot{a}^2 = \frac{8}{3}\pi G\rho_b a^2. \tag{11.4}$$

Since $\rho_b \propto a^{-3}$, the solution to this equation is

$$a \propto t^{2/3}, \qquad 6\pi G\rho_b t^2 = 1, \tag{11.5}$$

and equation (11.1) becomes

$$\frac{\partial^2\delta}{\partial t^2} + \frac{4}{3t}\frac{\partial\delta}{\partial t} = \frac{2}{3t^2}\delta. \tag{11.6}$$

This is homogeneous in t, so the solutions are powers of time. On trying $\delta \propto t^n$ in equation (11.6), one finds

$$\delta = A(\mathbf{x})t^{2/3} + B(\mathbf{x})t^{-1}. \tag{11.7}$$

B. Open model

If $\Omega - 1$ is not negligible, one way to proceed is to use equation (11.2) to change the independent variable in equation (11.1) from t to a (Mészáros 1974, Groth and Peebles 1975). With the new variable (eqs. 11.2, 97.20)

$$x = |\Omega^{-1}(t) - 1| = |\Omega_0^{-1} - 1|\frac{a(t)}{a_0} = \frac{3|R^{-2}|}{8\pi G\rho_b(t)a^2}, \tag{11.8}$$

one finds

$$\frac{d^2\delta}{dx^2} + \frac{(3 \pm 4x)}{2x(1 \pm x)}\frac{d\delta}{dx} - \frac{3\delta}{2x^2(1 \pm x)} = 0, \tag{11.9}$$

where the upper sign refers to an open model $\Omega < 1$, the lower to a closed model $\Omega > 1$. In the closed model $x = 1$ at maximum expansion, $\dot{a} = 0$ (eq. 11.2). In the open model $x = 1$ marks the transition from the early stage where the expansion is close to the Einstein-de Sitter model and the late stage of nearly free expansion.

The solutions to equation (11.9) can be expressed as hypergeometric functions, but it is easier to use equation (10.8) and the parametric solution for $a(t)$ to get the growing solution.

For an open model the parametric solution is

$$a = A(\cosh\eta - 1), \qquad t = B(\sinh\eta - \eta), \tag{11.10}$$

where A and B are constants. This expression in equation (11.2) gives

$$A = {}^4\!/_3\, \pi G \rho_b a^3 |R|^2, \qquad B = A|R| \propto |R|^3, \tag{11.11}$$

so equation (11.8) is (eq. 97.23)

$$x = \frac{a}{2A} = \frac{\cosh\eta - 1}{2} \equiv \Omega(\eta)^{-1} - 1, \tag{11.12}$$

where Ω is the density parameter at epoch η. With some manipulation one can express this equation in the forms

$$\cosh\eta = 2x + 1, \qquad \sinh\eta = 2x^{1/2}(1 + x)^{1/2},$$

$$\cosh(\eta/2) = (1 + x)^{1/2}, \qquad \sinh(\eta/2) = x^{1/2}, \tag{11.13}$$

$$\eta = 2\ln[(1 + x)^{1/2} + x^{1/2}] = -2\ln[(1 + x)^{1/2} - x^{1/2}].$$

The result of differentiating the second of equations (11.10) with respect to $|R|$ at fixed t and using equation (11.11) is

$$\frac{\partial\eta}{\partial|R|} = -\frac{3}{|R|}\frac{\sinh\eta - \eta}{\cosh\eta - 1}, \tag{11.14}$$

and so the first of equations (11.10) gives

$$\delta \propto \frac{1}{a}\frac{\partial a}{\partial|R|} \propto D_1 = \frac{3\sinh\eta\,(\sinh\eta - \eta)}{(\cosh\eta - 1)^2} - 2. \tag{11.15}$$

This is the growing solution in terms of the parameter η of equation (11.10). It was derived by Weinberg (1972) and Edwards and Heath (1976). With equations (11.13) the variable can be changed from η to $x \propto a(t)$ to obtain the form derived by Groth and Peebles (1975),

$$D_1(t) = 1 + \frac{3}{x} + \frac{3(1 + x)^{1/2}}{x^{3/2}}\ln[(1 + x)^{1/2} - x^{1/2}]. \tag{11.16}$$

The decaying solution follows directly from equation (10.12). From equations (11.2) and (11.8), we have

$$X = \dot{a}^2 \propto (1 + x)/x, \qquad a \propto x,$$

so

$$\delta \propto D_2 = (1 + x)^{1/2}/x^{3/2}. \tag{11.17}$$

This equation was derived by Guyot and Zel'dovich (1970).

At $x \ll 1$, where the expansion approximates the Einstein-de Sitter model, equations (11.16) and (11.17) are

$$D_1 = \frac{2x}{5} = \frac{2}{5}\left(\frac{3\tau}{2}\right)^{2/3}, \qquad D_2 = \frac{1}{x^{3/2}} = \frac{2}{3\tau}, \qquad x \ll 1, \tag{11.18}$$

where τ is the dimensionless time

$$\tau = \frac{3t}{8\pi G\rho a^3 |R|^3} = \frac{\sinh \eta - \eta}{2}. \tag{11.19}$$

Equations (11.18) agree with the Einstein-de Sitter case in equation (11.7), as expected. For $x \gg 1$ where the model is freely expanding, the solutions approach

$$D_1 = 1, \qquad D_2 = x^{-1} = \tau^{-1}; \qquad x \gg 1. \tag{11.20}$$

In this limit the density perturbation has stopped growing. Thus starting in the very early universe, where the density parameter is $\Omega_i \approx 1$ and $x_i = \Omega_i^{-1} - 1$ (eq. 11.8), one finds the total factor by which the growing mode increases is

$$\frac{\delta(\infty)}{\delta_i} = \frac{5}{2x_i} = \frac{5}{2(\Omega_i^{-1} - 1)}. \tag{11.21}$$

The universe expands by the factor $\sim(\Omega_i^{-1} - 1)^{-1}$ before Ω appreciably differs from unity and δ grows by a like factor.

It is interesting to compare the solutions in equations (11.16) and (11.17) to the power law solutions in equation (11.18) when $x \to 1$ and the model starts to deviate from the Einstein-de Sitter model. At $x = 1$, $\Omega = 0.5$ (eq. 11.12), half the Einstein-de Sitter density, $\eta = 1.763$ (eq. 11.12), $\tau = 0.533$ (eq. 11.19), and the small x approximations evaluated at this time give

$$D_1 \approx 0.344, \qquad D_2 \approx 1.25, \tag{11.22}$$

quite close to the exact solutions in equations (11.16) and (11.17) at the same τ,

$$D_1 = 0.261, \qquad D_2 = 1.41. \tag{11.23}$$

During the expansion from $x = 1$, $\Omega = 0.5$ to $x \to \infty$, $\Omega \to 0$, the growing solution D_1 increases by the additional factor $(0.261)^{-1} = 3.8$. At $x = 9$, $\Omega = 0.1$, the time is $\tau = 7.7$, and $D_1 = 0.69$, 30 percent of its final value. The growth of linear density fluctuation thus is quite strongly suppressed once Ω falls below 0.1.

C. Closed model

The equations describing a closed model are obtained by setting

$$\begin{gathered} \eta = i\eta', \qquad A = -A', \qquad B = iB', \\ R = iR', \qquad x = -x'. \end{gathered} \tag{11.24}$$

The first two lines of equation (11.13) give

$$\begin{aligned} \sin \eta &= \pm 2(x(1-x))^{1/2}; \quad \tan \eta/2 = \pm (x/(1-x))^{1/2}; \\ \eta &= 2 \tan^{-1}[x/(1-x)]^{1/2}; \quad 2\pi - 2\tan^{-1}[x/(1-x)]^{1/2}. \end{aligned} \tag{11.25}$$

In each equation the first case applies in the expanding phase, $0 < \eta < \pi$, the second in the contracting phase, $\pi < \eta < 2\pi$. On using equations (11.24) and (11.25) in equation (11.15) and reversing the sign to make the solution positive, one finds the growing solution (Groth and Peebles 1975)

$$\begin{aligned} \delta \propto D_1 &= -1 + \frac{3}{x} - \frac{3(1-x)^{1/2}}{x^{3/2}} \tan^{-1}\left(\frac{x}{1-x}\right)^{1/2}, \qquad 0 < \eta < \pi, \\ &= -1 + \frac{3}{x} - \frac{3(1-x)^{1/2}}{x^{3/2}} \left[\tan^{-1}\left(\frac{x}{1-x}\right)^{1/2} - \pi\right], \qquad \pi < \eta < 2\pi. \end{aligned} \tag{11.26}$$

An initially decaying solution follows directly from equations (11.17) and (11.24),

$$\delta \propto D_2 = \pm(1-x)^{1/2}/x^{3/2}, \tag{11.27}$$

where again the upper sign applies in the expanding phase, the lower sign in the contracting phase. A solution that decays to zero at the moment of the final singular collapse is

$$D_3 = D_1 + 3\pi D_2. \tag{11.28}$$

This is the first solution D_1 reversed in time.

The limiting value of D_1 at small x given in equation (11.18) applies here as well. At the epoch of maximum expansion, $x = 1$, $\tau = \pi/2$. This approximate solution gives

$$D_1 \approx \frac{2}{5}\left(\frac{3\tau}{2}\right)^{2/3} = 0.71, \tag{11.29}$$

while the exact solution in equation (11.26) is

$$D_1(x = 1) = 2. \tag{11.30}$$

During the collapsing phase as $x \longrightarrow 0$ and the universe approaches the singularity, D_1 diverges as

$$D_1 \approx \frac{3\pi}{x^{3/2}} \approx \frac{2\pi}{\pi - \tau}. \tag{11.31}$$

12. Solutions for $\delta(t)$: effect of a uniform radiation background

There is a very uniform microwave radiation background (presumably the primeval fireball radiation left over from the early universe, PC, Chapter V), and there could also be a uniform background of degenerate zero mass neutrinos or gravitational radiation. This background energy is very weakly coupled to matter, and the irregular matter distribution on scales $\ll ct$ causes negligible gravitational perturbation in it because the potential is $|\phi| \ll c^2$. However, it does affect the development of irregularities in the matter distribution because it speeds the expansion of the universe.

The effect is computed here under the simplifying assumption that the universe is cosmologically flat or close to it, R^{-2} negligible in equation (11.2). It is assumed also that the scale of irregularities is $\lambda \ll ct$, so the background energy can be taken to be accurately homogeneous. The case $\lambda \gtrsim ct$ is discussed in Chapter V (§ 86).

The expansion rate is

$$\frac{\dot{a}^2}{a^2} = \frac{8}{3}\pi G(\mu + \mathscr{E}),$$
$$\mu \propto a^{-3}, \qquad \mathscr{E} \propto a^{-4}. \tag{12.1}$$

Here μ is the mean density of the zero pressure matter and $\mathscr{E}$ is the mass density in the relativistic background. Because of the cosmological redshift, the energy of each photon (or zero mass neutrino or graviton) varies as $a(t)^{-1}$ and the number of quanta per unit proper volume varies as $a(t)^{-3}$, giving $\mathscr{E} \propto a^{-4}$ (eqs. 97.15, 97.16).

If the matter pressure is negligible and there is no drag force between matter and radiation, the linear perturbation equation is the same as before,

$$\frac{\partial^2 \delta}{\partial t^2} + 2\frac{\dot{a}}{a}\frac{\partial \delta}{\partial t} = 4\pi G\mu\delta, \tag{12.2}$$

where $\delta = \delta\mu/\mu$ is the fractional density contrast in the nonrelativistic matter. A convenient change of variables is

$$y = \mu/\mathscr{E} = a/a_{eq}, \tag{12.3}$$

where a_{eq} is the value of the expansion parameter at the epoch when $\mu = \mathscr{E}$. At $y \ll 1$, $\mathscr{E}$ fixes the expansion rate; at $y \gg 1$, μ dominates. On changing the independent variable in equation (12.2) from t to a to y, by using equations (12.1) and (12.3), one finds (Mészáros 1974, Groth and Peebles 1975)

$$\frac{d^2\delta}{dy^2} + \frac{2 + 3y}{2y(1 + y)}\frac{d\delta}{dy} - \frac{3\delta}{2y(1 + y)} = 0. \tag{12.4}$$

This equation cannot be solved by the trick used in Section 11 because the radiation does not follow the perturbation to the matter. (In the limit $\lambda \gg ct$ the trick does generate solutions; § 86). One can find a solution by trying $d^2\delta/dy^2 = 0$. The result is Mészáros' solution

$$\delta \propto D_1 = 1 + 3y/2. \tag{12.5}$$

This is the growing perturbation. One can then find the second solution by writing

$$\delta = f(y)D_1(y). \tag{12.6}$$

This reduces equation (12.4) to

$$\frac{f''}{f'} = -\frac{2}{D_1}\frac{dD_1}{dy} - \frac{2 + 3y}{2(1 + y)}, \tag{12.7}$$

or

$$f = -\int dy/[y(1 + 3y/2)^2(1 + y)^{1/2}], \tag{12.8}$$

which can be integrated, giving

$$\delta \propto D_2 = \left(1 + \frac{3y}{2}\right) \ln \left[\frac{(1 + y)^{1/2} + 1}{(1 + y)^{1/2} - 1}\right] - 3(1 + y)^{1/2}. \tag{12.9}$$

This is the decaying solution found by Groth and Peebles (1975).

The limiting values of D_2 are

$$\begin{aligned} D_2 &= \ln\left(\frac{4}{y}\right) - 3 = \frac{1}{2}\ln\left(\frac{8}{\tau}\right) - 3, \qquad y \ll 1; \\ D_2 &= \frac{4}{15y^{3/2}} = \frac{8}{45\tau}, \qquad y \gg 1, \\ \tau &= (8\pi G\mu^4/3\mathscr{E}^3)^{1/2}t. \end{aligned} \tag{12.10}$$

The limiting value of D_2 for $y \gg 1$ is most simply derived from equation (12.8).

The growing solution in equation (12.5) at large y is

$$D_1 = \frac{3y}{2} = \left(\frac{3}{2}\right)^{5/3} \tau^{2/3}, \tag{12.11}$$

which agrees with the Einstein-de Sitter model (eq. 11.7). At $y < 1$ the growing mode is nearly constant. This is because $\mu < \mathscr{E}$ so the time-scale for expansion $\sim(G\mathscr{E})^{-1/2}$ is less than the time-scale $\sim(G\mu)^{-1/2}$ for growth of irregularities in the matter distribution. The uniform background radiation thus stabilizes the matter distribution at high redshift, $\mathscr{E} > \mu$, even though matter and radiation are not coupled (as long as the scale of the irregularities is $\lambda \ll ct$; § 86).

We can treat an open cosmological model that makes the transition from radiation to matter dominated well before Ω appreciably deviates from unity by joining this solution to the one in Section 11B. In the interval when both are valid, equations (12.3) and (12.5) give

$$\delta = \left(1 + \frac{3y}{2}\right)\delta_i \simeq \frac{3}{2}\frac{a}{a_{eq}}\delta_i, \qquad (12.12)$$

and equations (11.8) and (11.18) give

$$\delta \propto D_1 = \frac{2x}{5} = \frac{2}{5}(\Omega_0^{-1} - 1)\frac{a}{a_0}. \qquad (12.13)$$

The solution valid in the matter-dominated phase is therefore

$$\delta = \frac{15}{4}\frac{1 + Z_{eq}}{\Omega_0^{-1} - 1} D_1(t)\delta_i, \qquad (12.14)$$

where (eq. 97.17)

$$1 + Z_{eq} = \frac{a_0}{a_{eq}} = 4.2 \times 10^4 \Omega_0 h^2. \qquad (12.15)$$

This is the redshift at equal densities of matter and radiation. The factor D_1 is given as a function of Ω_0 by equations (11.8) and (11.16). A rough lower limit to the range of values of Ω_0 now under discussion is $\Omega_0 = 0.03$, which gives $D_1 = 0.86$, and, with equations (12.14) and (12.15),

$$\delta \sim 130 h^2 \delta_i. \qquad (12.16)$$

If h is at the lower end of the range of current discussions, $h \sim 0.5$, the growth factor is only $\delta/\delta_i \sim 30$. If there is an appreciable contribution to the background density by diverse zero mass neutrinos left over from thermal equilibrium at high redshift, this number would be even smaller. Thus depending on cosmological parameters, the formation of galaxies and clusters of galaxies may have received only quite modest assistance from the linear growth of perturbations during the matter-dominated phase (Guyot and Zel'dovich 1970, Groth and Peebles 1975).

13. Solutions for $\delta(t)$: models with $\Lambda \neq 0$

Following the remark by Petrosian, Salpeter, and Szekeres (1967) and Shklovsky (1967) that the apparent peak in the distribution of quasar redshifts at $Z \sim 2$ might come from the quasi-static phase of a Lemaître model universe (§ 4C), there was a flurry of interest in the gravitational instability of this model (Byalko 1969, Brecher and Silk 1969, Nariai 1969, Rawson-Harris 1969). Two other cases of historical interest are the static Einstein model and the expanding Eddington model that asymptotically approaches the Einstein model at $t \rightarrow -\infty$.

It will be assumed that matter pressure may be neglected and that any radiation background is negligible. In the Eddington and Lemaître models the interesting question concerns the behavior of $\delta\rho/\rho$ through the quasi-static phase. Since it is thought that the redshift at this phase could not be much greater than unity, the energy density in the microwave-submillimeter background is negligible.

The Einstein model is a special case because the background is not expanding. Equation (10.3) becomes

$$\frac{\partial^2 \delta}{\partial t^2} = 4\pi G \rho_b \delta, \qquad \delta = \exp \pm (4\pi G \rho_b)^{1/2} t = \exp \pm t\, \Lambda^{1/2}. \tag{13.1}$$

As was pointed out by Tolman (1934a), this model is strongly unstable. The characteristic time for growth of fluctuations is the quantity $(4\pi G \rho_b)^{-1/2}$ supplied by the cosmological model.

For the Eddington and Lemaître models it is convenient to introduce some notation. In the Lemaître model there is an inflection point at expansion parameter $a = a_e$ where $\ddot{a} = 0$, $\dot{a} > 0$. Since

$$\frac{\ddot{a}}{a} = -\frac{4}{3}\pi G \rho + \frac{\Lambda}{3}, \tag{13.2}$$

the density at the inflection point is

$$\rho_e \equiv \Lambda/(4\pi G). \tag{13.3}$$

Convenient dimensionless quantities are

$$x = a/a_e, \qquad 1 - \epsilon = R^{-2}/\Lambda a_e^2. \tag{13.4}$$

If the model is always expanding, $\epsilon \geq 0$. With these definitions X (eq. 10.10) becomes

$$\dot{a} = X^{1/2}, \qquad X = \frac{\Lambda a_e^2}{3}\left[\frac{2}{x} + x^2 - 3(1 - \epsilon)\right], \tag{13.5}$$

so the solutions in equations (10.12) are

$$\begin{aligned} \delta \propto D_2 &= [x^3 - 3(1-\epsilon)x + 2]^{1/2}/x^{3/2}, \\ \delta \propto D_1 &= D_2(x) \int^x dx x^{3/2} [x^3 - 3(1-\epsilon)x + 2]^{-3/2}. \end{aligned} \tag{13.6}$$

The Eddington model is the limiting case $\epsilon = 0$. Here the growing solution is

$$D_2 = (x-1)(x+2)^{1/2} x^{-3/2}, \tag{13.7}$$

with the limiting values

$$\begin{aligned} D_2 &= 3^{1/2}(x-1) = 3^{1/2}(a-a_e)/a_e, \qquad a - a_e \ll a_e; \\ &= 1, \qquad a \gg a_e. \end{aligned} \tag{13.8}$$

In the solution D_1 the lower bound on the integral is $x_0 > 1$ because the expression diverges at $x = 1$. By admixing D_1 and D_2, we can get the new decaying solution

$$\begin{aligned} D_3(x) &= D_2(x) \int_x^\infty \frac{dx x^{3/2}}{(x-1)^3(x+2)^{3/2}} \\ &= \frac{1}{18}\left[\frac{4x^2 - 3x + 2}{x(x-1)} - 4D_2 \right. \\ &\quad \left. - \frac{D_2}{3^{1/2}} \ln\left(\frac{3^{1/2}-1}{3^{1/2}+1} \cdot \frac{3^{1/2} + (1+2/x)^{1/2}}{3^{1/2} - (1+2/x)^{1/2}}\right)\right]. \end{aligned} \tag{13.9}$$

This has limiting values

$$D_3 = \frac{1}{6(x-1)}, \qquad x \sim 1; \qquad D_3 = \frac{1}{2x^2}, \qquad x \gg 1. \tag{13.10}$$

Another form for these solutions is given by Edwards and Heath (1976).

In the Lemaître model $\epsilon > 0$. Here D_1 is a growing solution, while D_2 decays at $x \ll 1$ (and reaches a minimum at $x \approx 1$, grows to $D_2 = 1$ at $x \gg 1$). On setting the lower bound in the integral to $x = 0$, one finds the limiting expressions:

$$D_1 = x/5, \qquad D_2 = 2^{1/2}/x^{3/2}, \qquad x \ll 1;$$
$$D_1 = \int_0^\infty \frac{dx x^{3/2}}{[x^3 - 3(1-\epsilon)x + 2]^{3/2}}, \qquad D_2 = 1, \qquad x \gg 1; \tag{13.11}$$
$$D_2 = (3\epsilon)^{1/2}, \qquad x = 1.$$

If $\epsilon \ll 1$, corresponding to a long dwell time in the quasi-static phase, the main contribution to the integral is at $x \approx 1$. Then on writing $x = 1 + y$ and expanding the expression in brackets to order y^2, one finds

$$D_1 = \frac{2}{3^{3/2}\epsilon}, \qquad x \gg 1, \qquad \epsilon \ll 1. \tag{13.12}$$

If the density contrast in the growing mode is δ_i at some epoch $x_i \ll 1$ well before the quasi-static phase, then the final contrast at $x_f \gg 1$, well after, is (Byalko 1969)

$$\delta_f = \frac{10}{3^{3/2}\epsilon x_i}\,\delta_i. \tag{13.13}$$

This equation says the initial perturbation grows up to the quasi-static phase by the factor $\sim x_i^{-1} = a_e/a_i$, about the same as in the Einstein-de Sitter model, gains the factor ϵ^{-1} during the quasi-static period, then stops growing when $4\pi G\rho_b$ falls substantially below Λ.

One way to measure ϵ is in terms of the dwell time t_e in the quasi-static phase. The time elapsed from $x = 0$, at the time of the big bang, to $x = x_f \gg 1$ is (eq. 13.5)

$$t = (3/\Lambda)^{1/2} \int_0^{x_f} x^{1/2}\, dx/[x^3 - 3\,(1-\epsilon)\,x + 2]^{1/2}. \tag{13.14}$$

On differentiating this with respect to ϵ and then evaluating the integral in the same way as for equation (13.12), one finds

$$\partial t/\partial\epsilon = -(\Lambda^{1/2}\,\epsilon)^{-1}. \tag{13.15}$$

Thus a measure of the dwell time is

$$t_e = -(\ln \epsilon)/\Lambda^{1/2}. \tag{13.16}$$

By equation (13.13) the growth factor in the quasi-static phase is $\sim \epsilon^{-1}$, and equation (13.16) indicates

$$\epsilon^{-1} = \exp(\Lambda^{1/2} t_e), \tag{13.17}$$

which agrees with equation (13.1). This result was obtained by Brecher and Silk (1969).

As Lemaître pointed out, there can be substantial growth of irregularities during the quasi-static phase, and the growth is strongly suppressed once $\Lambda \gg 4\pi G\rho_b$. Opinions on whether the growth factor might reasonably be expected to be large may differ. If one supposes that the dwell time has been appreciably longer than $\Lambda^{-1/2}$, one certainly does find a large growth factor ϵ^{-1} (eq. 13.17). On the other hand, if ϵ^{-1} is a very large number, the space curvature Ra_e at the point of inflection of $a(t)$ must be arranged to differ from $\Lambda^{-1/2}$ by the exceedingly small fractional amount $\epsilon/2$ (eq. 13.4).

In all three of these models thermal fluctuations can develop into galaxies. For example, one could imagine starting with an Einstein model with the density as uniform as possible consistent with the discrete nature of matter. Gravitationally bound clumps of a few particles each would form; these would tend to move and initiate formation of larger clumps and so on to objects of galaxy mass after a time period that is a modest multiple of $\Lambda^{-1/2}$. According to the linear perturbation analysis, there is no reason why this process should initiate a general uniform expansion of the universe: to explain that it appears one must imagine that at the start x (eq. 13.4) was very slightly greater than unity (§ 3C).

An interesting problem with these models is that they present us with a characteristic density ρ_e (eq. 13.3). An object that forms during the static or quasi-static phase would have density on the order of ρ_e independent of size because only a modest collapse is needed to satisfy the virial theorem. Though the density in a bound system could increase through relaxation, it would seem to be difficult to account for the wide range of densities observed in the distribution of galaxies—from compact groups to clusters to superclusters—in a model where all the objects form at a particular density (§ 4B).

14. The peculiar velocity field

The peculiar velocity field $\mathbf{v}(\mathbf{x}, t)$ in the linear perturbation approximation with zero pressure satisfies the equations

$$\frac{\partial \mathbf{v}}{\partial t} + \frac{\dot{a}}{a}\mathbf{v} = \mathbf{g}, \qquad \frac{\nabla \cdot \mathbf{v}}{a} + \frac{\partial \delta}{\partial t} = 0, \tag{14.1}$$

where $\mathbf{g}$ is the peculiar gravitational acceleration,

$$\mathbf{g}(\mathbf{x}) = G\rho_b a \int d^3x'\delta(\mathbf{x}', t)(\mathbf{x}' - \mathbf{x})/|\mathbf{x}' - \mathbf{x}|^3. \tag{14.2}$$

These are obtained from equations (8.2) and (9.2), with (9.5), (9.6), and (9.7), for the streaming velocity in the free particle case, and from equations (9.17) for the ideal fluid model. The solution is

$$\mathbf{v} = a\frac{\partial}{\partial t}\left(\frac{\mathbf{g}}{4\pi G\rho_b a}\right) + \frac{\mathbf{F}(\mathbf{x})}{a(t)}, \qquad \nabla \cdot \mathbf{F} = 0, \tag{14.3}$$

where the second term is the homogeneous part. One sees by direct substitution that this satisfies the second of equations (14.1). The expression in the first of equations (14.1) gives

$$\frac{\partial}{\partial t}a^2\frac{\partial}{\partial t}\frac{\mathbf{g}}{\rho_b a} = 4\pi G a\mathbf{g}. \tag{14.4}$$

The perturbation δ is a sum of two terms, each of which vary as a solution $D_\alpha(t)$ to equation (10.3), so equation (14.2) indicates $\mathbf{g}$ is a sum of two terms that vary as

$$\mathbf{g}_\alpha \propto \rho_b a D_\alpha, \tag{14.5}$$

and this in equation (14.4) reproduces equation (10.3).

The velocity associated with each mode is, according to equations (14.3) and (14.5),

$$\mathbf{v}_\alpha = \frac{\mathbf{g}_\alpha}{4\pi G\rho_b}\frac{1}{D_\alpha}\frac{dD_\alpha}{dt}. \tag{14.6}$$

In the Einstein-de Sitter model the velocity fields in the growing and decaying modes are

$$\begin{aligned} D_1 &\propto t^{2/3}, & \mathbf{v} &= \mathbf{g}t \propto t^{1/3}; \\ D_2 &\propto t^{-1}, & \mathbf{v} &= -\tfrac{3}{2}\,\mathbf{g}t \propto t^{-4/3}. \end{aligned} \tag{14.7}$$

In the growing mode the peculiar velocity grows as $t^{1/3}$. In the decaying mode $\mathbf{v}$ is in the opposite direction to the peculiar acceleration, the matter moving so as to cancel the density irregularity.

A convenient form of equation (14.6) is

$$\mathbf{v} = \frac{Hf\mathbf{g}}{4\pi G\rho_b} = \frac{2f\mathbf{g}}{3\,H\Omega}, \qquad f = \frac{a}{D}\frac{dD}{da}, \tag{14.8}$$

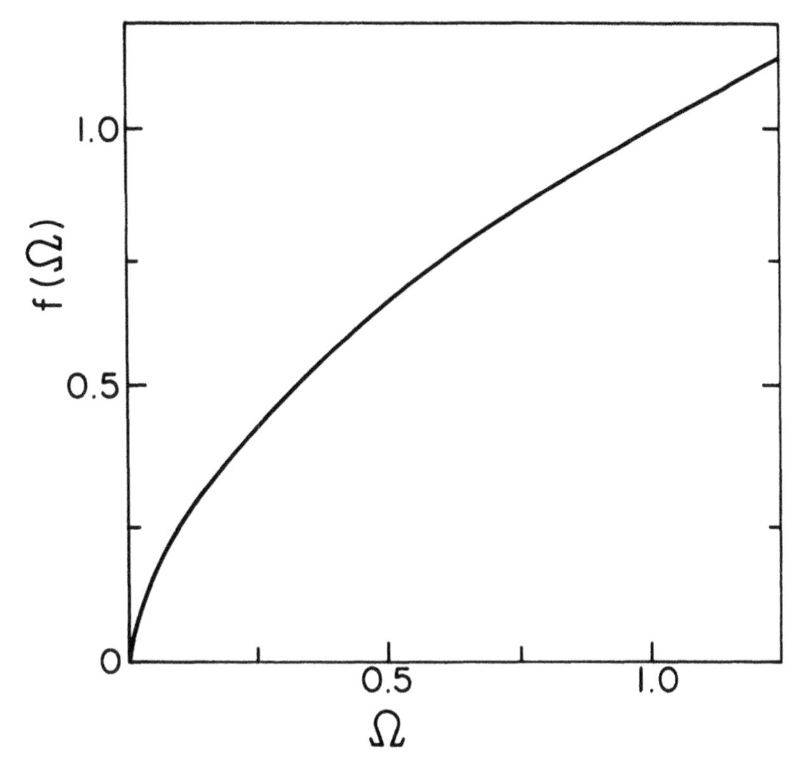

FIG. 14.1. Perturbation parameter f as a function of the density parameter Ω (from Peebles 1976a; published by The University of Chicago Press; copyright 1976 by the American Astronomical Society).

where H and Ω are the Hubble constant and density parameter (eqs. 97.3, 97.12). If the growing mode dominates and the relativistic background may be ignored, then f is given as a function of Ω by equations (11.8, 11.16, 11.26); the results are plotted in figure 14.1 (Peebles 1976a). A useful analytic approximation is $f(\Omega) = \Omega^{0.6}$. Finally, we note that equations (14.2) and (14.8) give

$$v_\alpha = \frac{Haf}{4\pi}\partial_\alpha \int d^3x'\delta(x')/|\mathbf{x}' - \mathbf{x}|. \tag{14.9}$$

These equations show the relation between the large-scale irregularities in density and the large-scale peculiar velocity field $\mathbf{v}$. With some improvement in data it may prove possible to test the relation and find a useful estimate of Ω. Preliminary applications of this concept have been discussed by van Albada (1960), Kantowski (1969), Silk (1974a), Peebles (1976a), and in Sections 76 to 78 below.

On squaring equation (14.9), averaging over position $\mathbf{x}$, and integrating by parts, one finds

$$\langle v^2 \rangle = (Haf)^2 \int_0^\infty y\, dy\, \xi(y), \tag{14.10}$$

where the dimensionless autocorrelation function is

$$\xi(y) = \langle \delta(\mathbf{x}) \delta(\mathbf{x} + \mathbf{y}) \rangle. \tag{14.11}$$

The generalization of this to nonlinear fluctuations is discussed in Section 24.

The shear of the peculiar velocity field is

$$\begin{aligned} \sigma_{\alpha\beta} &= \frac{(v^\alpha{}_{,\beta} + v^\beta{}_{,\alpha})}{2a} - \frac{\delta_{\alpha\beta}}{3} \frac{\nabla \cdot \mathbf{v}}{a} \\ &= \frac{Hf}{4\pi} \left(\frac{\partial}{\partial x^\alpha} \frac{\partial}{\partial x^\beta} - \frac{1}{3} \delta_{\alpha\beta} \nabla^2 \right) \int d^3x' \delta(\mathbf{x}')/|\mathbf{x} - \mathbf{x}'|. \end{aligned} \tag{14.12}$$

On squaring this, averaging over x, and integrating by parts, one finds

$$\langle \sigma_{\alpha\beta} \sigma_{\alpha\beta} \rangle = \tfrac{2}{3} (Hf)^2 \langle \delta(\mathbf{x}, t)^2 \rangle. \tag{14.13}$$

This was derived for an Einstein-de Sitter model by Silk (1974c). The general shear equation is discussed in Section 22 below.

15. Joining conditions for δ and $\mathbf{v}$

If the density contrast δ_i and peculiar velocity field $\mathbf{v}_i$ are given at some epoch t_i, perhaps the epoch of decoupling of matter and radiation, then in linear perturbation theory the density perturbation is carried forward in time by the equation

$$\delta = A(\mathbf{x}) D_1(t) + B(\mathbf{x}) D_2(t), \tag{15.1}$$

and $\mathbf{v}$ is carried forward by equation (14.3). The functions A and B are fixed by δ_i and $\mathbf{v}_i$ by the equations

$$\delta_i = AD_1(i) + BD_2(i),$$
$$-\nabla \cdot \mathbf{v}_i/a_i = A\frac{dD_1(i)}{dt} + B\frac{dD_2(i)}{dt}. \tag{15.2}$$

The argument i means that D and its derivative are evaluated at the starting time t_i. Equations (15.1) and (15.2) give

$$\delta = \frac{\delta_i}{E}\left[D_1(t)\frac{dD_2(i)}{dt_i} - D_2(t)\frac{dD_1(i)}{dt_i}\right]$$
$$+ \frac{\nabla \cdot \mathbf{v}_i}{a_i E}[D_1(t)D_2(i) - D_2(t)D_1(i)], \tag{15.3}$$
$$E = D_1(i)\frac{dD_2(i)}{dt_i} - D_2(i)\frac{dD_1(i)}{dt_i}.$$

This expression in (14.9) gives the velocity fields associated with the growing and decaying modes. In the result there is the factor

$$\int d^3x' \frac{\mathbf{x}' - \mathbf{x}}{|\mathbf{x}' - \mathbf{x}|^3}\nabla' \cdot \mathbf{v}_i(\mathbf{x}') = -4\pi\mathbf{v}_i(\mathbf{x}), \tag{15.4}$$

where the equation follows if $\mathbf{v}_i$ has zero circulation, $\nabla \times \mathbf{v}_i = 0$, because the curl of the left-hand side vanishes. It is convenient, therefore, to write the velocity as the sum of two parts that have zero curl and zero divergence, respectively

$$\mathbf{v}_i = \mathbf{v}_i^p + \mathbf{v}_i^r. \tag{15.5}$$

Then equation (15.4) applies to the part $\mathbf{v}^p$, and the part $\mathbf{v}_i^r$ is the homogeneous term $\mathbf{F}/a_i$ in equation (14.3). The result is

$$\mathbf{v} = \frac{a(t)}{4\pi}\int d^3x'\delta_i(\mathbf{x}')\frac{\mathbf{x}' - \mathbf{x}}{|\mathbf{x}' - \mathbf{x}|^3}\frac{1}{E}\left[\frac{dD_1(t)}{dt}\frac{dD_2(i)}{dt_i} - \frac{dD_2(t)}{dt}\frac{dD_1(i)}{dt_i}\right]$$
$$+ \mathbf{v}_i^p\frac{a(t)}{a_i}\frac{1}{E}\left[\frac{dD_2(t)}{dt}D_1(i) - \frac{dD_1(t)}{dt}D_2(i)\right] + \mathbf{v}_i^r a_i/a(t). \tag{15.6}$$

In the Einstein-de Sitter model these equations are

$$E = -\frac{5}{3}\frac{D_1(i)D_2(i)}{t_i},$$

$$\begin{aligned} \delta &= \frac{3}{5}\left(\frac{t}{t_i}\right)^{2/3}\left(\delta_i - t_i\frac{\nabla\cdot\mathbf{v}_i}{a_i}\right) + \frac{2}{5}\left(\frac{t_i}{t}\right)\left(\delta_i + \frac{3}{2}t_i\frac{\nabla\cdot\mathbf{v}_i}{a_i}\right), \\ \mathbf{v} &= \frac{2}{5}\left(\frac{t}{t_i}\right)^{1/3}\left[\mathbf{v}_i^p + \frac{a_i}{4\pi t_i}\int d^3x'\delta_i(\mathbf{x}')\frac{\mathbf{x}'-\mathbf{x}}{|\mathbf{x}'-\mathbf{x}|^3}\right] \\ &\quad + \frac{3}{5}\left(\frac{t_i}{t}\right)^{4/3}\left[\mathbf{v}_i^p - \frac{2}{3}\frac{a_i}{4\pi t_i}\int d^3x'\delta_i(\mathbf{x}')\frac{\mathbf{x}'-\mathbf{x}}{|\mathbf{x}'-\mathbf{x}|^3}\right] \\ &\quad + \left(\frac{t_i}{t}\right)^{2/3}\mathbf{v}_i^r. \end{aligned} \tag{15.7}$$

If at t_i the velocity is $\mathbf{v}_i = 0$, the first of these equations indicates that 3/5 of δ_i is in the growing mode. It may be that prior to decoupling the matter-radiation fluid was turbulent. Then assuming the turbulence was subsonic, one might expect that $\delta_i \approx 0$ and $\nabla \cdot \mathbf{v}_i \approx 0$ but that $\mathbf{v}_i$ is large. The amplitude of the growing mode in this case is given by equations (21.30, 21.31) below.

16. Critical Jeans Length

It is assumed here that matter can be approximated as an ideal fluid with pressure a function of density alone. Then to first order

$$p = p(\rho_b) + (c_s)^2\rho_b\delta, \qquad c_s^2 = dp/d\rho, \tag{16.1}$$

and equation (10.2) becomes

$$\frac{\partial^2\delta}{\partial t^2} + 2\frac{\dot a}{a}\frac{\partial\delta}{\partial t} = \left(\frac{c_s}{a}\right)^2\nabla^2\delta + 4\pi G\rho_b\delta. \tag{16.2}$$

Since the coefficients in this equation are independent of $\mathbf{x}$, the solution can be written as a sum of plane waves,

$$\delta = \delta(t)e^{i\mathbf{k}\cdot\mathbf{x}}, \qquad \lambda = 2\pi a(t)/k, \tag{16.3}$$

where λ is the proper wavelength. The differential equation for the amplitude $\delta(t)$ is

$$\frac{d^2\delta}{dt^2} + 2\frac{\dot{a}}{a}\frac{d\delta}{dt} = [4\pi G\rho_b - (c_s k/a)^2]\delta. \tag{16.4}$$

The right side of this equation shows the competing effects of gravity and the pressure gradient force. At very long wavelength, $k \rightarrow 0$, the equation reduces to the zero pressure case discussed in Sections 11 to 15 above. At very short wavelength, large k, the pressure term dominates and δ tends to oscillate as a sound wave. The pressure and gravity terms balance when the wavelength is equal to the Jeans length

$$\lambda_J = c_s(\pi/G\rho_b)^{1/2}. \tag{16.5}$$

Jeans' original 1902 discussion of gravitational instability included a general time-varying background, and equation (16.4) is a special case of this though not one he discussed. Jeans derived equation (16.5) under the assumption that the background is uniform and time independent. He recognized the problem with this: an unbounded uniform mass distribution has undefined Newtonian potential and a bounded distribution ought to collapse. The point that could be supplied only later is that the uniformly expanding case directly patches onto a Friedman–Lemaître cosmological model and that the potential ϕ is this case can be defined in a consistent way (§ 8).

Two special cases of equation (16.5) are of some interest. For an ideal monatomic gas with conserved entropy the velocity of sound is

$$c_s = (5\,kT/3m)^{1/2}, \tag{16.6}$$

so the Jeans length is

$$\lambda_J = (5\pi kT/3\,G\rho_b m)^{1/2}. \tag{16.7}$$

It can happen, as discussed in Chapter V, that the gas temperature is held very nearly uniform by the background radiation. Here $c_s^2 = kT/m$, and

$$\lambda_J = (\pi kT/G\rho_b m)^{1/2}. \tag{16.8}$$

In the limit $\lambda \ll \lambda_J$ the pressure term dominates gravity, and if in addition expansion can be neglected, equation (16.4) can be reduced to

$$\frac{d^2\delta}{dt^2} + \left(\frac{c_s k}{a}\right)^2 \delta = 0, \tag{16.9}$$

with the solution

$$\delta \propto e^{-ic_s kt/a}. \tag{16.10}$$

The effect of expansion on the amplitude of oscillation can be estimated in the adiabatic approximation. The expression

$$\delta \equiv A(t)e^{-i\phi(t)}, \tag{16.11}$$

in equation (16.4) gives

$$\begin{aligned}\frac{d^2A}{dt^2} - 2i\frac{dA}{dt}\frac{d\phi}{dt} - iA\frac{d^2\phi}{dt^2} - A\left(\frac{d\phi}{dt}\right)^2 \\ + 2\frac{\dot{a}}{a}\frac{dA}{dt} - 2iA\frac{\dot{a}}{a}\frac{d\phi}{dt} + \frac{c_s^2k^2A}{a^2} - 4\pi G\rho_b A = 0.\end{aligned} \tag{16.12}$$

According to equation (16.10), one expects

$$\frac{d\phi}{dt} \sim \frac{c_s k}{a} \sim \frac{1}{T}, \qquad \frac{1}{A}\frac{dA}{dt} \sim \frac{1}{t}, \tag{16.13}$$

where T is the vibration period of the wave and t is the expansion time for the universe. When $T \ll t$, the two dominant terms in equation (16.12) are on the order of T^{-2}, and these can be eliminated by choosing

$$d\phi/dt = c_s k/a. \tag{16.14}$$

The next largest terms in equation (16.12) are on the order of $(t\,T)^{-1}$. If the terms $\sim t^{-2}$ are dropped, it leaves

$$A\frac{d^2\phi}{dt^2} + 2\frac{dA}{dt}\frac{d\phi}{dt} + 2A\frac{d\phi}{dt}\frac{\dot{a}}{a} = 0, \qquad d\phi/dt \propto (Aa)^{-2}. \tag{16.15}$$

This fixes the time variation of the amplitude and equation (16.14) gives the phase, so the solution in the adiabatic approximation is

$$\delta \propto [c_s(t)a(t)]^{-1/2} \exp - i\int^t (c_s k/a)dt. \tag{16.16}$$

This shows how the expansion of the universe and the variation of the velocity of sound affects the amplitude of acoustic fluctuations in the density when $p \ll \rho c^2$. The case $p \sim \rho c^2$ is discussed in Section 88 below.

17. Primeval magnetic field as a source for $\delta\rho/\rho$

It has been suggested that the magnetic fields observed in the galaxy and in extragalactic objects may have existed before galaxies formed, perhaps originating in primeval turbulence at high redshift, perhaps present at the time of the big bang (§§ 4C, 95). (It has also been argued that the magnetic field of the Milky Way galaxy could not persist for 10^{10} years because it is dynamically unstable against leaving the galaxy in a shorter time. If so, a dynamo is needed to maintain the field in the galaxy and may be able to generate the field in the first place from a very weak seed, Parker 1975.) An interesting consequence of a weak tangled primeval field is discussed by Wasserman (1978): the field acts as a source of density irregularities that grow through ordinary gravitational instability. Thus if a tangled primeval field is postulated, it implies a minimum value of $\delta\rho/\rho$.

Very small residual ionization is needed to make Ohmic dissipation negligible, so it will be supposed that the field is tied to the matter. The magnetic force on the matter per unit volume is

$$\begin{aligned} \mathbf{F} &= \frac{1}{c}\mathbf{j} \times \mathbf{B} = \frac{1}{4\pi a}(\nabla \times \mathbf{B}) \times \mathbf{B} \\ &= \frac{1}{4\pi a}[(\mathbf{B} \cdot \nabla)\mathbf{B} - \nabla B^2/2]. \end{aligned} \tag{17.1}$$

If pressure is negligible, the motion of the matter is

$$\frac{\partial \mathbf{v}}{\partial t} + \frac{\dot{a}}{a}\mathbf{v} = \frac{\mathbf{F}}{\rho_b} + \mathbf{g}. \tag{17.2}$$

On taking the divergence of this equation and using (14.1) for δ, one finds

$$\frac{\partial^2 \delta}{\partial t^2} + 2\frac{\dot{a}}{a}\frac{\partial \delta}{\partial t} = 4\pi G\rho_b\delta + \frac{1}{4\pi a^2 \rho_b}\partial_\alpha\partial_\beta\left(\frac{B^2}{2}\delta^{\alpha\beta} - B^\alpha B^\beta\right). \tag{17.3}$$

The solution to equation (17.3) will be written only for the Einstein-de Sitter model. This simplifies the equations and is a reasonable approximation at moderately high redshift ($Z \gtrsim \Omega_0^{-1}$). Since flux is conserved, $B \propto$

a^{-2} and $\rho_b \propto a^{-3}$, so the source term varies as $a^{-3} \propto t^{-2}$ and equation (17.3) can be rewritten as

$$\frac{\partial^2 \delta}{\partial t^2} + \frac{4}{3t}\frac{\partial \delta}{\partial t} - \frac{2}{3}\frac{\delta}{t^2} = S_i \left(\frac{t_i}{t}\right)^2,$$
$$S_i = \frac{1}{4\pi a_i^2 \rho_b(i)} \partial_\alpha \partial_\beta \left(\frac{B_i^2}{2}\delta^{\alpha\beta} - B_i^\alpha B_i^\beta\right). \tag{17.4}$$

The source term has been referred to the starting time t_i, perhaps the epoch of decoupling of matter and radiation.

It will be assumed that at t_i, $\delta_i = 0$ and $d\delta_i/dt = 0$. Since the two homogeneous solutions are known, one can write down the Green's function,

$$G(t, t_0) = \frac{3}{5}\frac{t_0}{}\left[\left(\frac{t}{t_0}\right)^{2/3} - \frac{t_0}{t}\right], \qquad t > t_0$$
$$= 0, \qquad t < t_0, \tag{17.5}$$

and the wanted solution is (Wasserman 1978)

$$\delta(\mathbf{x}, t) = \int_{t_i} G(t, t_0) S_i t_i^2 / t_0^2 \, dt_0$$
$$= \frac{3}{5} S_i t_i^2 \left[\frac{3}{2}\left(\frac{t}{t_i}\right)^{2/3} + \frac{t_i}{t} - \frac{5}{2}\right] \tag{17.6}$$
$$\cong \frac{9}{10}\left(\frac{t}{t_i}\right)^{2/3} \frac{t_i^2}{4\pi a_i^2 \rho_b(i)} \partial_\alpha \partial_\beta \left(\frac{B_i^2}{2}\delta^{\alpha\beta} - B_i^\alpha B_i^\beta\right).$$

The last line assumes $t_i \ll t$. It says the primeval field generates an effective initial density perturbation.

$$\delta_i = \frac{9}{10} S_i t_i^2 \sim \frac{B_i^2 t_i^2}{\rho_i \lambda_i^2}, \tag{17.7}$$

where λ_i is the proper coherence length of the field, and of course it is assumed that $\mathbf{B}$ is not stress-free.

One can also compute the vorticity

$$\boldsymbol{\omega} = \nabla \times \mathbf{v}/a. \tag{17.8}$$

Taking the curl of equation (17.2), one finds (Wasserman 1978)

$$\frac{\partial}{\partial t} a^2 \omega = \frac{1}{4\pi \rho_b(t)} \nabla \times (\mathbf{B} \cdot \nabla)\mathbf{B},$$
$$t\omega(\mathbf{x}, t) = \frac{3\, t_i^2}{4\pi \rho_b(i)\, a_i^2} \nabla \times (\mathbf{B}_i \cdot \nabla)\, \mathbf{B}_i. \tag{17.9}$$

The second equation assumes an Einstein-de Sitter model and $t \gg t_i$. We see from equations (17.7) and (17.9) that

$$t\omega(t) \sim \delta_i. \tag{17.10}$$

The number $t\omega$ is a measure of the perturbation associated with circulation as it is the characteristic rotation in an expansion time. In the absence of a driving term, ω decays as a^{-2} (eq. 17.9) or $\omega t \propto t^{-1/3}$ in an Einstein-de Sitter model. The effect of the B term on the right side of equation (17.9) is to hold ωt constant at about the initial value of δ while of course gravity is making δ grow as $t^{2/3}$. Thus when $t \gg t_i$ the velocity associated with ω is small compared to the curl-free velocity (14.9) associated with $\delta(t)$.

The Alfvén speed is $\sim B\rho^{-1/2}$, so there is a magnetic Jeans length (eq. 16.5)

$$\lambda_B \sim B/\rho G^{1/2} \propto a(t), \tag{17.11}$$

and since $G\rho \sim t^{-2}$, the effective initial perturbation given by equation (17.7) is

$$\delta_i \sim (\lambda_B(i)/\lambda_i)^2. \tag{17.12}$$

This says the perturbation caused by B is appreciable if the coherence length λ is fairly close to the magnetic Jeans length.

Since $\lambda_B \propto a(t)$, we can find a useful measure of the size of B for an interesting perturbation in terms of present quantities assuming uniform expansion. If the universe is cosmologically flat or close to it, the growth factor since decoupling of matter and radiation in the linear approximation is $\sim 10^3$, so we would want $\delta_i \sim 10^{-3}$ or

$$\lambda_B(t_0) \sim 0.03\, \lambda_0. \tag{17.13}$$

Taking $\lambda_0 \sim 1$ Mpc, roughly the distance between small galaxies, one finds that the wanted present intergalactic magnetic field is $\sim 10^{-9}$ Gauss. By comparison the present interstellar magnetic field is $\sim 10^{-6}$ Gauss at ~ 1 proton cm^{-3}. If this were expanded isotropically to $\sim 10^{-6}$ cm^{-3}, it would amount to $\sim 10^{-10}$ Gauss, comparable to what is wanted. Thus if the magnetic field of the Galaxy is primeval and has not been strongly amplified (perhaps by differential rotation), then it could have had an important effect in generating initial density fluctuations on the scale of galaxies. Angular momentum transfer through the magnetic field, as measured by vorticity production, may be somewhat less interesting because this effect is not boosted by gravity. Finally, it is well to bear in mind that the effective perturbation δ_i computed here is a minimum value established under the assumption of no perturbation to the matter distribution at decoupling. It may well be that whatever produced B also produced some residual irregularity in the distribution of matter (see § 95).

18. Second order perturbation theory for $\delta\rho/\rho$

It is fairly easy to compute $\delta\rho/\rho = \delta(\mathbf{x}, t)$ in second order perturbation theory, and there is an interesting application in the growth of skewness in the distribution of values of δ. The velocity field in second order is discussed in Section 89. For general treatments see Tomita (1967, 1972) and Hunter (1964), and for an application to the angular momentum of a protogalaxy see Peebles (1969c).

It will be assumed that matter may be approximated as an ideal fluid with zero pressure. Equation (9.19) carried to second order in the perturbations δ and $\mathbf{v}$ is

$$\frac{\partial^2\delta}{\partial t^2} + 2\frac{\dot a}{a}\frac{\partial\delta}{\partial t} = 4\pi G\rho_b\delta(1+\delta) + \nabla\delta\cdot\nabla\phi/a^2 + \partial_\alpha\partial_\beta(v^\alpha v^\beta)/a^2, \tag{18.1}$$

where the potential is (eq. 8.1)

$$\phi = -G\rho_b a^2\Delta(\mathbf{x}), \qquad \Delta(\mathbf{x}) = \int \frac{d^3x'\delta(\mathbf{x}')}{|\mathbf{x}-\mathbf{x}'|}. \tag{18.2}$$

To calculate δ in second order, one can take the velocity in the last term in equation (18.1) from linear perturbation theory. If one assumes only the growing mode is present, this is (eq. 14.6)

$$v^\alpha = \frac{a}{4\pi D}\frac{dD}{dt}\Delta_{,\alpha}. \tag{18.3}$$

In the linear approximation the density perturbation is

$$\delta \approx \delta_0 \equiv A(\mathbf{x})D(t), \tag{18.4}$$

where again $D(t)$ is the growing mode. To estimate the deviation from this as δ approaches unity, we shall write

$$\begin{aligned} \delta &= \delta_0(\mathbf{x}, t)[1 + \epsilon(\mathbf{x}, t)], \\ \delta_0 &\ll 1, \qquad \epsilon \ll 1, \end{aligned} \tag{18.5}$$

in equation (18.1) and discard terms $\sim\delta_0{}^2\epsilon$ or smaller. The part $\sim\delta_0$ cancels because δ_0 is a solution to the linear perturbation equation, leaving

$$\begin{aligned} &\frac{d^2\epsilon}{dt^2} + 2\left(\frac{1}{D}\frac{dD}{dt} + \frac{\dot{a}}{a}\right)\frac{d\epsilon}{dt} \\ &\quad = 4\pi G\rho_b\delta_0 - G\rho_b\frac{\delta_{0,\alpha}}{\delta_0}\Delta_{,\alpha} + \frac{1}{16\pi^2 D^2\delta_0}\left(\frac{dD}{dt}\right)^2 \partial_\alpha\partial_\beta(\Delta_{,\alpha}\Delta_{,\beta}). \end{aligned} \tag{18.6}$$

In an Einstein-de Sitter model,

$$D \propto t^{2/3}, \qquad 6\pi G\rho_b t^2 = 1, \tag{18.7}$$

this can be integrated, giving

$$\begin{aligned} \delta &= \delta_0 + \epsilon\delta_0 \\ &= \delta_0 + \frac{5}{7}\delta_0{}^2 - \frac{1}{4\pi}\delta_{0,\alpha}\Delta_{,\alpha} + \frac{1}{56\pi^2}\Delta_{,\alpha\beta}\Delta_{,\alpha\beta}. \end{aligned} \tag{18.8}$$

The two constants of integration have been dropped, leaving only the most rapidly growing part of ϵ.

Equations (18.4) and (18.8) are the solution $\delta\rho/\rho$ as a function of $\mathbf{x}$ and t in second order perturbation theory. One sees that in this order the behavior is no longer local: the density perturbation at one spot depends on the initial perturbation at other places through Δ (eq. 18.2).

An interesting application is to the growth of skewness in the distribution of values of $\delta(\mathbf{x}, t)$ at given t. It will be supposed that the initial perturbation $\delta_i \ll 1$ is a random Gaussian process, so it is completely characterized by its autocorrelation function $\xi(x)$. Then the first moments of δ_i are:

$$\begin{aligned}
\langle \delta(\mathbf{x}) \rangle &= 0, \\
\langle \delta(\mathbf{x}_1)\delta(\mathbf{x}_2) \rangle &= \xi(|\mathbf{x}_1 - \mathbf{x}_2|), \\
\langle \delta(\mathbf{x}_1)\delta(\mathbf{x}_2)\delta(\mathbf{x}_3) \rangle &= 0, \\
\langle \delta(\mathbf{x}_1)\delta(\mathbf{x}_2)\delta(\mathbf{x}_3)\delta(\mathbf{x}_4) \rangle &= \xi(|\mathbf{x}_1 - \mathbf{x}_2|)\xi(|\mathbf{x}_3 - \mathbf{x}_4|) \\
&\quad + \xi(|\mathbf{x}_1 - \mathbf{x}_3|)\xi(|\mathbf{x}_2 - \mathbf{x}_4|) + \xi(|\mathbf{x}_1 - \mathbf{x}_4|)\xi(|\mathbf{x}_2 - \mathbf{x}_3|).
\end{aligned} \tag{18.9}$$

The frequency distribution of $\delta_i(\mathbf{x})$ is Gaussian, with variance $\xi_i(0)$, mean $\langle \delta_i \rangle = 0$. In linear perturbation theory δ varies as $D(t)$, and the variance varies as $\xi(0, t) \propto D(t)^2$.

The second and higher order corrections to δ_0 must preserve $\langle \delta \rangle = 0$ because mass is conserved. To check this for equation (18.8), we can use

$$\begin{aligned}
\langle \delta_{0,\alpha} \Delta_{,\alpha} \rangle &= -\langle \delta_0 \Delta_{,\alpha\alpha} \rangle = 4\pi \langle \delta_0^2 \rangle = 4\pi\xi(0), \\
\langle \Delta_{,\alpha\beta} \Delta_{,\alpha\beta} \rangle &= \langle \Delta_{,\alpha\alpha} \Delta_{,\beta\beta} \rangle = 16\pi^2 \xi(0),
\end{aligned} \tag{18.10}$$

which gives $\langle \epsilon\delta_0 \rangle = 0$, and so $\langle \delta \rangle = 0$ as expected. The interesting moment is the skewness

$$\langle \delta^3 \rangle = 3\langle \delta_0^3 \epsilon \rangle \tag{18.11}$$

to lowest order, because the distribution of δ_0 has zero skewness. To evaluate this we need the following means. The last of equations (18.9) indicates

$$\langle \delta_0^4 \rangle = 3\langle \delta_0^2 \rangle^2 = 3\xi(0)^2. \tag{18.12}$$

The next term from equation (18.8) is

$$\begin{aligned}
\langle \delta_0^2 \delta_{0,\alpha} \Delta_{,\alpha} \rangle &= 1/3 \langle \delta_{0,\alpha}^3 \Delta_{,\alpha} \rangle \\
&= -1/3 \langle \delta_0^3 \Delta_{,\alpha\alpha} \rangle = 4\pi\xi(0)^2.
\end{aligned} \tag{18.13}$$

The last term is, from the last of equations (18.9),

$$\langle \delta_0^2 \Delta_{,\alpha\beta} \Delta_{,\alpha\beta} \rangle = \xi(0) \langle \Delta_{,\alpha\beta} \Delta_{,\alpha\beta} \rangle + 2 \langle \Delta_{,\alpha\beta} \delta \rangle \langle \Delta_{,\alpha\beta} \delta \rangle. \tag{18.14}$$

So one sees that

$$\langle \Delta_{,\alpha\beta}\Delta_{,\alpha\beta}\rangle = \langle \Delta_{,\alpha\alpha}\Delta_{,\beta\beta}\rangle = 16\pi^2\xi(0),$$

$$\langle \delta_0\Delta_{,\alpha\beta}\rangle = \frac{1}{3}\delta^{\alpha\beta}\langle \delta_0\Delta_{,\gamma\gamma}\rangle = -\frac{4\pi}{3}\delta^{\alpha\beta}\xi(0), \tag{18.15}$$

$$\langle \delta_0{}^2\Delta_{,\alpha\beta}\Delta_{,\alpha\beta}\rangle = {}^{80}\!/_3\pi^2\xi(0)^2.$$

Collecting, one finds that the skewness of the density fluctuation is

$$\langle \delta^3\rangle = {}^{34}\!/_7\xi(0)^2, \tag{18.16}$$

where, to lowest order, the variance is

$$\langle \delta^2\rangle = \xi(0). \tag{18.17}$$

The positive skewness develops because as the fluctuations become non-linear δ in the dense spots tends to grow large while δ in the holes approaches the minimum value $\delta = -1$. When the rms fluctuation reaches 50 percent, $\delta\rho/\rho = 0.5$ or $\xi(0) = (0.5)^2$,

$$\left\langle \left(\frac{\delta\rho}{\rho}\right)^3\right\rangle^{1/3} = 0.67, \tag{18.18}$$

so the departure from a Gaussian distribution is considerable.

19. Spherical model

This is the simplest nontrivial model for the way an object like a galaxy or a cluster of galaxies breaks away from the general expansion. In the model the universe is spherically symmetric about one spot and the matter is an ideal fluid with zero pressure. This might give a realistic approximation to the early stages of fragmentation. However, except for very special cases (for example, § 25), it is not to be trusted once the model has stopped expanding and commenced the collapse, for by then the distribution is strongly unstable against the development of nonradial motion (§§ 20 and 21 below).

A. Energy and maximum radius

Suppose at some starting time t_i the spherical irregularity is moving with the general expansion, no peculiar velocity, and consider the mass shell initially at proper radius r_i. The kinetic energy per unit mass at r_i (relative to the center) is

$$K_i = \tfrac{1}{2} H_i^2 r_i^2. \tag{19.1}$$

In the homogeneous background model the potential energy per unit mass would be (Appendix, eq. 97.12)

$$|W_i^b| = \Omega_i K_i. \tag{19.2}$$

where Ω_i is the density parameter at t_i. If the mass within r_i is

$$M = \rho_b V_i (1 + \delta_i^*), \qquad V_i = \tfrac{4}{3}\pi r_i^3, \tag{19.3}$$

then equation (19.2) says the potential energy per unit mass at r_i is

$$|W_i| = \Omega_i K_i (1 + \delta_i^*), \tag{19.4}$$

and the total energy is

$$E = K_i - |W_i| = -\frac{|W_i|}{1 + \delta_i^*}\left[\delta_i^* - (\Omega_i^{-1} - 1)\right]. \tag{19.5}$$

In an open universe, where $\Omega_i < 1$, if δ_i^* is less than the critical value $\Omega_i^{-1} - 1$, the energy is positive and the shell never stops expanding (assuming no other shell moves inside it). If $\delta_i^* > (\Omega_i^{-1} - 1)$, the shell stops expanding at radius r_m, at which time the potential energy per unit mass is

$$|W_m| = \frac{r_i}{r_m}|W_i| = -E, \tag{19.6}$$

because $W \propto r^{-1}$ and $K_m = 0$. Then with equation (19.5) one sees that the shell radius at maximum expansion is

$$\frac{r_m}{r_i} = \frac{1 + \delta_i^*}{\delta_i^* - (\Omega_i^{-1} - 1)}. \tag{19.7}$$

This is the relation between the initial perturbation and the factor by which the shell expands.

B. Motion of a mass shell

To simplify the expressions in the general solution for $r(t)$, we will suppose that at t_i, $\Omega_i \approx 1$, $\delta_i^* \ll 1$ and only the growing density perturbation is present. Since δ_i^* (eq. 19.3) was defined for $\mathbf{v}_i = 0$, equation (15.7) says the amplitude of the growing mode is

$$\delta_i' = {}^3\!/_5\delta_i^*. \tag{19.8}$$

The decaying mode can be introduced in the equations below by adding a constant to the expression for t in equation (19.11).

The proper radius of the shell that encloses mass M is $r(t)$, and the acceleration of the shell is

$$d^2r/dt^2 = -GM/r^2. \tag{19.9}$$

It will be assumed that no mass shells cross this one, so M is constant. Then the first integral of equation (19.9), the energy equation, is

$$(dr/dt)^2 = 2GM/r + C, \tag{19.10}$$

where the constant of integration is $C > 0$ for a shell with positive energy, $C < 0$ for negative energy. The solution to equations (19.9) and (19.10) is the parametric form,

$$\begin{aligned} C > 0&: r = A(\cosh\theta - 1), \qquad t = B(\sinh\theta - \theta); \\ C < 0&: r = A(1 - \cos\theta), \qquad t = B(\theta - \sin\theta); \\ &A^3 = GMB^2. \end{aligned} \tag{19.11}$$

The homogeneous background model is described by like equations. On choosing the expansion parameter $a(t) \equiv r_b(t)$ so it contains the same mass M as in equation (19.11), one has

$$\begin{aligned} r_b = A_b(\cosh\eta - 1), \qquad t = B_b(\sinh\eta - \eta), \\ A_b{}^3 = GMB_b{}^2, \end{aligned} \tag{19.12}$$

for the open model and similarly for the closed case.

The behavior of the systems of equations (19.11) and (19.12) in the early universe is found by expanding the solutions in η and $\theta \ll 1$. The time in the first of equations (19.11) is

$$t \simeq \frac{B\theta^3}{6}\left(1 + \frac{\theta^2}{20}\right), \tag{19.13}$$

so

$$\frac{B\theta^3}{6} \simeq t\left[1 - \frac{1}{20}\left(\frac{6t}{B}\right)^{2/3}\right]. \tag{19.14}$$

The radius is

$$r \simeq \frac{A\theta^2}{2}\left[1 + \frac{\theta^2}{12}\right], \tag{19.15}$$

and with equations (19.11) and (19.14) this becomes

$$r \simeq \left(\frac{9}{2}GMt^2\right)^{1/3}\left[1 + \frac{1}{20}\left(\frac{6t}{B}\right)^{2/3}\right]. \tag{19.16}$$

The mean density within the shell r is then

$$\rho' = \frac{3M}{4\pi r^3} = \frac{1}{6\pi G t^2}\left[1 - \frac{3}{20}\left(\frac{6t}{B}\right)^{2/3}\right]. \tag{19.17}$$

Equation (19.17) shows the density as a perturbation from a cosmologically flat model. The background density can be described the same way; at t_i the density differs from a flat model with the same expansion rate by the fractional amount

$$(\delta\rho/\rho)_b = \Omega_i - 1 \simeq -(\Omega_i^{-1} - 1). \tag{19.18}$$

The first equation defines the density parameter, the second assumes $\Omega_i \approx 1$. The growing part is (eq. 19.8)

$$(\delta\rho/\rho)_b \equiv -\delta_c = -\tfrac{3}{5}(\Omega_i^{-1} - 1). \tag{19.19}$$

Then we see from equation (19.17) that the ratio of the mean density in the shell to the density in the background model at t_i is

$$\frac{\rho'}{\rho_b} = \left[1 + \delta_c - \frac{3}{20}\left(\frac{6t_i}{B}\right)^{2/3}\right] \equiv 1 + \delta_i'. \tag{19.20}$$

This fixes B in terms of δ_i', the fractional mass excess within r_i:

$$B = 6t_i/[{}^{20}\!/\!_3(\delta_c - \delta_i')]^{3/2}. \tag{19.21}$$

This in the last of equations (19.11) with (19.17) gives

$$A = \frac{3r_i}{10(\delta_c - \delta_i')}. \tag{19.22}$$

The values of A and B for the background model are found by setting $\delta_i' = 0$ in these two equations. One notes in particular that equations (19.12), (19.19), and (19.22) give (eq. 97.23)

$$2\Omega_i^{-1} - 1 = \cosh \eta_i. \tag{19.23}$$

These results fix the evolution of the mass shell relative to the background model. The epoch labeled by cosmological density parameter Ω has parameter η fixed by

$$\cosh \eta = 2\Omega^{-1} - 1. \tag{19.24}$$

The parameter θ for the mass shell is fixed by the conditions that the time t agree with the background,

$$B(\sinh \theta - \theta) = B_b(\sinh \eta - \eta), \tag{19.25}$$

or, by equation (19.21),

$$\sinh \theta - \theta = (1 - \delta_i'/\delta_c)^{3/2}(\sinh \eta - \eta). \tag{19.26}$$

Then the radius of the shell is (eq. 19.22)

$$r = \frac{3}{10} r_i \frac{(\cosh \theta - 1)}{\delta_c - \delta_i'}; \tag{19.27}$$

the mean density within the shell is

$$\rho' = 3M/4\pi r^3, \tag{19.28}$$

and the ratio of this to the density in the background model is

$$\frac{\rho'}{\rho_b} = \left[\frac{r_b(t)}{r(t)}\right]^3 = \left[\left(1 - \frac{\delta_i'}{\delta_c}\right)\frac{\cosh\eta - 1}{\cosh\theta - 1}\right]^3. \tag{19.29}$$

The velocity of the shell relative to the origin is

$$v = \frac{dr}{dt} = \frac{r}{t}\frac{\sinh\theta(\sinh\theta - \theta)}{(\cosh\theta - 1)^2}. \tag{19.30}$$

These equations describe a positive energy mass shell in an open universe. The equations for negative energy or a closed universe are obtained by the substitutions (see part E below).

$$\theta \rightarrow i\theta, \qquad A \rightarrow -A, \qquad B \rightarrow iB. \tag{19.31}$$

C. The density run

In equation (19.29) ρ' is the mean density within the shell. To get the mass per unit volume at r, one has to differentiate ρ'. The initial amplitude of the growing mode is supposed to be some given function of the initial radius, $\delta_i'(r_i)$. According to equation (19.26),

$$\frac{\partial\theta}{\partial r_i} = -\frac{3}{2(\delta_c - \delta_i')}\frac{d\delta_i'}{dr_i}\frac{\sinh\theta - \theta}{\cosh\theta - 1}, \tag{19.32}$$

at fixed t, hence fixed η. The derivative of r (equation 19.27) is then

$$\begin{aligned}\frac{r_i}{r}\frac{\partial r}{\partial r_i} &= 1 + \frac{r_i}{\delta_c - \delta_i'}\frac{d\delta_i'}{dr_i}\left[1 - \frac{3}{2}\frac{\sinh\theta(\sinh\theta - \theta)}{(\cosh\theta - 1)^2}\right] \\ &= 1 + \frac{r_i}{\delta_c - \delta_i'}\frac{d\delta_i'}{dr_i}\left[1 - \frac{3}{2}\frac{vt}{r}\right],\end{aligned} \tag{19.33}$$

where v is the expansion velocity of the shell (eq. 19.30). Since the mass between shells r_i and $r_i + dr_i$ at t_i is the same as the mass between r and $r + (\partial r/\partial r_i)\, dr_i$ at t, the mass density is

$$\rho(r, t) = \rho_i \frac{(r_i/r)^2}{(\partial r/\partial r_i)}, \tag{19.34}$$

and the ratio of this to the background density is

$$\frac{\rho(r, t)}{\rho_b(t)} = \left(\frac{r_b}{r}\right)^3 \left(\frac{r_i}{r}\frac{\partial r}{\partial r_i}\right)^{-1}. \tag{19.35}$$

The first factor is given in equation (19.29), the second in equation (19.33).

It will be recalled that δ_i' is the initial fractional mass excess within r_i;

$$\delta_i' = \frac{3}{r_i^3}\int_0^{r_i} r^2\,dr\,\delta_i(r), \tag{19.36}$$

where $\delta_i(r_i)$ is the initial density contrast at radius r_i. Thus

$$r_i\partial\delta_i'/\partial r_i = 3(\delta_i - \delta_i'). \tag{19.37}$$

This in equation (19.33) gives the final expression for the density,

$$\frac{\rho}{\rho_b} = \frac{\left(1 - \frac{\delta_i'}{\delta_c}\right)^3 \left(\frac{\cosh\eta - 1}{\cosh\theta - 1}\right)^3}{1 + 3\frac{\delta_i - \delta_i'}{\delta_c - \delta_i'}\left[1 - \frac{3}{2}\frac{\sinh\theta(\sinh\theta - \theta)}{(\cosh\theta - 1)^2}\right]}. \tag{19.38}$$

D. Behavior in the limit $\Omega \rightarrow 0$

In an open cosmological model at large t the density parameter Ω approaches zero and gravitational deceleration is unimportant, so small density fluctuations stop growing. In linear perturbation theory the net growth factor is (eqs. 11.18, 11.20)

$$\frac{\delta_f}{\delta_i} = \frac{5}{2(\Omega_i^{-1} - 1)} = \frac{3}{2\delta_c}, \qquad \delta_f \ll 1. \tag{19.39}$$

The spherical model can be used to compute this growth factor for somewhat larger fluctuations. In the limit $t \rightarrow \infty$ both θ and η are large, so equations (19.26) and (19.38) give

$$\frac{\rho_f}{\rho_b} = \frac{(1 - \delta_i'/\delta_c)^{-3/2}}{1 - \frac{3}{2}\frac{\delta_i - \delta_i'}{\delta_c - \delta_i'}}, \tag{19.40}$$

where it will be recalled that δ_i is the initial density contrast at radius r_i and δ_i' is the mean within r_i (eq. 19.36). In the limit $\delta_i, \delta_i' \ll \delta_c$, equation (19.40) is

$$\frac{\rho_f}{\rho_b} = 1 + \frac{3}{2}\frac{\delta_i}{\delta_c}, \tag{19.41}$$

in agreement with equation (19.39). If the density gradient across the fluctuation is small, so $\delta_i' \approx \delta_i$, equation (19.40) can be simplified to

$$\frac{\rho_f}{\rho_b} - 1 = \delta_f = \frac{1}{(1 - \delta_i/\delta_c)^{3/2}} - 1. \tag{19.42}$$

Equation (19.42) says the density reaches twice the background value, $\delta_f = 1$, if $\delta_i = 0.37\,\delta_c$. If $\delta_i > \delta_c$, the shell has negative energy and δ grows without limit. Thus there is rather a restricted range of initial density perturbations,

$$0.37\,\delta_c \lesssim \delta_i \lesssim \delta_c, \tag{19.43}$$

that can end up in the late stages of expansion, $\Omega_f \ll 1$, as density fluctuations with appreciable contrast, $\delta_f \gtrsim 1$, but yet freely expanding. It has been argued that some groups of galaxies are local density enhancements that are expanding with the general expansion (for example, Gott, Wrixon, and Wannier 1973). This certainly is possible if $\Omega_0 \sim 0.03$ perhaps and $\delta \ll 30$ in the group. However, equation (19.43) indicates that it is not so easy to arrange because it requires a very special choice for δ_i. Therefore in the gravitational instability picture, freely expanding groups could be a common phenomenon only if δ_i showed a strong tendency to be just slightly less than δ_c, which seems artificial. It is not clear whether this is a problem for the theory or the observations.

E. Closed and flat cosmological models

The result of applying the substitutions listed in equations (19.31) to equations (19.24) through (19.29) is

$$\begin{gathered} \cos\eta = 2\Omega^{-1} - 1, \\ \theta - \sin\theta = (1 + \delta_i'/\delta_c)^{3/2}(\eta - \sin\eta), \\ r = \frac{3}{10} r_i \frac{1 - \cos\theta}{\delta_i' + \delta_c}, \\ \frac{\rho'}{\rho_b} = \left[\left(1 + \frac{\delta_i'}{\delta_c}\right)\frac{1 - \cos\eta}{1 - \cos\theta}\right]^3, \qquad \delta_i' \geq -\delta_c, \end{gathered} \tag{19.44}$$

where δ_c has been redefined to

$$\delta_c = \tfrac{3}{5}(1 - \Omega_i^{-1}). \tag{19.45}$$

The solution for a cosmologically flat model is found by taking the limit $\Omega \rightarrow 1$, $\eta \rightarrow 0$ in equations (19.44). Equation (19.45) with the first of equations (19.44) gives in this limit

$$\eta_i = 2(1 - \Omega_i^{-1})^{1/2} = 2(5\delta_c/3)^{1/2}, \tag{19.46}$$

and since the expansion parameter varies as

$$a \propto 1 - \cos\eta \propto \eta^2, \tag{19.47}$$

equations (19.44) can be reduced to

$$\theta - \sin\theta = \frac{4}{3}\left(\frac{5}{3}\frac{a(t)}{a_i}\delta_i'\right)^{3/2},$$

$$r = \frac{3}{10}\frac{r_i}{\delta_i'}(1 - \cos\theta), \tag{19.48}$$

$$\frac{\rho'}{\rho_b} = \left[\frac{10}{3}\frac{a}{a_i}\frac{\delta_i'}{1 - \cos\theta}\right]^3 = \frac{9}{2}\frac{(\theta - \sin\theta)^2}{(1 - \cos\theta)^3}.$$

When $\theta \ll 1$, these equations yield

$$\frac{\rho'}{\rho_b} = 1 + \frac{a}{a_i}\delta_i', \tag{19.49}$$

as expected from linear perturbation theory.

The mass shell reaches maximum expansion at $\theta = \pi$, and equations (19.48) say that at maximum expansion the density relative to the background is

$$\frac{\rho_m'}{\rho_b} = \frac{9\pi^2}{16} = 5.6. \tag{19.50}$$

If all the matter were in such lumps, the filling factor would be $\rho_b/\rho_m' = 0.2$; in rough order of magnitude the lumps fill space when they fragment from the general expansion. For another derivation of equation (19.50) see Kihara (1968).

In the cosmologically flat model the time at maximum expansion is

$$t_m = [6\pi G\rho_b(t_m)]^{-1/2}, \tag{19.51}$$

so according to equation (19.50) the mean density within the shell at maximum expansion is

$$\rho_m' = \frac{3\pi}{32Gt_m{}^2}. \tag{19.52}$$

If the mass shell is taken to represent the outer boundary of a protogalaxy and if one has some estimate of ρ_m from the present structure of the galaxy, then equation (19.52) gives an estimate of the epoch t_m at which the object reaches maximum expansion (Partridge and Peebles 1967). The first of equations (19.48) with $\theta = \pi$ then gives (Peebles 1969a)

$$\delta_i' = \frac{3}{5}\left(\frac{3\pi}{4}\right)^{2/3}\frac{a_i}{a_m} = \frac{3}{5}\left(\frac{3\pi}{4}\frac{t_i}{t_m}\right)^{2/3}, \tag{19.53}$$

which fixes the wanted amplitude of the growing density perturbation at some starting time t_i. These results have been used in attempts to find quantitative scenarios for the development of galaxies and clusters of galaxies (for example, Partridge and Peebles 1967, Gunn and Gott 1972, Gott and Rees 1975).

Equations (19.50), (19.52), and (19.53) could have been obtained up to the numerical factors by dimensional analysis (with $\delta \propto t^{2/3}$; see eqs. 4.1–4.9). One might hope that numerical factors given by the model are in the appropriate direction, but it is doubtful that they should be trusted beyond factors of 2, for it is doubtful that there is a definite time of maximum expansion of a protogalaxy or protocluster, or even that the material originally at some chosen distance from an appropriate center comes to zero radial velocity at a roughly common time. The following sections deal with some aspects of the development of nonradial motions.

20. Homogeneous ellipsoid model

The spherical model in the last section assumes purely radial motion within the developing protoobject, which certainly is not realistic once it has stopped expanding. The simplest way to treat nonradial motion is to approximate the protoobject as an isolated homogeneous spheroid of ideal fluid at zero pressure. Almost all discussions of this model have dealt with oblate spheroids where two axes are equal as this greatly simplifies the equations and is a good general approximation to the final collapse of an ellipsoid. (Collapse to a spindle is a special case.) The collapse of a spheroid from rest was discussed by Lynden-Bell (1964) and Lin, Mestel, and Shu (1965). Icke (1973) considered the evolution of an initially expanding spheroid, and Nariai and Fujimoto (1972) analyzed a triaxial rotating ellipsoid. Zel'dovich (1964) discussed the analogy between the evolution of a homogeneous spheroid and of a section of a homogeneous anisotropic cosmological model: the local equations of motion can be expressed in the same form with the tidal field determined by the shape of the surface of the ellipsoid in the first case and by the topology of the model in the second.

The surface of the ellipsoid is given by the equations

$$\frac{x^2}{a^2} + \frac{y^2}{b^2} + \frac{z^2}{c^2} = 1, \tag{20.1}$$

where x, y, and z are proper Cartesian coordinates and a, b, c are the semiaxes. The gravitational potential inside the ellipsoid is (Kellogg 1953)

$$\Phi = \pi G\rho[Ax^2 + By^2 + Cz^2], \qquad \nabla^2\phi = 4\pi G\rho(t),$$

$$A = abc\int_0^\infty \frac{d\lambda}{(a^2+\lambda)f^{1/2}}, \qquad B = abc\int_0^\infty \frac{d\lambda}{(b^2+\lambda)f^{1/2}}, \tag{20.2}$$

$$C = abc\int_0^\infty \frac{d\lambda}{(c^2+\lambda)f^{1/2}}, \qquad f = (a^2+\lambda)(b^2+\lambda)(c^2+\lambda).$$

The three coefficients satisfy $A + B + C = 2$ from which it is apparent that Φ satisfies Poisson's equation. If two of the axes are equal, the integrals can be reduced to elementary functions. The interesting case is an oblate spheroid, where $a = b > c$. Here

$$\begin{aligned} A = B &= \frac{(1-e^2)^{1/2}}{e^2}\left[\frac{\sin^{-1} e}{e} - (1-e^2)^{1/2}\right], \\ C &= \frac{2(1-e^2)^{1/2}}{e^2}\left[\frac{1}{(1-e^2)^{1/2}} - \frac{\sin^{-1} e}{e}\right], \end{aligned} \tag{20.3}$$

where the eccentricity is defined as

$$e = (1 - c^2/a^2)^{1/2}. \tag{20.4}$$

The gravitational potential energy of the spheroid is

$$W = \frac{3}{5}\frac{GM^2}{a}\frac{\sin^{-1} e}{e}, \tag{20.5}$$

where the mass is

$$M = \tfrac{4}{3}\pi\rho a^2 c = \tfrac{4}{3}\pi\rho a^3(1 - e^2)^{1/2}. \qquad (20.6)$$

It is convenient to assign to each mass element fixed comoving coordinates $\mathbf{x}$. If the motion is homogeneous, as is consistent with the potential, one can assign a linear relation between x^α and the proper position r^α of the mass element measured in an inertial Cartesian coordinate system:

$$r^\alpha = A^{\alpha\beta}(t)x^\beta, \qquad (20.7)$$

where $A^{\alpha\beta}$ is a function of time alone. The potential in the ellipsoid can be written as

$$\Phi = \tfrac{1}{2}\Phi_{\alpha\beta}(t)r^\alpha r^\beta. \qquad (20.8)$$

This is the result of rotating the form for Φ in equation (20.2) from coordinate axes along the axes of the ellipsoid. The acceleration of a fluid element is

$$\frac{d^2 r^\alpha}{dt^2} = \frac{d^2 A^{\alpha\beta}}{dt^2} x^\beta = -\Phi_{\alpha\beta} r^\beta, \qquad (20.9)$$

giving

$$\frac{d^2 A^{\alpha\beta}}{dt^2} = -\Phi_{\alpha\gamma} A^{\gamma\beta}. \qquad (20.10)$$

If the initial velocity field is homogeneous, it can be put in the form

$$u^\alpha = \frac{dA^{\alpha\beta}}{dt} x^\beta, \qquad (20.11)$$

which defines the initial values of $dA^{\alpha\beta}/dt$.

For the initial values of $A^{\alpha\beta}$ we might take the ellipsoid to be along the coordinate axes and then set

$$A^{11} = a, \qquad A^{22} = b, \qquad A^{33} = c, \qquad (20.12)$$

with the off-diagonal components equal to zero. Then the surface of the ellipsoid (eq. 20.1) expressed in comoving coordinates is

$$(x^1)^2 + (x^2)^2 + (x^3)^2 = 1. \tag{20.13}$$

As the fluid moves, the equation for the surface, expressed in proper Cartesian coordinates, is

$$\sum_\alpha (A^{-1\alpha\beta} r^\beta)^2 = 1, \tag{20.14}$$

representing an ellipsoid rotated relative to the coordinate axes. This fixes the $\Phi_{\alpha\beta}$ through rotation of equation (20.2), and then equation (20.10) with the initial values for $dA^{\alpha\beta}/dt$ gives the homogeneous motion of the fluid.

The velocity field is

$$u^\alpha = \frac{dr^\alpha}{dt} = H^{\alpha\beta} r^\beta, \qquad H^{\alpha\beta} = \frac{dA^{\alpha\gamma}}{dt} A^{-1\gamma\beta}. \tag{20.15}$$

This is the generalization of $\mathbf{u} = (\dot{a}/a)\mathbf{r}$ for homogeneous isotropic motion. The volume of the ellipsoid is $4\pi/3$ in x-units, according to equation (20.13), so the proper volume is

$$V = \tfrac{4}{3}\pi A, \qquad A = \det|A^{\alpha\beta}|. \tag{20.16}$$

By using the matrix identity

$$\frac{dA}{dt} = A \frac{dA^{\alpha\beta}}{dt} A^{-1\beta\alpha}, \tag{20.17}$$

one finds

$$\frac{1}{V}\frac{dV}{dt} = H^{\alpha\alpha}, \tag{20.18}$$

as expected because the trace of H is the divergence of $\mathbf{u}$. The kinetic energy is

$$K = \frac{1}{2}\rho \int u^2 d^3r = \frac{1}{2}\rho A \frac{dA^{\alpha\beta}}{dt}\frac{dA^{\alpha\gamma}}{dt} \int x^\beta x^\gamma d^3x = \frac{1}{10} M \frac{dA^{\alpha\beta}}{dt}\frac{dA^{\alpha\beta}}{dt}. \tag{20.19}$$

The second step follows on changing variables from $\mathbf{r}$ to $\mathbf{x}$, the third if the surface of the ellipsoid is given by equation (20.13).

The result of differentiating $H^{\alpha\beta}$ is

$$\frac{dA^{-1\alpha\beta}}{dt} = -A^{-1\alpha\gamma}\frac{dA^{\gamma\delta}}{dt}A^{-1\delta\beta},$$
$$\frac{dH^{\alpha\beta}}{dt} + H^{\alpha\gamma}H^{\gamma\beta} = -\Phi_{\alpha\beta}. \tag{20.20}$$

The first equation is the expression for the derivative of an inverse of a matrix, the second follows from equation (20.10). It is convenient to reduce $H^{\alpha\beta}$ to

$$H^{\alpha\beta} = \Theta\frac{\delta^{\alpha\beta}}{3} + \sigma_{\alpha\beta} - \omega_{\alpha\beta},$$
$$\sigma_{\alpha\beta} = \sigma_{\beta\alpha}, \qquad \sigma_{\alpha\alpha} = 0, \qquad \omega_{\alpha\beta} = -\omega_{\beta\alpha}. \tag{20.21}$$

The antisymmetric part is the vorticity $\omega_{\alpha\beta}$; the trace-free symmetric part is the shear $\sigma_{\alpha\beta}$, and the trace is the divergence or expansion $\Theta = \nabla \cdot \mathbf{u}$. The result of substituting equation (20.21) into equation (20.20) and separating symmetric and antisymmetric parts is

$$\frac{d\Theta}{dt} + \frac{1}{3}\Theta^2 + \sigma^2 - \omega^2 = -\Phi_{\alpha\alpha} = -4\pi G\rho(t),$$

$$\sigma^2 = \sigma_{\alpha\beta}\sigma_{\alpha\beta}, \qquad \omega^2 = \omega_{\alpha\beta}\omega_{\alpha\beta}, \qquad \frac{d\rho}{dt} = -\rho\Theta,$$

$$\frac{d\sigma_{\alpha\beta}}{dt} + \frac{2}{3}\Theta\sigma_{\alpha\beta} + \sigma_{\alpha\gamma}\sigma_{\gamma\beta} - \frac{1}{3}\delta^{\alpha\beta}\sigma^2 + \omega_{\alpha\gamma}\omega_{\gamma\beta} + \frac{1}{3}\delta^{\alpha\beta}\omega^2 = -\Phi_{\alpha\beta} + \frac{1}{3}\delta^{\alpha\beta}\Phi_{\gamma\gamma}, \tag{20.22}$$

$$\frac{d\omega_{\alpha\beta}}{dt} + \frac{2}{3}\Theta\omega_{\alpha\beta} + \sigma_{\alpha\gamma}\omega_{\gamma\beta} + \omega_{\alpha\gamma}\sigma_{\gamma\beta} = 0.$$

The gravitational source of $\sigma_{\alpha\beta}$ is the trace-free part of $\Phi_{\alpha\beta}$, which is the homogeneous part of the solution to Poisson's equation. This is the tidal field determined by the shape of the surface of the ellipsoid.

A spheroid has a fixed axis of symmetry, which can be taken to be the coordinate axis x^3. Then the rotational symmetry about x^3 implies that $\sigma_{\alpha\beta}$ is diagonal with $\sigma_{11} = \sigma_{22}$, and the only nonzero components of $\omega_{\alpha\beta}$ are

$$\omega_{12} = -\omega_{21} = \Omega, \tag{20.23}$$

the angular velocity of rotation. The radial component of the velocity evaluated at the surface of the ellipsoid at the pole and equator gives the rates of change of the semiaxes,

$$\begin{aligned} \frac{da}{dt} &= H^{11}a = \left(\frac{1}{3}\Theta + \sigma_{11}\right)a, \\ \frac{dc}{dt} &= H^{33}c = \left(\frac{1}{3}\Theta + \sigma_{33}\right)c, \end{aligned} \tag{20.24}$$

and on differentiating these expressions and using equation (20.22), one finds the equations of motion

$$\frac{1}{a}\frac{d^2a}{dt^2} = \Omega^2 - 2\pi G\rho A, \qquad \frac{1}{c}\frac{d^2c}{dt^2} = -2\pi G\rho C. \tag{20.25}$$

The last of equations (20.22) with equation (20.23) gives

$$\frac{d\Omega}{dt} + \frac{2\Omega}{a}\frac{da}{dt} = 0, \qquad \Omega \propto a^{-2}. \tag{20.26}$$

As expected, the acceleration of the radius has the centrifugal term $a\Omega^2$, and Ω varies inversely as the square of the radius.

The kinetic energy of the spheroid is, according to equation (20.19),

$$K = \frac{M}{10}\left[2\left(\frac{da}{dt}\right)^2 + \left(\frac{dc}{dt}\right)^2 + 2a^2\Omega^2\right], \tag{20.27}$$

and the potential energy is given by equation (20.5).

Figures 20.1 and 20.2 show examples of the evolution of an oblate spheroid with no rotation. In both cases the initial expansion rate is isotropic,

$$\frac{1}{a}\frac{da}{dt} = \frac{1}{c}\frac{dc}{dt}, \qquad t = t_i, \tag{20.28}$$

the initial axis ratio is $c/a = 0.8$, and the unit on the vertical axes is the initial value of a for the ellipsoid. In Figure 20.1 the spheroid has negative energy: the initial kinetic energy is 0.9 times the magnitude of the initial

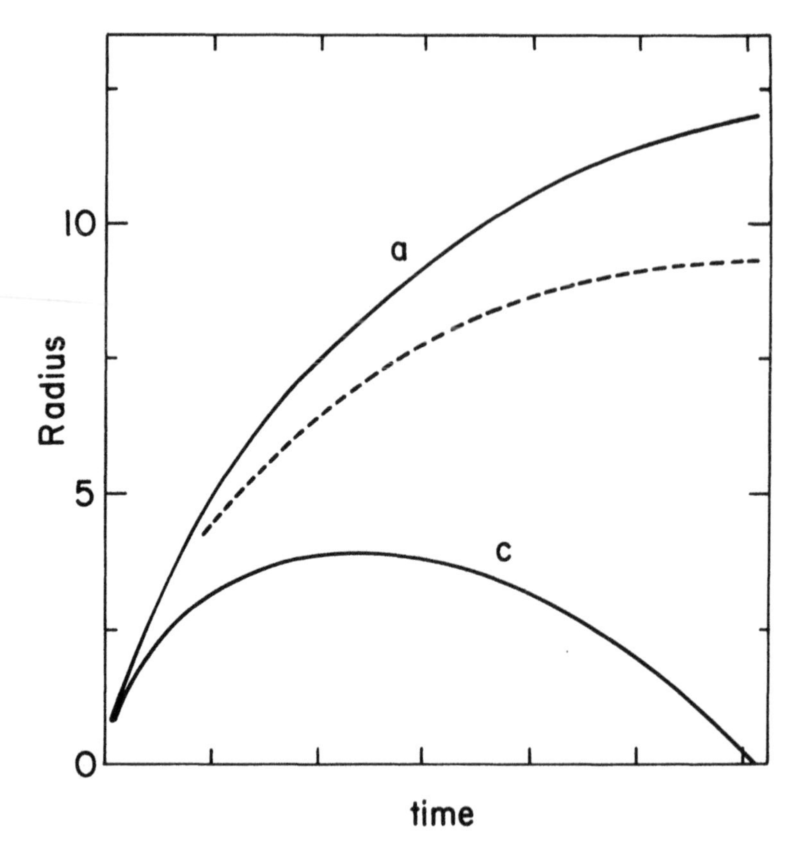

FIG. 20.1. Evolution of an oblate spheroid. The initial axis ratio is $c/a = 0.8$ and the initial ratio of kinetic to potential energy is $K/W = 0.9$. The dotted line shows the radius of a sphere with the same initial volume, mass and K/W.

potential energy. The dashed curve shows the radius of a homogeneous sphere with the same initial volume, mass, and ratio of kinetic to potential energy as for the spheroid. In Figure 20.2 the spheroid has positive energy: the initial kinetic energy is 1.01 times the magnitude of the initial potential energy. It is interesting that this system nevertheless collapses in the narrow direction.

As the spheroids approach the pancake singularity $c \to 0$, equations (20.3) and (20.25) give

$$\frac{d^2 a}{dt^2} = a\Omega^2 - \frac{3\pi}{4}\frac{GM}{a^2}, \qquad \frac{d^2 c}{dt^2} = -3\frac{GM}{a^2}. \tag{20.29}$$

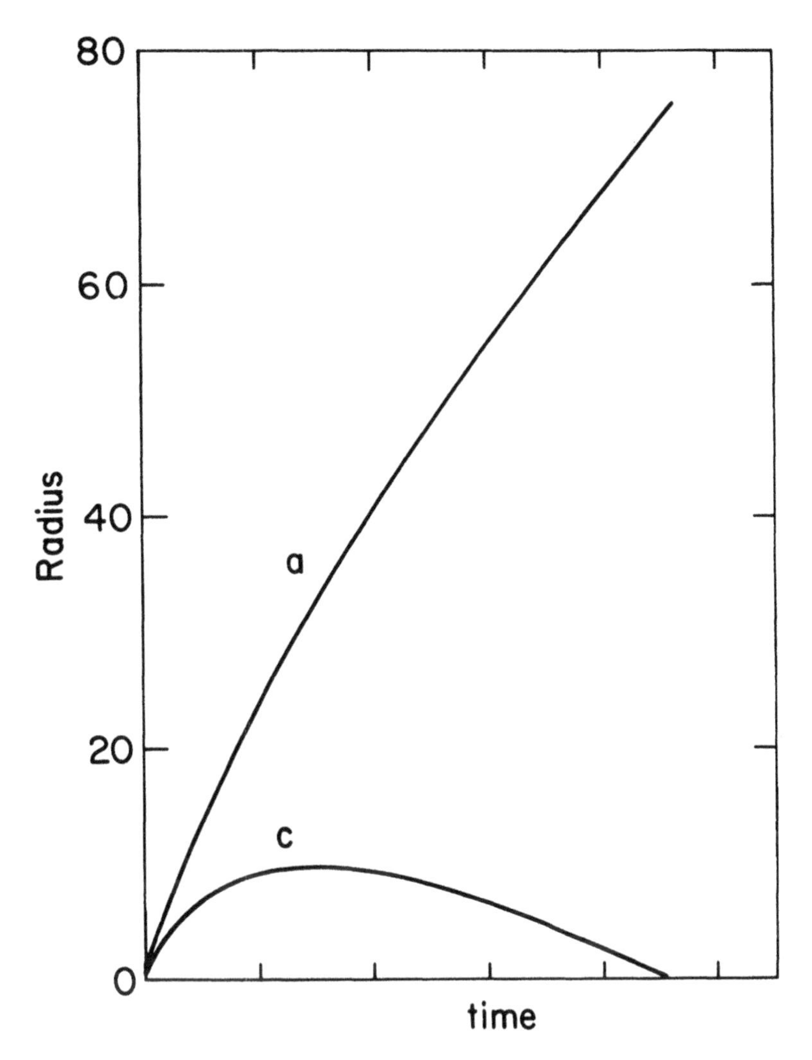

FIG. 20.2. Evolution of an oblate spheroid. The initial axis ratio is $c/a = 0.8$ and $K/W = 1.01$.

The first equation says the gravitational acceleration of the radius of the pancake is close to that of a sphere with the same mass and radius. The second equation says the acceleration of c is comparable to that of a: the mass per unit area of the pancake is $\sigma = 3M/2\pi a^2$ at the axis, and since it approximates a uniform sheet, the acceleration is nearly uniform at

$$g = 2\pi G\sigma = 3GM/a^2 \qquad (20.30)$$

just above the slab. Thus one sees in the figures that the graphs of $a(t)$ and $c(t)$ have similar curvature. Since $c \rightarrow 0$ at nearly constant speed, one can say the pancake singularity is kinematic. Rotation tends to slow the collapse by increasing a, but once $dc/dt < 0$ the approach to the singularity is little affected.

If the three axes are different, for example, $a > b > c$, then $A < B < C$ in equation (20.2), which tends to make $c(t)$ reach zero first, producing an elliptical pancake. Of course, this is prevented if the expansion rate along the c axis is large, but then one of the other axes assumes the role of c. Only with very special initial conditions would two axes reach zero at the same time (hot dog singularity).

The ellipsoidal model does not give a good approximation to the behavior of a small density fluctuation, $\delta\rho/\rho \ll 1$ because it assumes there is no matter outside the ellipsoid. Consider for example an ellipsoid with uniform mass density ρ surrounded by matter at uniform density ρ_b, with $|\rho/\rho_b - 1| \ll 1$. The gravitational field of the matter inside the ellipsoid tends to make it promptly collapse to a pancake. The tidal field of the matter outside almost exactly cancels this, with the result that $\delta\rho/\rho$ grows slowly, as described by linear perturbation theory, the spheroid preserving its shape. Only when the density contrast has become appreciable, perhaps $\rho/\rho_b - 1 \sim 1$, is it reasonable to ignore the matter outside and apply the spheroid model. Prior to that the spherical model is a better approximation because there is no tidal field due to the outside matter.

Figures 20.1 and 20.2 show that once the contrast reaches $\rho/\rho_b - 1 \sim 1$ the distribution tends to develop strongly nonradial motions, which of course would vitiate the spherical model. It is occasionally argued that once a protogalaxy or protocluster breaks away from the general expansion it has little kinetic energy, so it must collapse a factor of about 2 in radius to convert enough potential to kinetic energy to satisfy the virial theorem (for example, Gott and Rees 1975, Sargent and Turner 1977, Press and Lightman 1978). This could be an important effect for it would indicate that a stable virialized cluster must have a density well above the background. However, the spheroid model shows that the argument is doubtful. In the example shown in Figure 20.1, when the equivalent sphere has reached the point of maximum expansion, the spheroid has collapsed to a pancake and in the process developed strongly nonradial motion. At this time the ratio of kinetic to potential energy in the sphere is of course $K/W = 0$ while the shear of the spheroid makes $K/W = 0.19$. The model still is highly simplified, of course, and it seems reasonable to expect that adding to the complexity would tend to increase the kinetic energy at the nominal point of maximum expansion and perhaps even eliminate any collapse phase. The relation to the shape of the galaxy two-point correlation function is discussed in Section 71.

21. Caustics and pancakes

In some cases the development of the first generation of mass concentrations in an expanding universe can be likened to the formation of caustic surfaces in ray optics—the matter piles up in sheets or pancakes. The central assumption is that the initial density and velocity fields are smoothly varying functions of position. A small patch of matter thus behaves like a homogeneous distribution in the tidal field of the matter nearby, and in the manner discussed in the last section an initially spherical section of the fluid evolves into an ellipsoid and may end up as a flat pancake. This concept has been used in discussions of the nature of the singularity at the moment of the big bang in inhomogeneous cosmological models (Lemaître 1933b, § 12; Lifshitz and Khalatnikov 1963); its importance as a way to picture the nature of the first generation of mass concentrations in an expanding universe was pointed out by Zel'dovich (1970). Reviews of the implications for scenarios of galaxy and cluster formation are given by Zel'dovich (1978) and Doroshkevich, Shandarin, and Saar (1978).

The following is a formal treatment of the formation of pancakes or caustic surfaces. The assumptions are that matter can be approximated as an ideal fluid with initial values of ρ and $\mathbf{v}$ smoothly varying functions of position and that nongravitational forces are quite negligible and orbits can intersect. This last point, of course, is not realistic if the matter is a gas: the results would have to be modified once the first pancakes form. The pancake formation is discussed first in general terms, then in a convenient kinematic approximation.

A. General treatment

Consider an observer moving with one of the fluid elements. The position of a neighboring mass element measured in an inertial frame with origin at the observer is $\mathbf{s}(t)$, where t is the observer's proper time. The local gravitational potential mapped by the observer is $\Phi(\mathbf{s}, t)$. Since the mass distribution is supposed to be smooth, we can expand Φ in a series in $\mathbf{s}$ and keep the lowest nontrivial part. The first derivative of Φ vanishes because the fluid and observer are freely moving, so

$$\Phi = \tfrac{1}{2}\Phi_{\alpha\beta}(t)s^{\alpha}s^{\beta}, \qquad (21.1)$$

and Poisson's equation gives

$$\Phi_{\alpha\alpha} = 4\pi G\rho, \tag{21.2}$$

where ρ is the mass density at the observer. The equations of motion of the neighboring mass element are then

$$d^2 s^\alpha / dt^2 = -\Phi_{\alpha\beta} s^\beta. \tag{21.3}$$

Since this equation is linear in **s** and the first nontrivial term in the expansion of the velocity field is also linear, the motion of the matter in the neighborhood of the observer may be written as

$$s^\alpha(t) = A^{\alpha\beta}(t) s_i^\beta, \tag{21.4}$$

where s_i^β is the position of the mass element at the starting time t_i. This expression in equation (21.3) gives

$$d^2 A^{\alpha\beta} / dt^2 = -\Phi_{\alpha\gamma} A^{\gamma\beta}. \tag{21.5}$$

The local velocity field is

$$\begin{aligned} u^\alpha &= \frac{ds^\alpha}{dt} = \frac{dA^{\alpha\beta}}{dt} s_i^\beta = H^{\alpha\beta}(t) s^\beta, \\ H^{\alpha\beta} &= \frac{dA^{\alpha\gamma}}{dt} A^{-1\gamma\beta}, \end{aligned} \tag{21.6}$$

and the result of differentiating $H^{\alpha\beta}$ and using equation (21.5) is

$$\frac{d}{dt} H^{\alpha\beta} + H^{\alpha\gamma} H^{\gamma\beta} = -\Phi_{\alpha\beta}. \tag{21.7}$$

Equations (21.3) to (21.7) can be compared to equations (20.7) to (20.20) for a homogeneous ellipsoid. There is the important difference that here the trace-free part of $\Phi_{\alpha\beta}$, which represents the tidal stress due to nearby matter, would be known only if the full three-dimensional mass distribution were computed. However, some useful results may be obtained even if the tidal force is not known.

The variable s_i^α is fixed to a fluid element, so a volume element fixed to the fluid varies as

$$d^3s = P(t)d^3s_i, \qquad P = \det|A^{\alpha\beta}(t)|, \tag{21.8}$$

where P is the Jacobian of the transformation from $\mathbf{s}_i$ to $\mathbf{s}$. Since the mass in the element is fixed, the density is

$$\rho(t) = \rho_i/P(t). \tag{21.9}$$

Using equation (20.17), one can verify that this is consistent with the usual equation of mass conservation,

$$\frac{1}{\rho}\frac{d\rho}{dt} = -\nabla \cdot \mathbf{u} = -H^{\alpha\alpha}. \tag{21.10}$$

One can see from equation (21.9) that where the Jacobian P vanishes the density is infinite because neighboring orbits are intersecting. The two-dimensional surface $P = 0$ at fixed t defines sheets $\rho \longrightarrow \infty$ analogous to caustic surfaces in ray optics (Landau and Lifshitz 1979, p. 142).

If at $t = t_1$ the observer moving with the particle $x_i(1)$ is on one of these caustic surfaces, then

$$A^{\alpha\beta}(1)d^\beta = 0 \tag{21.11}$$

has a nontrivial solution d^β because the determinant of $A^{\alpha\beta}$, which is P, vanishes. According to equation (21.4), the particles with initial positions $\mathbf{s}_i$ along the (infinitesimal) line segment $\mathbf{s}_i \propto \mathbf{d}$ are at $\mathbf{s} = 0$ at t_1; the orbits of these particles have intersected at the common point $\mathbf{x}_1 = \mathbf{x}(\mathbf{x}_i(1), t_1)$.

The density near $\mathbf{x}_1$ at t_1 is estimated as follows. The family of particles initially along the line

$$\mathbf{x}_i = \mathbf{x}_i(1) + s\mathbf{d}, \tag{21.12}$$

where the parameter s measures the position along d, is at time $t = t_1$ along the line

$$x^\alpha = x_1{}^\alpha + \frac{1}{2}\frac{\partial A^{\alpha\beta}}{\partial x_i{}^\gamma} d^\beta d^\gamma s^2 + \cdots, \tag{21.13}$$

the term linear in s vanishing, according to equation (21.11). The Jacobian P can be expanded along the line (21.12) by using equation (20.17):

$$P = P(1) + \frac{\partial P}{\partial x_i{}^\mu} sd^\mu = PA^{-1\alpha\beta}\frac{\partial A^{\beta\alpha}}{\partial x_i{}^\mu} sd^\mu. \tag{21.14}$$

The first term vanishes because $P = 0$ at the caustic. In the second term $PA^{-1\alpha\beta}$ generally is finite because it is the matrix of the subdeterminants of $A^{\alpha\beta}$. The density along the line (21.12) thus varies with distance $|\mathbf{x} - \mathbf{x}_1|$ (eq. 21.13) from the caustic as

$$\rho \propto |\mathbf{x} - \mathbf{x}_1|^{-1/2}. \tag{21.15}$$

The local gravitational field measured by the observer at $\mathbf{x}_1$ varies with distance from the caustic as the mass per unit area,

$$g \propto |\mathbf{x} - \mathbf{x}_1|^{1/2}. \tag{21.16}$$

Although the density diverges, the gravitational acceleration is finite varying smoothly across the caustic.

Next let us consider the relative velocities of the particles along the intersecting orbits as $t \rightarrow t_1$. For the observer $x_i(1)$,

$$A^{\alpha\beta}(t) \simeq A^{\alpha\beta}(t_1) + (t - t_1)\frac{dA^{\alpha\beta}}{dt}, \tag{21.17}$$

and so for the family of particles along the line (21.12), the relative positions at t close to t_1 are

$$s^\alpha(t) = (t - t_1)\frac{dA^{\alpha\beta}}{dt} s d^\beta \equiv (t - t_1) q^\alpha s, \tag{21.18}$$

where q^α gives the direction of the line of converging particles as $t \rightarrow t_1$. The relative velocity along this line for particles separated by proper distance r is

$$\mathbf{u} = -\mathbf{r}/(t_1 - t), \qquad \mathbf{r} \propto \mathbf{q}. \tag{21.19}$$

Since the time derivative of P at $(\mathbf{x}_i(1), t_1)$ generally does not vanish, the density measured by the observer at $x_i(1)$ varies with time as $t \rightarrow t_1$ as

$$\rho \propto |t - t_1|^{-1}. \tag{21.20}$$

This agrees with equations (21.10) and (21.19).

The particles initially on the small sphere

$$\sum_\beta (s_i^\beta)^2 = \text{constant} \tag{21.21}$$

centered on $\mathbf{x}_i(1)$ later are on the surface of the ellipsoid (eq. 21.4)

$$\sum_{\alpha} (A^{-1\alpha\beta} s^{\beta})^2 = \text{constant}, \tag{21.22}$$

and at $t \rightarrow t_1$ have flattened to a pancake along the caustic. If $\mathbf{s}_i$ is not parallel to $\mathbf{d}$ (eq. 21.11), then $A^{\alpha\beta} s_i^{\beta}$ ends up in the plane of the pancake, so the normal n^{α} to the caustic satisfies

$$n^{\alpha} A^{\alpha\beta}(t_1) s_i^{\beta} = 0. \tag{21.23}$$

Since this is true for any s_i^{β}, the normal is fixed by

$$n^{\alpha} A^{\alpha\beta}(t_1) = 0. \tag{21.24}$$

If $H^{\alpha\beta}$ (eq. 21.6) is symmetric, $\mathbf{n}$ is parallel to $\mathbf{q}$ (eq. 21.18). To see this, note that the symmetry of $H^{\alpha\beta}$ gives

$$\frac{dA^{\alpha\gamma}}{dt} A^{-1\gamma\beta} = \frac{dA^{\beta\gamma}}{dt} A^{-1\gamma\alpha}, \qquad A^{\alpha\delta} \frac{dA^{\alpha\gamma}}{dt} A^{-1\gamma\beta} = \frac{dA^{\beta\delta}}{dt}. \tag{21.25}$$

When $t \rightarrow t_1$, $A^{-1\gamma\beta}$ diverges as P^{-1}, so generally (unless the determinant of $PA^{-1\gamma\beta}$ vanishes)

$$A^{\alpha\delta}(1) \frac{dA^{\alpha\gamma}}{dt}(1) = 0, \qquad q^{\alpha} A^{\alpha\delta}(1) = 0, \tag{21.26}$$

where the second equation follows from equation (21.18). Comparing this with equation (21.24), one sees that generally

$$q_{\alpha} = n_{\alpha}, \tag{21.27}$$

that is, the relative velocities of the particles moving toward intersection at t_1 are normal to the pancake. One can understand this by noting that if $H^{\alpha\beta}$ is symmetric, coordinates can be oriented so it is diagonal with $H^{11} = -(t_1 - t)^{-1}$ (eq. 21.19) and H_{22} and H_{33} finite, representing motion along the (1) axis to a pancake in the (2, 3) plane.

Equation (21.27) assumes $H^{\alpha\beta}$ is symmetric. This means the flow has no vorticity, $\nabla \times \mathbf{u} = 0$ (eq. 21.6). One sees from equation (21.7) that if $H^{\alpha\beta}$ is symmetric to begin with, it remains symmetric because $\Phi^{\alpha\beta}$ is symmetric. Vorticity is discussed further in Section 22.

B. Kinematic approximation

The discussion can be made more direct by noting that once the density fluctuations have grown large, the velocities are large and in the final stages of collapse gravitational acceleration can be ignored (Zel'dovich, 1970). Accordingly let us start at epoch t_i, when the density contrast is roughly $|\delta_i| \sim 1$, and let $u^\alpha(\mathbf{r}_i)$ be the velocity of the mass element at $\mathbf{r}_i$ at t_i. Both $\mathbf{r}$ and $\mathbf{u}$ will be proper inertial coordinates relative to some convenient origin.

The subsequent gravitational acceleration will be ignored so at time $t_i + \tau$ the particle originally at $\mathbf{r}_i$ is at

$$r^\alpha = r_i^\alpha + u^\alpha(r_i^\beta)\tau. \tag{21.28}$$

For the neighboring mass elements originally at $\mathbf{r}_i$ and $\mathbf{r}_i + \mathbf{s}_i$, the relative position as a function of time is formed by differentiating this,

$$\begin{gathered} s^\alpha(\tau) = A^{\alpha\beta}(\tau)s_i^\beta, \qquad A^{\alpha\beta}(\tau) = \delta^{\alpha\beta} + \tau\Delta^{\alpha\beta}, \\ \Delta^{\alpha\beta} = \partial u^\alpha/\partial r_i^\beta. \end{gathered} \tag{21.29}$$

Equation (21.28) shows how $\mathbf{r}_i$ maps to $\mathbf{r}$, so the density at the element labeled $\mathbf{r}_i$ is

$$\rho(\tau) = \rho_i/P(\mathbf{r}_i, \tau), \qquad P = \det|A^{\alpha\beta}|. \tag{21.30}$$

This is a special case of equations (21.4) and (21.8).

On multiplying out the determinant and collecting powers of τ, one finds (Peebles 1971a)

$$\begin{aligned} P &= \det|\delta^{\alpha\beta} + \tau\partial u^\alpha/\partial r_i^\beta| \\ &= 1 + \tau\nabla\cdot\mathbf{u}^\alpha + \frac{1}{2}\tau^2\left[(\nabla\cdot\mathbf{u})^2 + (\nabla\times\mathbf{u})^2 - \sum_{\alpha\beta}(\Delta^{\alpha\beta})^2\right] \\ &\qquad + \tau^3\det|\Delta^{\alpha\beta}|. \end{aligned} \tag{21.31}$$

At small τ this gives $\rho \propto 1 - \tau\nabla\cdot\mathbf{u}$, in agreement with linear perturbation theory. If the initial velocity field has zero divergence (as might be assumed if t_i is the epoch of decoupling of matter and radiation and if prior to decoupling the matter-radiation fluid were in turbulent incompressible flow), the contrast starts to grow as τ^2. In each of the three terms in powers of τ the characteristic time for P to change by a factor of 2 is $\sim\lambda/v$, where λ is the coherence length for $\mathbf{u}$; this is simply the characteristic crossing time.

As pointed out by Zel'dovich, equation (21.28) offers a way to extrapolate beyond perturbation theory to the development of the first strong mass concentrations. The discussion closely follows part (A) above. For fixed mass element $\mathbf{r}_i$ the Jacobian P is a cubic polynomial in τ. Suppose it has a positive zero, and let τ_1 be the smallest positive zero. Then the density at $\mathbf{r}_i$ diverges at $\tau \rightarrow \tau_1$. Since the determinant of $A^{\alpha\beta}$ vanishes, there is a nontrivial solution to

$$A^{\alpha\beta}(\tau_1) d^\beta = 0. \tag{21.32}$$

Then since $A = I + \tau\Delta$ (eq. 21.29), $\mathbf{d}$ is an eigenvector of Δ,

$$\Delta^{\alpha\beta} d^\beta = -\tau_1^{-1} d^\alpha, \tag{21.33}$$

and so $\mathbf{d}$ also is an eigenvector of $A^{\alpha\beta}(\tau)$,

$$A^{\alpha\beta}(\tau) d^\beta = (1 - \tau/\tau_1) d^\alpha. \tag{21.34}$$

If the initial relative position vector $\mathbf{s}_i$ for neighboring mass elements is chosen to be along d^α, then equations (21.29) and (21.34) indicate the separation varies with time as

$$\mathbf{s}(\tau) = (1 - \tau/\tau_1)\mathbf{s}_i, \qquad \mathbf{s} \propto \mathbf{d}, \tag{21.35}$$

so these particles are moving toward intersection at $\tau = \tau_1$.

The relative velocity field within a small patch of matter around $\mathbf{r}_i$ is, according to equations (21.28) and (21.29),

$$\delta u^\alpha = \frac{\partial u^\alpha}{\partial r_i^\beta} s_i^\beta = \Delta^{\alpha\beta} A^{-1\beta\gamma} s^\gamma = \tau^{-1}(\delta^{\alpha\beta} - A^{-1\alpha\beta}) s^\beta. \tag{21.36}$$

If $\mathbf{s}$ is parallel to $\mathbf{d}$, then equation (21.34) gives

$$A^{-1\alpha\beta} d^\beta = (1 - \tau/\tau_1)^{-1} d^\alpha, \tag{21.37}$$

and with equation (21.36)

$$\delta\mathbf{u} = \frac{-\mathbf{s}}{\tau_1 - \tau}, \qquad \mathbf{s} \propto \mathbf{d}. \tag{21.38}$$

This describes the motion toward intersection of orbits. If $\mathbf{u}(\mathbf{r}_i)$ is irrotational, a reasonable assumption, then $\Delta^{\alpha\beta}$ is symmetric and so has three orthogonal eigenvectors $d_n^{\ \alpha}$, with eigenvalues $-\tau_n^{\ -1}$, τ_1 being the smallest positive time. With coordinates along the d_n^{α}, the $A^{\alpha\beta}$ and $H^{\alpha\beta}$ tensors are diagonal,

$$A^{\alpha\beta} = \delta^{\alpha\beta}(1 - \tau/\tau_\alpha),$$
$$H^{\alpha\beta} = -\delta^{\alpha\beta}/(\tau_\alpha - \tau). \tag{21.39}$$

C. Remarks

This calculation rests on two assumptions, that ρ and $\mathbf{v}$ are smooth fields to begin with and that pressure may be neglected. It is doubtful that there is any scale in the present universe for which the first is a useful approximation, for the mass density has no coherence length: the density autocorrelation function rises fairly smoothly with decreasing r all the way down to atomic distances. It is a matter of speculation whether in the past ρ had a useful coherence length. As discussed in Chapter VI this does happen, and the Jeans length is much less than the coherence length, if prior to decoupling of matter and radiation, the entropy per baryon is constant and the density irregularities consist of adiabatic acoustic waves. The short wavelength part is strongly dissipated during decoupling by photon diffusion; the resulting coherence length contains a mass somewhere between that of a giant galaxy and that of a cluster of galaxies, depending on the assumed cosmological density parameter. The development of mass concentrations in this picture has been discussed in detail by Zel'dovich and his colleagues (Zel'dovich 1978 and references therein). The first dense spots would be pancakes with matter entering the pancake normal to its surface (since $\nabla \times \mathbf{v}$ would be small), at first with low relative velocity. If orbits were allowed to cross, the caustic surfaces would move about as different orbits intersect. Since gas clouds cannot appreciably interpenetrate on the scale of a galaxy, once the first sheets form matter tends to pile up in them at steadily increasing velocity until a shock forms parallel to the pancake. In the later stages of accumulation of matter in the pancake the shock perhaps would tend to become oblique. The pancake once formed is unstable; it tends to fragment into lumps that might be protogalaxies. The pancake surface has a radius of curvature comparable to the original coherence length of $u(\mathbf{r}_i)$, so the sheet-like distribution of objects would tend to be washed out in a crossing time $\sim\lambda/u$. It is not clear whether it could be arranged that this mixing time is comparable to the present age of the universe, so that some trace of the sheet-like distribution might remain. It is interesting that Einasto (1978) has found observational evidence of a sheet-like character in the large-scale distribution of galaxies.

22. Expansion, vorticity, and shear

The equations describing the behavior of a small patch of matter were obtained in the last section under the assumption that pressure may be neglected. If the matter is approximated as an ideal fluid with pressure $p \ll \rho c^2$, one can take account of the pressure gradient force by noting that in a locally Minkowski coordinate frame the acceleration of the fluid element at r^α is

$$\frac{d^2 r^\alpha}{dt^2} = -\frac{\partial \Phi}{\partial r^\alpha}(\mathbf{r}(t), t) - \frac{1}{\rho}\frac{\partial p}{\partial r^\alpha}, \tag{22.1}$$

so the relative acceleration of elements separated by the small distance s^α is

$$\frac{d^2 s^\alpha}{dt^2} = -\frac{\partial^2 \Phi}{\partial r^\alpha \partial r^\beta} s^\beta - s^\beta \frac{\partial}{\partial r^\beta}\left(\frac{1}{\rho}\frac{\partial p}{\partial r^\alpha}\right). \tag{22.2}$$

The calculation leading to equation (21.7) then gives the relative velocity of the two fluid elements as

$$\begin{gathered} u^\alpha = H^{\alpha\beta} s^\beta, \qquad \Phi_{,\alpha\alpha} = 4\pi G \rho, \\ \frac{dH^{\alpha\beta}}{dt} + H^{\alpha\gamma} H^{\gamma\beta} = -\Phi_{,\alpha\beta} - \frac{\partial}{\partial r^\beta}\left(\frac{1}{\rho}\frac{\partial p}{\partial r^\alpha}\right). \end{gathered} \tag{22.3}$$

If the flow conserves entropy (or more generally if p can be written as a function of ρ), this last equation may be rewritten as

$$\frac{dH^{\alpha\beta}}{dt} + H^{\alpha\gamma} H^{\gamma\beta} = -\Phi_{,\alpha\beta} - h_{,\alpha\beta}, \tag{22.4}$$

where h is the enthalpy per unit mass, $dh = dp/\rho$.

As in Section 20 it is convenient to break $H^{\alpha\beta}$ into the parts

$$\begin{gathered} H^{\alpha\beta} = \tfrac{1}{3}\Theta\delta^{\alpha\beta} + \sigma_{\alpha\beta} - \omega_{\alpha\beta}, \\ \omega_{\alpha\beta} = \tfrac{1}{2}\epsilon_{\alpha\beta\gamma}\omega_\gamma, \qquad \omega_\alpha = \epsilon_{\alpha\beta\gamma} H^{\gamma\beta}, \qquad \boldsymbol{\omega} = \nabla \times \mathbf{u}, \end{gathered} \tag{22.5}$$

where $\epsilon_{\alpha\beta\gamma}$ is the completely antisymmetric tensor ($\epsilon_{123} = 1$), Θ is the expansion, the trace-free symmetric part $\sigma_{\alpha\beta}$ is the shear, and the antisymmetric part $\boldsymbol{\omega}$ is the vorticity.

Mass conservation (eq. 21.10) gives

$$\frac{1}{\rho}\frac{d\rho}{dt} = -H^{\alpha\alpha} = -\Theta. \tag{22.6}$$

The trace of equation (22.4) with equation (22.5) yields

$$\frac{d\Theta}{dt} + \frac{\Theta^2}{3} + \sigma^2 - \omega^2 = -4\pi G\rho - \nabla^2 h, \tag{22.7}$$

where the square of the shear and vorticity are

$$\sigma^2 = \sum_{\alpha,\beta} (\sigma_{\alpha\beta})^2, \qquad \omega^2 = \sum_{\alpha,\beta} (\omega_{\alpha\beta})^2 = \frac{1}{2}(\nabla \times \mathbf{u})^2. \tag{22.8}$$

If $h = 0$, equation (22.7) with equation (22.6) is the relation derived in general relativity theory by Raychaudhuri (1955). Though the calculation is framed here in Newtonian terms, the derivation is general because the Newtonian approximation accurately describes the internal motions $v \ll c$ in a small enough patch of matter (§ 6). As in Section 10, equation (22.7) is local because the gravitational field of the matter outside the patch affects only the tidal field, which has zero divergence.

The antisymmetric part of equation (22.4) with equation (22.5) is the vorticity equation (compare eq. 20.22)

$$\begin{aligned} \frac{d\omega_{\alpha\beta}}{dt} + \frac{2}{3}\Theta\omega_{\alpha\beta} + \sigma_{\alpha\gamma}\omega_{\gamma\beta} + \omega_{\alpha\gamma}\sigma_{\gamma\beta} &= 0, \\ \frac{d\omega_\alpha}{dt} + \frac{2}{3}\Theta\omega_\alpha - \sigma_{\alpha\beta}\omega_\beta &= 0. \end{aligned} \tag{22.9}$$

If $\omega = 0$ to begin with, this equation says it remains zero: no vorticity is created. Of course vorticity is created if the matter stress represented by the last term in equation (22.3) is not symmetric, as in an oblique shock or a strongly viscous medium. For incompressible flow $\rho \propto a^{-3}$ as the universe expands so $\Theta = 3\dot{a}/a$ and equation (22.9) becomes

$$\frac{d}{dt} a^2\omega_\alpha = a^2\sigma_{\alpha\beta}\omega_\beta. \tag{22.10}$$

The trace-free symmetric part of equation (22.4) yields the shear equation (20.22).

These equations describe the behavior of a fixed patch of matter. It is also useful to write down the time derivatives at fixed coordinate positions: convenient variables are the expanding cosmological coordinates x^α, cosmic time t, and proper peculiar velocity v^α. In these variables the equations describing an ideal nonrelativistic fluid are (eqs. 9.15, 9.17)

$$\frac{\partial \rho}{\partial t} + 3\frac{\dot{a}}{a}\rho + \frac{1}{a}\nabla \cdot \rho \mathbf{v} = 0,$$
$$\frac{\partial \mathbf{v}}{dt} + \frac{1}{a}(\mathbf{v} \cdot \nabla)\mathbf{v} + \frac{\dot{a}}{a}\mathbf{v} = -\frac{1}{\rho a}\nabla p - \frac{1}{a}\nabla\phi \tag{22.11}$$

The source for ϕ is the departure from homogeneity, $\rho - \rho_b$.

The analog of equation (22.6) is

$$\frac{1}{\rho}\left(\frac{\partial}{\partial t} + \frac{1}{a}\mathbf{v} \cdot \nabla\right)\rho = -\Theta,$$
$$\Theta = \theta + 3\frac{\dot{a}}{a}, \qquad \theta = \frac{1}{a}\nabla \cdot \mathbf{v}. \tag{22.12}$$

The analog of equation (22.7) is found by taking the divergence of the second of equations (22.11) and using the definitions of shear and vorticity in terms of these variables,

$$v^\alpha{}_{,\beta}/a = {}^1\!/\!_3\theta\delta^{\alpha\beta} + \sigma_{\alpha\beta} - {}^1\!/\!_2\epsilon_{\alpha\beta\gamma}\omega_\gamma. \tag{22.13}$$

The result is

$$\left(\frac{\partial}{\partial t} + 2\frac{\dot{a}}{a} + \frac{1}{a}\mathbf{v} \cdot \nabla\right)\theta + \frac{\theta^2}{3} + \sigma^2 - \frac{1}{2}\omega \cdot \omega$$
$$= -4\pi G(\rho - \rho_b) - \nabla^2 h/a^2. \tag{22.14}$$

The vorticity equation is found by taking the curl of the second of equations (22.11). The only complicated term is the second,

$$[\nabla \times (\mathbf{v} \cdot \nabla)\mathbf{v}]_\alpha = a(\mathbf{v} \cdot \nabla)\omega_\alpha + A_\alpha,$$
$$A_\alpha = \epsilon_{\alpha\beta\gamma}v^\delta{}_{,\beta}v^\gamma{}_{,\delta} = {}^1\!/\!_2\epsilon_{\alpha\beta\gamma}(v^\delta{}_{,\beta} + v^\beta{}_{,\delta})(v^\gamma{}_{,\delta} - v^\delta{}_{,\gamma}). \tag{22.15}$$

Using the identities (for $\alpha = 1, 2, 3$)

$$\epsilon_{\alpha\beta\gamma}\epsilon_{\beta\gamma\delta} = 2\delta_{\alpha\delta}, \qquad \epsilon_{\alpha\beta\gamma}\epsilon_{\gamma\delta\mu} = \delta_{\alpha\delta}\delta_{\beta\mu} - \delta_{\alpha\mu}\delta_{\beta\delta}, \tag{22.16}$$

with the definitions in equation (22.13), one finds

$$A_\alpha = a^2[\tfrac{2}{3}\theta\omega_\alpha - \sigma_{\alpha\beta}\omega_\beta]. \tag{22.17}$$

If p may be expressed as a function of ρ, the pressure term on the right hand side of equation (22.11) is a gradient, so the curl vanishes and the vorticity equation becomes

$$\frac{\partial\omega_\alpha}{\partial t} + \frac{1}{a}\mathbf{v}\cdot\nabla\omega_\alpha + \left(2\frac{\dot a}{a} + \frac{2}{3}\theta\right)\omega_\alpha = \sigma_{\alpha\beta}\omega_\beta, \tag{22.18}$$

in agreement with equations (22.9) and (22.12). By using

$$\omega_\beta(v^\beta{}_{,\alpha} - v^\alpha{}_{,\beta}) = a\omega_\beta\epsilon_{\alpha\beta\gamma}\omega_\gamma = 0, \tag{22.19}$$

we can rewrite this equation as

$$\frac{\partial\boldsymbol\omega}{\partial t} + \frac{1}{a}(\mathbf{v}\cdot\nabla)\boldsymbol\omega + \left(2\frac{\dot a}{a} + \theta\right)\boldsymbol\omega = \frac{1}{a}(\boldsymbol\omega\cdot\nabla)\mathbf{v}. \tag{22.20}$$

The dot product of equation (22.18) with $\boldsymbol\omega$ gives the rate of change of the magnitude of the vorticity,

$$\frac{\partial\omega^2}{\partial t} + \frac{1}{a}\mathbf{v}\cdot\nabla\omega^2 + \left(4\frac{\dot a}{a} + \frac{4}{3}\theta\right)\omega^2 = 2\sigma_{\alpha\beta}\omega_\alpha\omega_\beta. \tag{22.21}$$

The result of averaging this over space and integrating the second term by parts is

$$\frac{d}{dt}\langle\omega^2\rangle + 4\frac{\dot a}{a}\langle\omega^2\rangle + \frac{1}{3}\langle\theta\omega^2\rangle = 2\langle\sigma_{\alpha\beta}\omega_\alpha\omega_\beta\rangle. \tag{22.22}$$

The uniform expansion tends to reduce the vorticity as $\omega \propto a^{-2}$ while the shear term can serve to amplify the vorticity.

Equations (22.17) to (22.21) are the usual hydrodynamic vorticity equations (for example, Tennekes and Lumley 1972) modified by the expansion term $\dot a/a$ (Olson and Sachs 1973, Silk 1974c).

Kelvin's circulation theorem says the integral

$$\Gamma = a\oint \mathbf{v}\cdot d\mathbf{x} \tag{22.23}$$

around a closed path fixed to elements of the fluid is independent of time if the fluid is ideal with p a single-valued function of ρ. Of course, the general expansion of the cosmological model does not alter this (for example, Kihara and Saki 1970).

23. Origin of the rotation of galaxies

In the standard gravitational instability picture the angular momentum of a galaxy is not primeval but rather the result of exchange of angular momentum among protogalaxies. One imagines that a protogalaxy is highly asymmetric; the surface containing the material destined to end up in a single galaxy could have a very irregular shape. Therefore, the interaction among neighboring clumps would cause an appreciable torque. The final angular momentun of a galaxy thus would have come from the rotational and relative orbital angular momenta of its original neighbors. This process was first discussed by Hoyle (1949a), who considered a collapsing protogalaxy, and Peebles (1969c), who considered the process in the context of an expanding world model.

It is not clear whether this effect can account for the rotation of galaxies. The wanted angular momentum transfer is uncertain because the typical mass distribution in a galaxy is uncertain and so is the amount of the spin-up due to the collapse after formation. Furthermore, a realistic theoretical treatment proves difficult, as is illustrated by the following perturbation theory approach.

It will be assumed that the density and velocity fields are only slightly perturbed from the uniformly expanding background model and that ρ and $\mathbf{v}$ are smooth functions with coherence length λ. We shall consider the angular momentum in a sphere of radius r about the center of the sphere, and to compute we will expand $\rho\mathbf{v}$ in a power series about the center, which is valid if the radius is small compared to the coherence length. Since the angular momentum of a protoobject is wanted, the sphere will be centered on a local maximum of ρ.

The momentum density is

$$\rho v^\alpha(r) = (\rho v^\alpha) + (\rho v^\alpha)_{,\beta} r^\beta + \tfrac{1}{2}(\rho v^\alpha)_{,\beta\gamma} r^\beta r^\gamma + \cdots . \qquad (23.1)$$

The first nontrivial term in this expression is the second, which gives

$$\begin{aligned} L_\alpha &= \epsilon_{\alpha\beta\gamma}(\rho v^\gamma)_{,\delta} \int r^\beta r^\delta d^3 r, \\ \mathbf{L} &= \frac{4\pi}{15} r^5 \nabla \times \rho\mathbf{v} = \frac{4\pi}{15} r^5 \rho \nabla \times \mathbf{v}. \end{aligned} \qquad (23.2)$$

The last step follows because the sphere is centered on a maximum of ρ. Since vorticity is conserved, if $\nabla \times \mathbf{v}$ was negligible to begin with, this contribution to L is negligible. The next nontrivial term in equation (1) is the fourth, which gives

$$L_\alpha = \frac{1}{6}\epsilon_{\alpha\beta\gamma}(\rho v^\gamma)_{,\sigma\mu\nu}\int d^3r\,r^\beta r^\sigma r^\mu r^\nu,$$
$$\int d^3r\,r^\beta r^\sigma r^\mu r^\nu = \frac{4\pi}{105}r^7(\delta^{\beta\sigma}\delta^{\mu\nu} + \delta^{\beta\mu}\delta^{\sigma\nu} + \delta^{\beta\nu}\delta^{\sigma\mu}), \tag{23.3}$$

and, with $\nabla \times \mathbf{v} = 0$,

$$\mathbf{L} = \frac{2\pi}{105}r^7\rho_b\nabla^2(\nabla\delta \times \mathbf{v}), \tag{23.4}$$

where $\delta = \delta\rho/\rho$. One can see the size of this by comparing it to the angular momentum of a sphere uniformly rotating at angular velocity Ω:

$$L = \frac{8\pi}{15}\rho_b r^5\Omega. \tag{23.5}$$

The effective angular velocity of equation (23.4) is then

$$\Omega \sim r^2|\nabla^2(\nabla\delta \times \mathbf{v})| \sim v\delta r^2/\lambda^3, \tag{23.6}$$

where λ is the coherence length, and (14.9) says

$$v \sim \lambda\delta/t, \tag{23.7}$$

so

$$t\Omega \sim \delta^2(r/\lambda)^2. \tag{23.8}$$

The protoobject breaks away from the general expansion when $\delta \sim 1$, and its collapse time is on the order of the expansion time t for the universe at that epoch. With $r \sim \lambda \sim$ the size of the object, equation (23.8) indicates that the rotation period is on the order of the collapse time, which is what is wanted for an object supported by rotation. Of course, this result is not surprising because $\Omega t \sim 1$ is demanded by dimensional analysis. It is also apparent from equation (23.8) that the typical size of the number Ωt is difficult to estimate with any accuracy because it critically depends on the contribution to **L** by the matter on the outer fringes of the protoobject, $r \sim \lambda$, where the expansion of equation (23.1) is not reliable.

A further apparent difficulty might be mentioned. In the instability picture the circulation (eq. 22.23) is conserved and small, while the gas in the disc of a spiral galaxy evidently moves in fairly smooth circular orbits with substantial circulation. However, it is doubtful that this is a serious problem because the collapse of a protogalaxy or protocluster would be expected to be violent, hence strongly dissipative, which would violate the conditions of the circulation theorem (Chernin, 1970). As one example, consider a fluid of stars where the cross sections are small, so the fluid can mix with negligible scattering. If this fluid has no vorticity to begin with and if one could isolate and follow those stars that originated in a small common patch, one would find that the vorticity remains zero even in a well-mixed object like a rotating elliptical galaxy. But, of course, that is not relevant: the stars one finds in a small patch in the galaxy are a mixture of components that arrive there by different paths, and the best one can do is measure the streaming velocity

$$\mathbf{v}(\mathbf{r}) = \Sigma \rho_a(\mathbf{r}) \mathbf{v}_a(\mathbf{r}) / \Sigma \rho_a(\mathbf{r}), \tag{23.9}$$

where ρ_a and v_a are the density and velocity fields of the stars that originated in patch a, and

$$\nabla \times \mathbf{v}_a = 0, \tag{23.10}$$

if there was no initial vorticity. If the denominator of equation (23.9) is independent of position, the vorticity of the streaming velocity is

$$\nabla \times \mathbf{v} = \Sigma \nabla \rho_a \times \mathbf{v}_a / \Sigma \rho_a. \tag{23.11}$$

As is to be expected, orbit mixing can produce a smooth galaxy with streaming velocity that has circulation. In a similar way formation of a uniformly rotating disc of gas in a spiral galaxy would require strong dissipation, not only to produce vorticity, but also to dissipate the original strongly noncircular motions of the gas. An example illustrating motion without vorticity of a gas in a protogalaxy is given by Peebles (1973a). The motion soon becomes quite complicated, giving one the impression that it would be difficult to see how such motions could persist for more than a crossing time without strong dissipation leading to circular motion with circulation.

An interesting potential test of the angular momentum transfer picture was discussed by Jones (1976). In close pairs of galaxies one ought to be able to devise a way to detect the residuum of the detailed balance of angular momentum exchange. Any such effect certainly is not very strong: in close pairs of galaxies the orientation of the long axes relative to the line joining the pair is quite close to isotropic (Hawley and Peebles 1975).

The significance of the rotation of galaxies in the gravitational instability picture is not at all well understood. It is obvious from dimensional analysis that the angular momentum tranfer in this picture is the right order of magnitude, but a check beyond the factors of ten proves difficult. Perhaps the most reliable estimates are from N-body experiments (Peebles 1971b, Haggerty and Janin 1974, Efstathiou and Jones 1979). They suggest that the gas in spiral galaxies collapsed a factor ~ 10 in radius before it could be supported by rotation. One must bear in mind that the angular momentum transfer critically depends on the material in the outer fringes of the object, and it is not clear that even the N-body models correctly take this into account. In view of the uncertainty in the size of the theoretical effect and in what is needed to account for the observations, it is perhaps not surprising that quite variable opinions have been expressed on whether angular momentum is a serious problem for the instability picture. One can also find diverse arguments on whether the circulation theorem is a problem for any scenario where primeval vorticity is negligible. The spread of opinion on these points can be seen by comparing the discussions of Peebles (1969c, 1971b, 1973a), Hunter (1970), Chernin (1970), Oort (1970), Harrison (1971), Doroshkevich (1973), Tomita (1973), Binney (1974), Field (1975), Doroshkevich, Sunyaev, and Zel'dovich (1974, § 5), and Thuan and Gott (1977).

24. Cosmic energy equation

The ordinary Newtonian energy conservation equation when expressed in expanding cosmological coordinates assumes an interesting and useful form. This was discovered by Irvine (1961, 1965) and Layzer (1963, 1964) and independently by Dmitriev and Zel'dovich (1963) and Haggerty (1970). Two forms of the equation are derived here: one applies to point-like particles that interact only through gravity, the other to an ideal fluid that conserves entropy.

A. Free particle model

The particles in a finite region of the universe will be considered, and it will be supposed that the region is large enough so that the perturbation due to matter outside it has a negligible effect on the behavior of a typical particle inside. The mass of the sample is

$$M = \Sigma m_j, \tag{24.1}$$

and a Lagrangian for the system is (eqs. 8.6)

$$\begin{aligned} \mathcal{L} &= \Sigma \tfrac{1}{2} m_j a^2 \dot{x}_j^2 - MW, \\ MW &= -\tfrac{1}{2} G a^2 \int d^3x_1 d^3x_2 [\rho(\mathbf{x}_1) - \rho_b][\rho(\mathbf{x}_2) - \rho_b]/x_{12}. \end{aligned} \tag{24.2}$$

Here ρ is a sum over delta functions (in $\mathbf{r}$) and the integral is supposed to exclude $x_1 = x_2$. The Hamiltonian is

$$\begin{aligned} \mathrm{H} &= \Sigma\, \mathbf{p}_j \cdot \dot{x}_j - \mathcal{L} \\ &= \sum \frac{p_j^2}{2\, m_j a^2} + MW, \qquad \mathbf{p}_j = m_j a^2 \dot{\mathbf{x}}_j, \end{aligned} \tag{24.3}$$

and the energy equation is

$$\frac{dH}{dt} = \frac{\partial H}{\partial t}. \tag{24.4}$$

The kinetic energy term in this equation is

$$MK = \sum \frac{p_j^2}{2\, m_j a^2} = \frac{1}{2} \sum m_j v_j^2, \qquad \frac{\partial K}{\partial t} = -2 \frac{\dot{a}}{a} K, \tag{24.5}$$

since K varies as a^{-2} at fixed $\mathbf{p}$. The potential energy term varies as $MW \propto a^{-1}$ at fixed $\mathbf{x}_i$ (because ρ varies as $\delta(\mathbf{r} - \mathbf{r}_i) = a^{-3}\delta(\mathbf{x} - \mathbf{x}_i)$), so

$$\frac{\partial W}{\partial t} = -\frac{\dot{a}}{a} W. \tag{24.6}$$

The energy equation (24.4) is then

$$\frac{dK}{dt} + \frac{dW}{dt} + \frac{\dot{a}}{a}(2K + W) = 0. \tag{24.7}$$

On using the density autocorrelation function

$$\langle (\rho(\mathbf{x}_1) - \rho_b)(\rho(\mathbf{x}_2) - \rho_b) \rangle = \rho_b{}^2 \xi(|\mathbf{x}_1 - \mathbf{x}_2|), \tag{24.8}$$

we can rewrite the second of equations (24.2) as

$$
\begin{aligned}
W &= -\frac{Ga^5\rho_b^2}{2M}\int d^3x_1 d^3x_2 \xi(x_{12})/x_{12} \\
&= -\frac{1}{2}Ga^2\rho_b \int d^3x\xi(x)/x,
\end{aligned}
\tag{24.9}
$$

since the region is supposed to be large compared to the maximum correlation length for ξ.

Equation (24.9) defines the cosmic potential energy W per unit mass; equation (24.5) defines the cosmic kinetic energy K per unit mass, and the energy equation (24.7) gives the relation between the two. If $W \equiv 0$, the equation indicates $K \propto a^{-2}$, in agreement with the usual decay of peculiar motion of a freely moving particle, $v \propto a^{-1}$. If there is no substantial time variation in the clustering so dK/dt and dW/dt are negligibly small, the equation becomes $2K + W = 0$, in agreement with the virial theorem.

The energy equation can be rewritten as

$$
\frac{d}{dt}a(K + W) = -K\dot{a} < 0, \tag{24.10}
$$

in an expanding universe, $\dot{a} > 0$. This means $a(K + W)$ is decreasing, and if $K = 0$, $W \leq 0$ to begin with (or more generally the initial values of K and W are negligibly small), $a(K + W)$ is negative, so

$$
K < -W. \tag{24.11}
$$

This inequality was derived by Zel'dovich (1965a.)

Unlike the standard Newtonian formalism the potential W need not be negative: if $\xi < 0$, representing anticorrelation of positions, it makes a positive contribution to W (eq. 24.9). Zel'dovich's inequality shows that if the galaxy distribution is the result of gravitational interactions alone, W must be negative, that is, the possible positive contribution to W by anticorrelation at fairly large scales must be overbalanced by the negative contribution due to clustering on smaller scales.

A second inequality is obtained by writing equation (24.7) as

$$
\frac{d}{dt}a^2\left(K + \frac{1}{2}W\right) = -\frac{a^2}{2}\frac{dW}{dt}. \tag{24.12}
$$

The function $\rho_b\xi$ is the number of neighbors in excess of random per unit proper volume, and this number evaluated at fixed proper distance r is expected to increase with time as clusters develop. If so, $-dW/dt > 0$. Thus if $K + W/2$ is small to begin with, $a^2(K + W/2) > 0$. With equation (24.11) we have then

$$\tfrac{1}{2}|W| < K < |W|. \tag{24.13}$$

The relation between K and W in linear perturbation theory was derived in Section 14. Equation (14.10) is

$$K = \frac{2}{3}\frac{f^2}{\Omega}|W|. \tag{24.14}$$

It is left as an exercise to verify that this is consistent with equation (24.7). Another relation between K and W is derived in Section 74 below (eq. 74.6).

B. Ideal fluid model

In the early universe the matter might approximate an ideal fluid that conserves entropy. The equation for the internal energy of the fluid, expressed in expanding coordinates, is

$$\frac{\partial}{\partial t}\rho u + \frac{1}{a}\mathbf{v}\cdot\nabla\rho u = -(\rho u + p)\left(\frac{1}{a}\nabla\cdot\mathbf{v} + 3\frac{\dot{a}}{a}\right). \tag{24.15}$$

Here u is the internal energy per unit mass, ρu the internal energy per unit proper volume. The right-hand side gives the rate of change of this quantity in a fixed fluid element due to expansion and pdV work. The usual fluid energy conservation equation is found by writing down the time derivative of $\rho v^2/2$, using the mass and momentum equations to evaluate the time derivatives of ρ and $\mathbf{v}$, and adding the derivative of ρu. The result of repeating this calculation in expanding coordinates (eqs. 22.11) is the energy equation

$$\frac{\partial}{\partial t}\rho\left(u + \frac{1}{2}v^2\right) + \frac{1}{a}\nabla\cdot\rho\left(u + \frac{1}{2}v^2\right)\mathbf{v}$$

$$= -\frac{1}{a}\nabla\cdot p\mathbf{v} - 3\frac{\dot{a}}{a}(\rho u + p) - \frac{5}{2}\frac{\dot{a}}{a}\rho v^2 - \frac{\rho}{a}\mathbf{v}\cdot\nabla\phi, \tag{24.16}$$

$$\phi = -G\rho_b a^2 \int d^3x'\delta(\mathbf{x}')/|\mathbf{x} - \mathbf{x}'|.$$

The cosmic energy relation is found by averaging this equation over position. This eliminates the terms that are total divergences, and the gravity term with equation (9.17) becomes

$$\begin{aligned}
-\frac{1}{a\rho_b}\langle\rho\mathbf{v}\cdot\nabla\phi\rangle &= -\langle\phi\partial\delta/\partial t\rangle \\
&= G\rho_b a^2\int d^3x'\,\langle\delta(\mathbf{x}')\partial\delta(\mathbf{x})/\partial t\rangle/|\mathbf{x}'-\mathbf{x}| \\
&= \frac{1}{2}G\rho_b a^2\frac{\partial}{\partial t}\int d^3x\,\xi(x)/x \\
&= -\left(\frac{\partial}{\partial t}+\frac{\dot a}{a}\right)W,
\end{aligned} \tag{24.17}$$

where W is defined as in equation (24.9). The cosmic kinetic energy per unit mass and the mean internal energy per unit mass are

$$K=\langle\rho v^2/2\rangle/\rho_b,\qquad U=\langle\rho u\rangle/\rho_b,\qquad \rho_b=\langle\rho\rangle. \tag{24.18}$$

Since $\rho_b\propto a^{-3}$, one finds from equations (24.16) to (24.18)

$$\frac{d}{dt}(K+U+W)+\frac{\dot a}{a}\left(2K+\frac{3}{\rho_b}\langle p\rangle+W\right)=0. \tag{24.19}$$

Another form is obtained by using the space average of the internal energy equation (24.15),

$$\frac{\partial U}{\partial t}+\frac{3}{\rho_b}\frac{\dot a}{a}\langle p\rangle+\frac{1}{\rho_b a}\langle p\nabla\cdot\mathbf{v}\rangle=0. \tag{24.20}$$

The last term is

$$\langle\mathbf{v}\cdot\nabla p\rangle=\langle\rho\mathbf{v}\cdot\nabla h\rangle=-\langle h\nabla\cdot\rho\mathbf{v}\rangle=a\rho_b\langle h\partial\delta/\partial t\rangle, \tag{24.21}$$

where $h=u+p/\rho$ is the enthalpy per unit mass. With these expressions we can rewrite equation (24.19) as

$$\frac{d}{dt}(K+W)+\frac{\dot a}{a}(2K+W)=\frac{1}{\rho_b a}\langle p\nabla\cdot\mathbf{v}\rangle=-\langle h\partial\delta/\partial t\rangle. \tag{24.22}$$

If $p=0$, the first of equations (24.22) reduces to the energy equation (24.7). It is amusing to note that the free particle case described in equation (24.7) can be considered a statistically uniform ideal gas with $K = 0$ and $p = 2\rho u/3$, for which equation (24.19) with equation (24.18) can be reduced to

$$\frac{d}{dt}(U + W) + \frac{\dot{a}}{a}(2U + W) = 0. \tag{24.23}$$

For incompressible flow, as in subsonic turbulence, δ is nearly zero so $W \approx 0$ and equation (24.22) gives

$$K \propto a^{-2}. \tag{24.24}$$

This decay law is the same as for freely moving particles, $v \propto a^{-1}$. The generalization of equation (24.24) when $p \sim \rho c^2$ is discussed in Section 90 below.

25. Spherical accretion model

This is a special model for the development of clustering around a mass concentration. The key assumption is that at some starting epoch t_i matter is fairly uniformly distributed overall while at some spot there happens to be an unusually strong mass excess around which more matter tends to collect. This process was first discussed by Hoyle and Narlikar (1966), who suggested that it might account for the star distribution around the nucleus in an elliptical galaxy. They worked through the calculation leading to equation (25.8) below, but because they used some unconventional cosmological conditions they arrived at a different power law, $\rho \propto r^{-8/3}$. Gunn and Gott (1972) discussed spherical accretion of extracluster gas by a cluster of galaxies. Ryan (1972) and Gribbin (1974) discussed accretion by primeval black holes as a way to make galaxies. The 9/4 power law in equation (25.8) is derived by Gott (1975) and Gunn (1977).

Matter is treated here as an ideal zero pressure fluid, which, depending on the circumstances, might be a gas of atoms, or stars, or galaxies.

It is assumed that at some starting time t_i there is a mass concentration m_0, with the matter everywhere else uniformly distributed to high accuracy, all with zero peculiar velocity. Then the spherical model applies (§ 19). At the mass shell with initial radius r_i the initial kinetic energy per unit mass is

$$K_i = \frac{1}{2} H_i^2 r_i^2 = \frac{4\pi}{3} G\rho_i \Omega_i^{-1} r_i^2, \tag{25.1}$$

where Ω_i is the cosmological density parameter at t_i. The potential energy per unit mass is

$$|W_i| = \frac{4}{3}\pi G\rho_i r_i^2 + \frac{Gm_0}{r_i}. \tag{25.2}$$

If no other mass shell crosses this one, its maximum radius r_m is given by (eq. 19.6)

$$\frac{|W_i| - K_i}{|W_i|} = \frac{r_i}{r_m}. \tag{25.3}$$

After the shell collapses, its time average radius is

$$r = fr_m, \tag{25.4}$$

where $f \sim 0.5$, depending on how shells cross. This factor might depend somewhat on the value of Ω, but the effect should be small and will be ignored. The relaxed density run around m_0 is estimated by imagining that the material originally at r_i ends up at radius r:

$$\rho(r)r^2dr = \rho_i r_i^2 dr_i. \tag{25.5}$$

It will be assumed that $\Omega_i \approx 1$, so to begin with the expansion is close to Einstein-de Sitter, and that $\rho_i r_i^3 \gg m_0$, so the mass concentration at first is a small perturbation.

Equations (25.1) to (25.4) give

$$\frac{r_i}{r} = \frac{\Omega_i^{-1} - 1}{f}\left[\left(\frac{r_0}{r_i}\right)^3 - 1\right], \qquad r_0^3 = \frac{3\,m_0}{4\pi\rho_i(\Omega_i^{-1} - 1)}, \tag{25.6}$$

and this with equation (25.5) yields

$$\rho(r) = \rho_i \frac{(\Omega_i^{-1} - 1)^3}{f^3} \frac{[(r_0/r_i)^3 - 1]^4}{[4(r_0/r_i)^3 - 1]}. \tag{25.7}$$

These equations describe the halo $\rho(r)$ that develops around m_0.

In an open model $\Omega_i < 1$, $r_0 > 0$, the mass shell at initial radius $r_i = r_0$ has zero net energy; only the matter at $r_i < r_0$ eventually falls toward m_0. In the limit $r_i \ll r_0$ equations (25.6) and (25.7) become

$$\rho = \frac{\rho_i}{4}\left(\frac{\Omega_i^{-1} - 1}{f}\right)^{3/4}\left(\frac{r_0}{r}\right)^{9/4}. \tag{25.8}$$

This part forms while $\Omega(t) \sim 1$. If $r_0 \sim r_i$,

$$\rho = \frac{\rho_i f}{3\,(\Omega_i^{-1} - 1)} \left(\frac{r_0}{r}\right)^4. \tag{25.9}$$

This part forms when $\Omega \ll 1$. The transition between these two limiting cases is at

$$r \simeq \frac{f r_0}{\Omega_i^{-1} - 1}, \qquad \rho(r) \simeq \frac{\rho_i}{3}\left(\frac{\Omega_i^{-1} - 1}{f}\right)^3. \tag{25.10}$$

It will be noted that relaxation (through two-body interactions or the fluctuating potential) has been ignored; if important this could erase the steep r^{-4} part that develops when Ω drops well below unity.

Equations (25.8) and (25.9) assume also that the material has been mixed through several crossing times. The initial rise in density as the material first falls toward m_0 is given by equations (19.21) through (19.26), (19.29), and (19.38). The initial fractional mass excess within r_i is (eq. 19.8)

$$\delta_i' = \frac{9}{20\,\pi} \frac{m_0}{\rho_i r_i^3}, \qquad \delta_i = 0, \tag{25.11}$$

for the only perturbation is the mass m_0 at $r = 0$. Figure 25.1 shows an example of the shape of the halo given by these equations. The density parameter is $\Omega = 0.03$. The coordinate axes both are linear. The lower curve is the mass density at radius r divided by the density in the background cosmological model,

$$\rho(r)/\rho_b = 1 + \delta(r). \tag{25.12}$$

This curve has been plotted only to $\delta_i' = \delta_c$, at the shell that has zero energy. Somewhat below this radius the model is no longer to be trusted because the infalling mass shells would be crossed by rebounding ones. The upper curve shows the ratio of the mass within r to that expected for a homogeneous mass distribution (eq. 19.29),

$$\frac{M(<r)}{\rho_b V} = \frac{\rho'}{\rho_b} = 1 + \delta'(r). \tag{25.13}$$

In the part of the halo pictured here there has been appreciable accretion: the mass within r is well above the value for a uniform distribution.

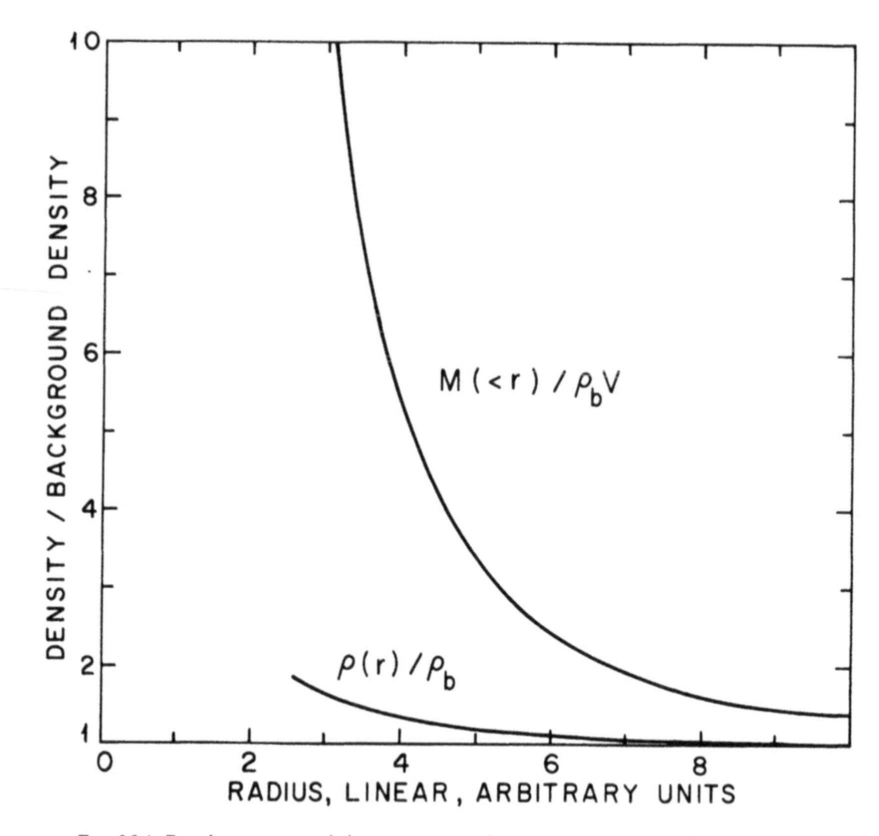

FIG. 25.1. Density run around the outer parts of a mass concentration in the spherical accretion model. The density parameter is $\Omega = 0.03$. The lower curve is the mass density in units of the background density. It is plotted for the shells that have positive energy. The upper curve is the total mass within r divided by the volume.

However, the matter has moved so as to hold ρ very nearly constant. This happens also at smaller radius. For example, the shell that has just stopped expanding has a radius slightly more than half that of the zero energy shell. The mean density within this shell is 90 ρ_b while the density at the shell is 3.5 ρ_b. The spherical accretion process thus is efficient at pulling in mass but not so efficient at building a smooth mass distribution at densities a modest multiple of the mean.

A similar but smaller imbalance is found in denser cosmological models. For example, equations (19.48) and (19.38) with equation (25.11) describe accretion in an Einstein-de Sitter model. On expanding this solution in a power series in the parameter θ and keeping only the first nontrivial part, one finds

$$\frac{\rho'(r)}{\rho_b} = 1 + \frac{3}{20}\theta^2, \qquad \frac{\rho(r)}{\rho_b} = 1 + \frac{3}{700}\theta^4,$$
$$\theta^2 = \frac{3}{\pi}\frac{m_0}{\rho_b(t)}\frac{a(t)}{a_i}\frac{1}{r^3}. \tag{25.14}$$

The mass density thus falls very sharply, the perturbation varying as r^{-6}. The perturbation to the mean density within r varies as r^{-3}, that is as the volume, which just says that there is a mass concentration well inside r amounting to

$$M_c = \frac{3}{5} m_0 \frac{a(t)}{a_i}. \tag{25.15}$$

At the mass shell that has just stopped expanding, the parameter is $\theta = \pi$, and the densities are (eq. 19.50)

$$\frac{\rho'}{\rho_b} = \frac{9\pi^2}{16} = 5.6, \qquad \frac{\rho}{\rho_b} = \frac{9\pi^2}{64} = 1.4. \tag{25.16}$$

In an object formed by this spherical accretion process one would find that the density $\rho(r)$ in the outskirts is substantially less than the mass within r divided by the volume within r. This is not observed in the run of galaxy number density around an Abell cluster (§ 77), but conceivably it has been washed out by the average over clusters. A test for the effect might be possible when we have detailed maps of galaxy positions around individual large clusters.

One should bear in mind that all these results depend on the fundamental assumption that m_0 is the dominant perturbation. The mass distribution around m_0 presumably is somewhat irregular, so the model is unrealistic if r_i is too large. For example, suppose the initial density $\rho(\mathbf{x}, t_i)$ around m_0 is a random Gaussian process, and let the rms fluctuation in density after smoothing through a window of radius r be

$$\delta\rho_i = \alpha r^{-\beta}, \tag{25.17}$$

where α and β are positive constants. For white noise $\beta = 3/2$. Then if r_i satisfies

$$\delta\rho_i = \alpha r_i^{-\beta} \lesssim m_0/r_i^3, \tag{25.18}$$

the spherical accretion model is a reasonable approximation. If the limit is violated the dominant perturbation is the fluctuating mass outside m_0. A model for the development of clustering in this case is discussed next.

26. Hierarchical clustering model

The idea here is that at some epoch t_i the fluctuation in the matter distribution approximated a random Gaussian process. This means the mass found within a randomly placed sphere of fixed radius has a roughly Gaussian frequency distribution with dispersion fixed by the assumed mass density power spectrum (or autocorrelation function). Where this mass happens to be high the matter tends eventually to stop expanding and form a stable system. Within such a patch there were at t_i fluctuations on smaller scales that would have had larger initial amplitude and so would have stopped expanding sooner to form denser subsystems: thus a sort of continuous clustering hierarchy develops.

The idea that a clustering hierarchy might develop was implicit in the early discussions of Lemaître (1933b, 1934), at least to the extent of galaxies forming within clusters, but it seems to have been first clearly stated by Layzer (1954). The scaling argument below was given by Peebles (1965, 1974a).

The assumed initial conditions can be stated in terms of the Fourier transform of the mass distribution. It seems to be simplest to take the universe to be periodic in some large rectangular volume V_u. (Of course this has no physical significance and no effect on the final results if V_u is large compared to the maximum correlation length for the distribution.) Then the mass density can be expanded as the sum

$$\rho(\mathbf{x}) = \rho_b \left[1 + \sum_{k \neq 0} \delta_{\mathbf{k}} e^{-i\mathbf{k}\cdot\mathbf{x}} \right]. \tag{26.1}$$

The assumption is that the $\delta_{\mathbf{k}}$ have random phases and that the power spectrum has some prescribed shape, which will be supposed to be a power law,

$$|\delta_{\mathbf{k}}|^2 \propto k^n, \qquad -3 < n < 4. \tag{26.2}$$

This assumption like that of Section 25 is only a guess, that will be justified if the results are found to approximate the present situation in some detail (and perhaps ultimately if it is found to be consistent with some deeper theory of the early universe and the big bang). The random phase condition rules out the dominant mass concentrations assumed in the last section or at least makes them rare exceptions. The power law form of equation (26.2) is adopted because it is convenient and seems not contrived. If the index n were $n \leq -3$, then for convergence (eq. 26.4 below) the spectrum would have to be cut off at some wavelength λ_c and the dominant effect would be density fluctuations of size $\sim\lambda_c$ on which are superposed relatively minor ripples $\lambda < \lambda_c$ from the power spectrum at $k > k_c$. This would vitiate the clustering hierarchy argument that assumes the density fluctuations on small scales are substantially larger than the fluctuations on large scales. The calculation given here fails if $n > 1$, but, as described in Section 28, can be adjusted to apply to $n \leq 4$.

The mass found within a sphere of radius x centered at $\mathbf{x}_0$ is

$$M(\mathbf{x}_0) = \rho_b a^3 V + \rho_b a^3 \sum_{\mathbf{k}} \delta_{\mathbf{k}} e^{-i\mathbf{k}\cdot\mathbf{x}_0} \int_x d^3x' e^{-i\mathbf{k}\cdot\mathbf{x}'}. \tag{26.3}$$

Averaging the fluctuation $(M - \rho_b a^3 V)^2$ over position $\mathbf{x}_0$ eliminates the cross terms, leaving

$$\begin{aligned} \left(\frac{\delta M}{M}\right)^2 &= \sum_{k\neq 0} |\delta_{\mathbf{k}}|^2 W(kx) = \frac{V_u}{(2\pi)^3} \int d^3k\, |\delta_{\mathbf{k}}|^2 W(kx), \\ W &= \left[\int d^3x' e^{-i\mathbf{k}\cdot\mathbf{x}'} \Big/ \int d^3x'\right]^2 \\ &= \frac{9}{y^6}[\sin y - y\cos y]^2, \qquad y = kx. \end{aligned} \tag{26.4}$$

The window function is $W \sim 1$, $y \lesssim 1$, and $W \sim y^{-4}$, $y \gg 1$.

If the power spectrum is not growing as rapidly as k or decreasing as rapidly as k^{-3}, then equation (26.4) yields

$$\begin{aligned} (\delta M/M)^2 &\sim V_u k^3 |\delta_{\mathbf{k}}|^2, \qquad k \sim x^{-1}, \\ (\delta M/M)^2 &\propto x^{-(3+n)}, \end{aligned} \tag{26.5}$$

in the power law model. This result is independent of the phases of the $\delta_{\mathbf{k}}$. If the phases are random, the distribution of $M(\mathbf{x}_0) - \rho_b V a^3$ is Gaussian, so we have the picture that the density measured on the scale x fluctuates up and down by the amount $\propto x^{-(3+n)/2}$.

Now consider the course of evolution of these fluctuations in a universe that is close to the Einstein-de Sitter model. Linear perturbation theory establishes that $\delta M/M$ at fixed x grows at $t^{2/3}$ until $\delta M/M \sim 1$, at time $t \sim t_x$. At this time the mass distribution has fragmented into more or less stable lumps of size $r \sim xa(t_x)$, density $\sim(Gt_x^2)^{-1}$ (§ 19E). Of course, each lump contains a hierarchy of sublumps that formed earlier. The time t_x is fixed by

$$(\delta M/M)^2 \propto x^{-(3+n)} t_x^{4/3} \sim 1, \tag{26.6}$$

whence the radius and density are

$$\begin{aligned} r &\sim xa(t_x) \propto x t_x^{2/3} \propto x^{(5+n)/2}, \\ n &\propto t_x^{-2} \propto x^{-(9+3n)/2}, \end{aligned} \tag{26.7}$$

yielding the radius-density relation

$$n \propto r^{-\gamma}, \qquad \gamma = (9 + 3n)/(5 + n). \tag{26.8}$$

Equation (26.8) describes a clustering hierarchy that is continuous: there is no natural dividing line between levels of clustering on scales smaller than the sizes of the clusters just in the process of forming. The clusters forming at epoch t will have characteristic size $r_0(t)$ and characteristic density $n(r_0)$ that is on the order of the mean background density $n_b(t)$. By equation (26.7), this transition radius scales with time as

$$r_0(t) \propto t^{2/\gamma}. \tag{26.9}$$

One interpretation of this model is in terms of how the mass distribution would appear if viewed with linear resolution r. For $r < r_0$ the matter would appear in patches of size $\sim r$ and typical density $\sim n \propto r^{-\gamma}$. Viewed on a smaller resolution $r' < r$ these patches would tend to resolve into smaller patches, size $\sim r'$ and density $\sim n(r')$. At resolution $r > r_0$ the mass distribution would appear close to uniform, and one would see fluctuations of size $\sim r$ and amplitude $\delta\rho \propto r^{-(3+n)/2}$, representing some future generation of the clustering hierarchy. An example is shown in Figure 26.1. The brightness pattern is white noise, $n = 0$ in equation (26.2), viewed with resolution comparable to the size of the coherent patches. Though there is equal power at all wavelengths longer than the cutoff, the shortest wavelengths appear most prominent because of the $k^3 \propto \lambda^{-3}$ factor in equation (26.5). If the brightness pattern in this figure were viewed at lower resolution it would again appear mottled, but the amplitude $\delta i/i$ of the fluctuations would diminish (as r^{-1} in two dimensions, $r^{-3/2}$ in three dimensions).

In this clustering hierarchy picture each particle is in a nested set of clusters, so another interpretation of equation (26.8) is that the mean

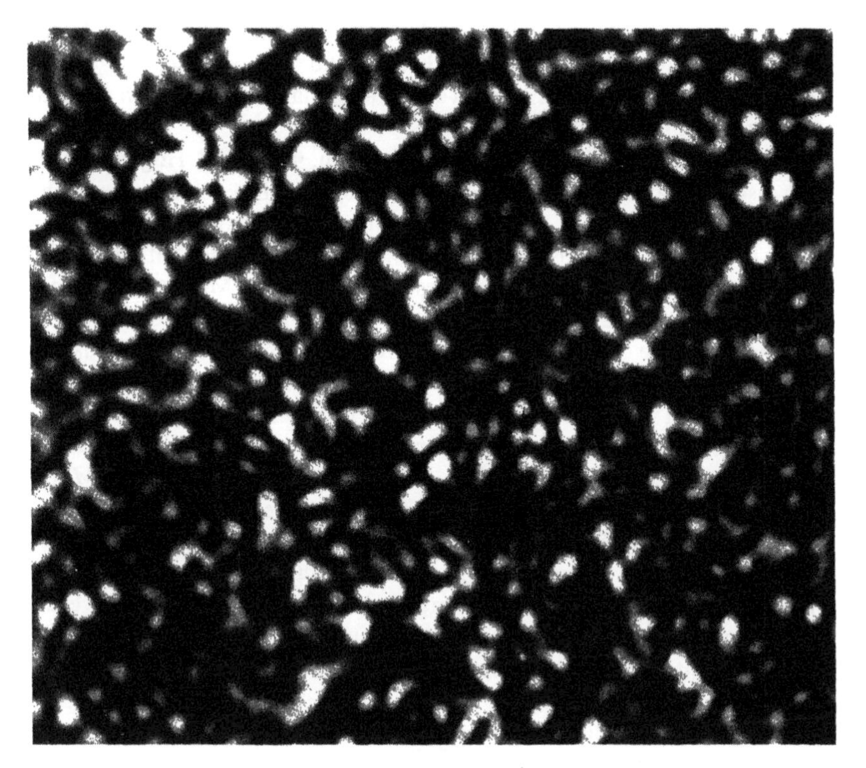

FIG. 26.1. Illustration of white noise. The spectrum is cut off at a wavelength ~ the coherence length in this mottled pattern. This is a pinhole photograph of a spot illuminated by a laser; the size of the pinhole determined the resolution (from Peebles and Dicke 1968; published by The University of Chicago Press; copyright 1968 by The University of Chicago).

(ensemble average) density at distance r from randomly chosen particle varies as $r^{-\gamma}$. This appears similar to the result of the spherical accretion model discussed in Section 25, but the situation is quite different. Suppose, for example, that at t_i the matter is in particles each of mass m_0, randomly distributed, $n = 0$. Each particle represents a mass concentration: because the positions are uncorrelated, the mean mass found within distance r of a randomly chosen particle is larger than the mean mass found within a randomly placed sphere of the same size by the amount m_0. One might then try to compute the development of the mean clustering of matter around randomly chosen particles by using the spherical accretion model. However, this is not valid when x is larger than the interparticle distance because under the initial conditions assumed here the dominant perturbation on scale x comes not from the bias m_0 but rather from the random fluctuations in density on scale x. The chosen particle ends up in a cluster of size $\sim x$, but the cluster forms not around the particle, but rather wherever the density averaged over scale x happens to reach a local maximum.

Press and Schechter (1974) and Efstathiou, Fall, and Hogan (1979) have discussed an interesting extension of the model to an estimate of the frequency distribution of the masses of the objects that are fragmenting out of the general expansion at epoch t. They observe that if the initial density $\rho(\mathbf{x}) - \rho_b$ is a Gaussian process, then the mass $M(\mathbf{x}_0)$ (in a sphere of radius x centered at $\mathbf{x}_0$; eq. 26.3) has a Gaussian frequency distribution, and they assume that the positive (overdense) half of this Gaussian distribution is the frequency distribution of initial density contrasts in protoclusters of size x. It is doubtful that this is a good approximation, however, for as noted above protoobjects of size $\sim x$ would tend to form not around randomly placed centers $\mathbf{x}_0$, but rather where the function $M(\mathbf{x}_0)$ happens to reach a local maximum. Apparently one would prefer instead of the distribution of $M(\mathbf{x}_0)$ for randomly placed $\mathbf{x}_0$, the distribution of $M(\mathbf{x}_0)$ for $\mathbf{x}_0$ placed at local maxima of the function $M(\mathbf{x}_0)$ (Jones 1976). Even this would be a crude approximation to a difficult problem, however, for the mass of a protoobject is a sensitive function of the radius; a region with unusually large mean density would tend to sweep in more of the surroundings.

The evidence of a clustering hierarchy in the general galaxy distribution is discussed in Chapter III below; some formal developments in the theory of the formation of a clustering hierarchy are presented in Chapter IV, and some details of possible scenarios are reviewed in Chapter VI.

27. Fourier transform of the equations of motion

For some purposes it is useful to study how the Fourier transform of the mass density varies with time. Some relevant general relations are given here for two models; an ideal zero pressure fluid and a gas of particles that interact only by gravity.

As in the last section a discrete rather than continuous transform seems somewhat more convenient. The universe is imagined to be periodic in V_u fixed in $\mathbf{x}$ coordinates, with V_u much larger than the (assumed) maximum correlation length. The periodic boundary conditions mean the potential ϕ can be expanded as a sum over plane waves periodic in V_u,

$$\phi = \Sigma \phi_{\mathbf{k}} e^{-i\mathbf{k}\cdot\mathbf{x}}. \tag{27.1}$$

On substituting this in the field equation

$$\nabla^2\phi = 4\pi Ga^2[\rho(\mathbf{x}) - \rho_b] \tag{27.2}$$

and using

$$\int e^{i(\mathbf{k}-\mathbf{k}')\cdot\mathbf{x}}d^3x = V_u\delta_{\mathbf{k},\mathbf{k}'}, \tag{27.3}$$

one finds

$$-k^2V_u\phi_{\mathbf{k}} = 4\pi Ga^2\int[\rho(\mathbf{x}) - \rho_b]e^{i\mathbf{k}\cdot\mathbf{x}}d^3x. \tag{27.4}$$

The transform of the density contrast is defined as

$$\delta_{\mathbf{k}} = \int d^3x\rho e^{i\mathbf{k}\cdot\mathbf{x}}/\rho_bV_u, \tag{27.5}$$

and this in equation (27.4) with equation (27.1) gives

$$\phi(\mathbf{x}) = -4\pi Ga^2\rho_b\sum_{k\neq 0}e^{-i\mathbf{k}\cdot\mathbf{x}}\delta_{\mathbf{k}}/k^2. \tag{27.6}$$

Since $\langle\rho(\mathbf{x})\rangle = \rho_b$, the term $\mathbf{k} = 0$ vanishes on both sides of equation (27.4), and we can simply drop this term from equation (27.6).

Let us suppose first that matter can be represented as a collection of point particles, the jth particle at position $\mathbf{x}_j(t)$ having mass m_j. The particles interact only by gravity, and as usual the Newtonian approximation is used, but this will be described in full, not in the Vlasov model of Section 9.

The density is a sum of delta functions

$$\rho(\mathbf{x}, t) = \Sigma m_j\delta^3(\mathbf{x} - \mathbf{x}_j(t))/a(t)^3, \tag{27.7}$$

so the transform of equation (27.5) is

$$\delta_{\mathbf{k}} = \Sigma e^{i\mathbf{k}\cdot\mathbf{x}_j}m_j/M_u, \qquad M_u = \Sigma m_j = \rho_bV_ua^3, \tag{27.8}$$

where M_u is the total mass in V_u. The first two time derivatives of this are

$$\begin{aligned}\frac{d\delta_{\mathbf{k}}}{dt} &= \sum(i\mathbf{k}\cdot\dot{\mathbf{x}}_j)e^{i\mathbf{k}\cdot\mathbf{x}_j}m_j/M_u,\\ \frac{d^2\delta_{\mathbf{k}}}{dt^2} &= \sum[i\mathbf{k}\cdot\ddot{\mathbf{x}}_j - (\mathbf{k}\cdot\dot{\mathbf{x}}_j)^2]e^{i\mathbf{k}\cdot\mathbf{x}_j}m_j/M_u.\end{aligned} \tag{27.9}$$

On using the equations of motion (7.10)

$$\ddot{\mathbf{x}} + 2\frac{\dot{a}}{a}\dot{\mathbf{x}} = -\nabla\phi/a^2 \tag{27.10}$$

in the second of these equations, one finds

$$\begin{gathered}\frac{d^2\delta_{\mathbf{k}}}{dt^2} + 2\frac{\dot{a}}{a}\frac{d\delta_{\mathbf{k}}}{dt} = -\sum\frac{m_j i\mathbf{k}\cdot\nabla\phi_j}{M_u a^2}e^{i\mathbf{k}\cdot\mathbf{x}_j} - C,\\ C = \Sigma(\mathbf{k}\cdot\dot{\mathbf{x}}_j)^2 e^{i\mathbf{k}\cdot\mathbf{x}_j}m_j/M_u.\end{gathered} \tag{27.11}$$

The potential for the jth particle is given by equations (27.6) and (27.8):

$$\phi_j = -\frac{4\pi G}{aV_u}\sum_{\substack{k'\neq 0\\ l\neq j}} m_l \exp\left[i\mathbf{k}'\cdot(\mathbf{x}_l - \mathbf{x}_j)\right]/(k')^2. \tag{27.12}$$

On using this in the first term on the right side of equation (27.11) and adding the term $l = j$ (which vanishes in the sum below), one finds

$$\begin{aligned}&\frac{4\pi G}{M_u V_u a^3}\sum_{k'\neq 0} m_j m_l \frac{\mathbf{k}\cdot\mathbf{k}'}{k'^2}\exp i\left[\mathbf{x}_l\cdot\mathbf{k}' + \mathbf{x}_j\cdot(\mathbf{k}-\mathbf{k}')\right]\\ &\qquad = 4\pi G\rho_b\sum_{k'\neq 0}\frac{\mathbf{k}\cdot\mathbf{k}'}{k'^2}\delta_{\mathbf{k}'}\delta_{\mathbf{k}-\mathbf{k}'}.\end{aligned} \tag{27.13}$$

The term $\mathbf{k}' = \mathbf{k}$ is

$$4\pi G\rho_b\delta_{\mathbf{k}}, \tag{27.14}$$

because $\delta_0 = 1$, and this should be written separately since it is larger than any of the others. The remainder of the sum is unchanged if the index is changed from $\mathbf{k}'$ to $\mathbf{k}'' = \mathbf{k} - k'$, so it can be rewritten as

$$\begin{gathered}A = 2\pi G\rho_b\sum_{k'\neq 0,\mathbf{k}} D\delta_{\mathbf{k}'}\delta_{\mathbf{k}-\mathbf{k}'},\\ D = \frac{\mathbf{k}\cdot\mathbf{k}'}{k'^2} + \frac{\mathbf{k}\cdot(\mathbf{k}-\mathbf{k}')}{|\mathbf{k}-\mathbf{k}'|^2}.\end{gathered} \tag{27.15}$$

This form is convenient for the discussion in Section 28 because it shows an important cancellation of terms; when $k \ll k'$, $D \sim (k/k')^2$.

The final result is (Peebles 1974c)

$$\frac{d^2\delta_{\mathbf{k}}}{dt^2} + 2\frac{\dot{a}}{a}\frac{d\delta_{\mathbf{k}}}{dt} = 4\pi G\rho_b\delta_{\mathbf{k}} + A - C. \qquad (27.16)$$

This is the linear perturbation theory equation (10.3) with two nonlinear terms added on A (eq. 27.15) and C (eq. 27.11).

An equation very similar to (27.16) is found in the ideal fluid model (Wickes and Peebles 1976, Vishniac and Press 1978). When the pressure vanishes, equation (9.19) is

$$\frac{\partial^2\delta}{\partial t^2} + 2\frac{\dot{a}}{a}\frac{\partial\delta}{\partial t} = \frac{1}{a^2}\nabla\cdot(1+\delta)\nabla\phi + \frac{1}{a^2}\partial_\alpha\partial_\beta[(1+\delta)v^\alpha v^\beta]. \qquad (27.17)$$

On multiplying this equation by $\exp i\mathbf{k}\cdot\mathbf{x}/V_u$ and integrating over $\mathbf{x}$, and then using equation (27.5) for $\delta_{\mathbf{k}}$ and equation (27.6) for ϕ, one finds

$$\frac{d^2\delta_{\mathbf{k}}}{dt^2} + 2\frac{\dot{a}}{a}\frac{d\delta_{\mathbf{k}}}{dt} = 4\pi G\rho_b\delta_{\mathbf{k}} + A - C, \qquad (27.18)$$

where A is given by equation (27.15) and the velocity term is

$$C = (a/M_u)k_\alpha k_\beta \int d^3x\,\rho v^\alpha v^\beta e^{i\mathbf{k}\cdot\mathbf{x}}, \qquad (27.19)$$

which can be compared to equation (27.11).

The Fourier transform of the velocity field is

$$v_{\mathbf{k}}{}^\alpha = \int v^\alpha e^{i\mathbf{k}\cdot\mathbf{x}}d^3x/V_u, \qquad (27.20)$$

and this in (9.17) is

$$\frac{dv_{\mathbf{k}}{}^\alpha}{dt} + \frac{\dot{a}}{a}v_{\mathbf{k}}{}^\alpha = -4\pi iGa\rho_b\delta_{\mathbf{k}}k^\alpha/k^2 + E_\alpha,$$
$$E_\alpha = (i/a)\Sigma k'_\beta v_{\mathbf{k}-\mathbf{k}'}{}^\beta v_{\mathbf{k}'}{}^\alpha. \qquad (27.21)$$

In the linear perturbation approximation the term E_α is dropped. It is left as an exercise to verify that if $\delta_{\mathbf{k}} \propto D(t)$ where D is a solution to equation

(10.3), then the inhomogeneous solution to equation (27.21) with $E = 0$ is

$$v_{\mathbf{k}}^{\alpha} = -i\frac{a}{D}\frac{dD}{dt}\frac{k^{\alpha}}{k^{2}}\delta_{\mathbf{k}}, \tag{27.22}$$

and that this agrees with equation (14.9) for $v^{\alpha}(\mathbf{x})$.

Equations (27.18) to (27.21) offer a possible way to compute the development of nonlinear density fluctuations in an expanding world model (Wickes and Peebles 1976): replace the functions of four variables $\rho(\mathbf{x}, t)$, $\mathbf{v}(\mathbf{x}, t)$ with the infinite set of coupled functions of one variable, $\delta_{\mathbf{k}}(t)$, $v_{\mathbf{k}}^{\alpha}(t)$, and then truncate the set at short wavelength. If, as is argued in the next section, the existence of strongly nonlinear clumps has little effect on the growth of new levels of the clustering hierarchy, the truncation of the spectrum might be a reasonable approximation, and numerical integration of the truncated set of equations might be feasible. So far, however, nothing much has come of this.

28. Coupling of density fluctuations

The linear perturbation calculation in Section 10 assumes there are only slight departures from homogeneity, but nevertheless we can use it to describe the behavior of large-scale density irregularities in the present universe by treating each bound cluster of galaxies as a diffuse particle. This is because momentum conservation says the strongly nonlinear motions balance out within the cluster, the center of mass responding only to the gravitational attraction of neighboring clusters: there is no strong coupling of substantially different wavelengths even when the clustering is nonlinear. As is described here, this idea can be made more formal by using equation (27.16) for the Fourier transform of the mass distribution. The analysis follows Peebles (1974c).

It has been suggested that what mode coupling is present might play a significant dynamic role; the strongly nonlinear behavior on small scales might force the development of clustering on larger scales and that in turn force the growth of still larger clusters, and so on (Carlitz, Frautschi, and Nahm 1973, Press and Schechter 1974). Thus one might speculate that at the time of the Big Bang the universe was as close to homogeneous and isotropic as is allowed by the discrete nature of matter and radiation. The minimal perturbations forced by the fact that energy appears in quanta would grow under gravity and perhaps develop into galaxies and clusters of galaxies. The scheme seems attractive because it offers a particularly simple and natural prescription for initial conditions (if one accepts that the universe at the time of the Big Bang naturally attempts to be as uniform as possible). However, it appears from the following analysis that the effect is too weak to be interesting.

Let us consider first how to define the density fluctuations on a given length scale. Once this is done we can find the minimal density fluctuations forced by discrete clumping of matter. This is discussed in parts (A) and (B) below. The role of relaxed clusters of matter is analyzed in part (C), and a model for the effect of the process of formation of clusters is analyzed in part (D).

A. Miminal fluctuations

A measure of the typical density fluctuation on scale x is (eq. 26.5)

$$(\delta\rho/\rho)_x^2 = \frac{k^3 V_u}{(2\pi)^3} |\delta_{\mathbf{k}}|^2, \qquad k = 2\pi/x. \tag{28.1}$$

As before the Fourier component belonging to wave number $\mathbf{k}$ is (eq. 27.5)

$$\delta_{\mathbf{k}} = \int d^3x \rho(\mathbf{x}) e^{i\mathbf{k}\cdot\mathbf{x}}/\rho_b V_u, \tag{28.2}$$

where the distribution is periodic in V_u fixed in x coordinates. The variance of the density is (eq. 26.4)

$$\langle(\rho(\mathbf{x}) - \rho_b)^2\rangle/\rho_b^2 = \sum_{\mathbf{k}\neq 0} |\delta_{\mathbf{k}}|^2 = \frac{V_u}{(2\pi)^3}\int d^3k\, |\delta_{\mathbf{k}}|^2, \tag{28.3}$$

so one sees that equation (28.1) is the contribution to the variance of ρ by the Fourier components with wavelengths $\lambda \sim x$ in the range $\delta\lambda \sim x$.

It will be supposed that the matter is in point-like particles, mass $\sim m_0$. The goal is to discover the minimal fluctuations in density on large scale forced by the discreteness of the matter. If the particles were randomly distributed, equation (28.2) would be a random walk of N_u steps, N_u being the number of particles in V_u, and since the length of each step is

$$m_0/\rho_b \mathrm{V_u} a^3 = 1/N_u, \tag{28.4}$$

equation (28.1) gives

$$(\delta\rho/\rho)_x^2 = V_u/N_u x^3 = (x_0/x)^3, \qquad x_0 = (V_u/N_u)^{1/3}. \tag{28.5}$$

This is the usual result: the variance is the reciprocal of the mean number of particles in x^3. The density fluctuation is larger than is given by equation (28.5) if the particles are clustered, and it is smaller than equation (28.5) if the particles are anticorrelated (§ 36 below). To see how small the fluctuations reasonably might be, let us start with the density precisely uniformly distributed, $\rho(\mathbf{x}) = \rho_b$. Then $\delta_{\mathbf{k}} = 0, \mathbf{k} \neq 0$. To make the particles one must shift elements of mass, the minimum typical shift being on the order of the interparticle distance x_0 (eq. 28.5). For $kx_0 \ll 1$ this perturbs the exponential in equation (28.2) by the amount $\sim kx_0$, perturbing the spectrum from zero by the amount

$$|\delta_{\mathbf{k}}|^2 \sim (kx_0)^2/N_u, \qquad (\delta\rho/\rho)_x^{\,2} \sim (x_0/x)^5, \qquad x \gg x_0, \tag{28.6}$$

if the contributions from forming each particle add like a random walk. We can reduce this by observing that the particle building ought to conserve momentum locally: for each mass element shift there should be an equal and opposite shift of an element a distance $\sim x_0$ away. This cancels the first derivative of the exponential in equation (28.2) leaving the second and

$$|\delta_{\mathbf{k}}|^2 \sim (kx_0)^4/N_u, \qquad (\delta\rho/\rho)_x^{\,2} \sim (x_0/x)^7, \qquad x \gg x_0. \tag{28.7}$$

One could go further and arrange each pair of mass element shifts to cancel a nearby pair of opposite shifts (zero quadrupole moments). However, this is not something that would naturally follow in ordinary dynamics, and the arrangement would not be conserved if neighboring particles interact, for that conserves only the dipole moment through momentum conservation. Thus equation (28.7) represents the minimal fluctuations forced by the assumption of discrete particles. This k^4 minimal power spectrum was pointed out by Zel'dovich (1965a).

Another treatment of the argument goes as follows. Suppose again mass is exactly uniformly distributed to begin with, space is divided up into cells of size $\sim x_0$, and the mass found in each cell is gathered up into a point particle at the cell center of mass. Since it would seem artificial to place the cells in a regular lattice, we shall specify that there is no correlation among positions of cell centers at separations greater than a few times the cell size. The Fourier transform of the original distribution vanishes. This can be written as a sum over the cells c_j:

$$0 = \sum_j V_u^{-1} \int_{c_j} d^3x e^{i\mathbf{k}\cdot\mathbf{x}} = \sum_j V_u^{-1} e^{i\mathbf{k}\cdot\mathbf{x}_j} \int_{c_j} d^3y e^{i\mathbf{k}\cdot\mathbf{y}}, \tag{28.8}$$

where y is referred to the position x_j where the particle will be placed in c_j.

For $\lambda \gg x_0$ this last integral can be expanded in a rapidly converging series:

$$0 = \sum_j e^{i\mathbf{k}\cdot\mathbf{x}_j}\left[m_j/M_u + ik_\alpha \int_{c_j} y^\alpha d^3y/V_u \right. \\ \left. - \frac{1}{2}k_\alpha k_\beta \int_{c_j} y^\alpha y^\beta d^3y/V_u + \cdots \right]. \tag{28.9}$$

The first term is the Fourier transform δ_k of the particle distribution (eq. 27.8). If the particles were placed at random in the cells the next term would indicate $\delta_\mathbf{k} \propto \mathbf{k}$, as in equation (28.6), but to conserve momentum locally we have placed $\mathbf{x}_j$ at the cell center of mass so the integral vanishes. This leaves the third term. Since the cell positions are uncorrelated beyond near neighbors, this third sum approximates a random walk of $\sim N_u$ steps, step length $\sim (x_0 k)^2/N_u$, so

$$|\delta_\mathbf{k}|^2 \sim (x_0 k)^4/N_u, \qquad (\delta\rho/\rho)_x^{\ 2} \sim (x_0/x)^7, \tag{28.10}$$

as before.

There is another measure of density fluctuations that appears to give a different answer. A sphere of fixed radius x is placed at random and the mass within the sphere recorded. A measure of the density fluctuations on scale x is the variance of this mass,

$$(\delta\rho/\rho)_x^{\ 2} \equiv (\delta M/M)^2 = \langle (M - \rho_b V)^2 \rangle/(\rho_b V)^2. \tag{28.11}$$

In the above minimal construction of particles the surface of the sphere cuts $\sim (x/x_0)^2$ cells. Since the particle in each cell independently ends up either inside or outside the sphere (apart from a minor correction for the anticorrelation among positions of neighboring cells),

$$(\delta M)^2 \sim m_0^{\ 2}(x/x_0)^2, \qquad M \sim m_0(x/x_0)^3, \tag{28.12}$$

or

$$(\delta M/M)^2 \sim (x_0/x)^4, \qquad x \gg x_0. \tag{28.13}$$

This is only one power of x_0/x smaller than for a random distribution (eq. 28.5), so by this measure minimal fluctuations are not much less than for a random Poisson distribution of particles (Press and Schechter 1974,

Carlitz, Frautschi, and Nahm 1973). However, equation (28.13) is three powers of (x_0/x) larger than equation (28.10).

The reason for the discrepancy between equations (28.10) and (28.13) is apparent from equations (26.4). If the power spectrum $|\delta_{\mathbf{k}}|^2$ does not vary with k too rapidly, then the two measures defined in equations (28.1) and (28.11) are equivalent (eq. 26.5). However, if the spectrum increases faster than k, as here, then the main contribution to $(\delta M)^2$ is at large k from the spectrum that enters the integral through the sidelobes of the window function because the window function at large k varies as

$$k^3 W \propto k^{-1}. \tag{28.14}$$

Indeed one finds that if $|\delta_{\mathbf{k}}|^2$ is given by equation (28.10) at $kx_0 < 1$ and is constant at $kx_0 > 1$, then equation (26.4) yields equation (28.13). Clearly in this case $(\delta M/M)^2$ is not a useful measure of the fluctuations on scale x because it is responding to the high frequency (short wavelength) noise at the surface of the sphere, that is to the accidents of whether particles end up just inside or just outside the surface. Equation (28.10) is the preferred measure here.

B. k^4 Spectrum from gravity dynamics

The shape k^4 for the minimal spectrum can be derived also from the equation of motion (27.16) for the Fourier transform of the mass distribution. As discussed above one can arrange the initial conditions for the particle distribution so that $|\delta_{\mathbf{k}}|^2$ is cut off more sharply than λ^{-4} at $\lambda > x_0$, the mean interparticle distance. Then at long wavelength the dominant source terms on the right side of (27.16) are the two nonlinear ones A and C. The gravity term A is given by equation (27.15). If the spectrum cuts off at $k < x_0^{-1}$ $(\lambda > x_0)$, then the shape of the gravity source at long wavelength is determined by the factor D:

$$D = \frac{\mathbf{k}\cdot\mathbf{k}'}{k'^2} + \frac{\mathbf{k}\cdot(\mathbf{k}-\mathbf{k}')}{|\mathbf{k}-\mathbf{k}'|^2} \sim \frac{k^2}{k'^2} \sim (kx_0)^2, \qquad k \ll k' \gtrsim x_0^{-1}. \tag{28.15}$$

The gravity term thus acts as a source for a $k^4 \propto \lambda^{-4}$ tail of $|\delta_{\mathbf{k}}|^2$. The second nonlinear term is (eq. 27.11)

$$C = k_\alpha k_\beta \Sigma \dot{x}_j^\alpha \dot{x}_j^\beta e^{i\mathbf{k}\cdot\mathbf{x}_j} m_j / M_u. \tag{28.16}$$

If the particle velocities are uncorrelated, the sum is independent of k and $C \propto k^2$, as for A.

While this shows the spectrum grows a k^4 tail, it does not give the amplitude. Some aspects of this problem are discussed next.

C. Cancellation of nonlinear effects from stable clusters

When the particles are in tight massive clusters, it makes the velocities large and so makes the source term C large. It might seem that this would tend to drive the growth of fluctuations at wavelengths much greater than the cluster size (with a k^4 spectrum), and it might seem that the tighter the cluster the higher the velocities and the stronger the source of new fluctuations at long wavelength: an example of a mode-coupling effect of the sort discussed at the beginning of this section. However, we know this cannot be because the violent motions within a cluster cannot affect the motion of the cluster center of mass: the contributions to A and C by stable compact clusters of matter must cancel. We can see how this comes about by using the virial theorem.

The quantity

$$\frac{d^2}{dt^2}\sum_{j\epsilon\nu} m_j r_j^\alpha r_j^\beta \tag{28.17}$$

for the particles in the νth cluster has zero time average if the cluster is stable. The sum is over the particles in the cluster and the r_j^α are proper Cartesian coordinates relative to the center of mass. The result of differentiating this expression and using

$$\frac{d^2 r_j^\alpha}{dt^2} = \sum_{l\epsilon\nu} G m_l \frac{(r_l^\alpha - r_j^\alpha)}{|\mathbf{r}_l - \mathbf{r}_j|^3}, \tag{28.18}$$

is the tensor virial theorem,

$$\sum_{j\epsilon\nu} m_j \frac{dr_j^\alpha}{dt}\frac{dr_j^\beta}{dt} = \frac{1}{2} G \sum_{j\neq l\epsilon\nu} \frac{m_j m_l}{|\mathbf{r}_j - \mathbf{r}_l|^3}(r_j^\alpha - r_l^\alpha)(r_j^\beta - r_l^\beta). \tag{28.19}$$

The A term as it appears in equation (27.11) is

$$\begin{aligned} A &= \frac{ik_\alpha}{M_u a^2}\sum m_j(\phi_j)_{,\alpha} e^{i\mathbf{k}\cdot\mathbf{x}_j}, \\ -(\phi_j)_{,\alpha} &= \frac{G}{a}\sum_l \frac{x_l^\alpha - x_j^\alpha}{|\mathbf{x}_l - \mathbf{x}_j|^3} m_l. \end{aligned} \tag{28.20}$$

The contribution to A from the interactions within the νth group is found by restricting both sums to j, l in ν. On using

$$e^{i\mathbf{k}\cdot\mathbf{x}_j} = e^{i\mathbf{k}\cdot\mathbf{x}_\nu}[1 + i\mathbf{k}\cdot(\mathbf{x}_j - \mathbf{x}_\nu) + \cdots], \tag{28.21}$$

where $\mathbf{x}_\nu$ is the position of the center of mass of the group,

$$\Sigma m_j(\mathbf{x}_j - \mathbf{x}_\nu) = 0, \tag{28.22}$$

one finds that the contribution to A is, to lowest order,

$$\begin{aligned} A_\nu &= -\frac{Gk_\alpha k_\beta}{M_u a^3}\sum \frac{m_j m_l}{|\mathbf{x}_j - \mathbf{x}_l|^3}(x_l^\alpha - x_j^\alpha)(x_j^\beta - x_\nu^\beta)e^{i\mathbf{k}\cdot\mathbf{x}_\nu} \\ &= \frac{Gk_\alpha k_\beta}{2M_u a^3}\sum \frac{m_j m_l}{|\mathbf{x}_j - \mathbf{x}_l|^3}(x_l^\alpha - x_j^\alpha)(x_l^\beta - x_j^\beta)e^{i\mathbf{k}\cdot\mathbf{x}_\nu} \\ &= \frac{k_\alpha k_\beta}{a^2 M_u}\sum m_j \frac{dr_j^\alpha}{dt}\frac{dr_j^\beta}{dt}e^{i\mathbf{k}\cdot\mathbf{x}_\nu}. \end{aligned} \tag{28.23}$$

The first term in the expansion of equation (28.21) gives zero because for this term the summand is antisymmetric in j and l. The second term in the expansion gives the first line of equation (28.23) and with the virial theorem (28.19) this can be reduced to the third of equations (28.23), which is seen to be the same as the contribution to C (eq. 28.16) by the motions within the cluster. That is, the strong nonlinear motions within the cluster cancel out of A–C, as is to be expected.

D. Model for newly forming clusters

A rough model for the clusters just in the process of forming is obtained as follows. The newly forming generation is made up of clusters that can be treated as particles, each of mass m_0 say. If these particles were randomly distributed the Fourier components $\delta_\mathbf{k}$ would have random phases and power spectrum (eq. 28.5)

$$|\delta_\mathbf{k}|^2 = m_0/M_u. \tag{28.24}$$

We are interested in a sub-random distribution, where the large-scale fluctuations are suppressed as much as possible, so the power spectrum will be taken to be

$$\begin{aligned} |\delta_\mathbf{k}|^2 &= m_0/M_u, \qquad k > x_0^{-1}, \\ &= (m_0/M_u)(kx_0)^n, \qquad k \le x_0^{-1}, \\ n &> 0, \qquad \rho_b(ax_0)^3 \sim m_0. \end{aligned} \tag{28.25}$$

The $\delta_{\mathbf{k}}$ will be assumed to have random phases. The characteristic interparticle distance is x_0. On scales $x \lesssim x_0$ the spectrum is similar to random, and this is reasonable because the chance of finding more than one particle inside a volume $x^3 < x_0^3$ is small in any case. (Since the particles are distributed masses with size $\sim x_0$, the spectrum ought to decrease with increasing k at $k > x_0^{-1}$, but this detail will be ignored.) The fall-off of the power spectrum at long wavelength corresponds to anticorrelation of particle positions at $x \gtrsim x_0$. The density fluctuations (eq. 28.1) on scale x are

$$(\delta\rho/\rho)_x^2 \sim (x_0/x)^{3+n}, \qquad x \gtrsim x_0. \tag{28.26}$$

The fluctuations thus are just becoming nonlinear at scale $\sim x_0$, signaling the incipient formation of a new generation of clusters on this scale. Since this generation has not yet formed, we shall suppose the velocity term C may be neglected, and we shall estimate the size of the gravity term A.

By the random phase assumption we can write the mean of the square of A as (eq. 27.15)

$$\begin{aligned}\langle |A|^2\rangle &= (2\pi G\rho_b)^2 \sum_{\mathbf{k}'\neq 0,\mathbf{k}} D(\mathbf{k}')D(\mathbf{k}'')\langle \delta_{\mathbf{k}'}\delta_{\mathbf{k}-\mathbf{k}'}\delta_{-\mathbf{k}''}\delta_{\mathbf{k}''-\mathbf{k}}\rangle \\ &= 2(2\pi G\rho_b)^2 \frac{V_u}{(2\pi)^3}\int d^3k' D^2 |\delta_{\mathbf{k}'}|^2 |\delta_{\mathbf{k}-\mathbf{k}'}|^2,\end{aligned} \tag{28.27}$$

because only the terms with $\mathbf{k}'' = \mathbf{k}'$ or $\mathbf{k}'' = \mathbf{k} - \mathbf{k}'$ are not zero. The factor D^2 is

$$\begin{aligned} D^2 &\approx (k/k')^2, \qquad k' < k; \\ &\approx (k/k')^4, \qquad k' > k. \end{aligned} \tag{28.28}$$

For a random particle distribution where the spectrum is constant the factor D^2 makes the integral in equation (28.27) converge at large and small k', and one sees from equation (28.24) that the rms value of A in this case is

$$A_k \sim G\rho_b(m_0/M_u)(V_u k^3)^{1/2} \sim G\rho_b(m_0/M_u)^{1/2}(kx_0)^{3/2}. \tag{28.29}$$

The linear gravity term in equation (27.16) is

$$4\pi G\rho_b|\delta_{\mathbf{k}}| = 4\pi G\rho_b(m_0/M_u)^{1/2}, \tag{28.30}$$

and this is larger than the nonlinear term if $kx_0 \ll 1$: as expected, for a random particle distribution the linear approximation applies at wavelengths larger than the interparticle distance.

If the spectrum is given by equation (28.25) then with $kx_0 \ll 1$ and $n > \frac{1}{2}$ the main contribution to the integral (28.27) is at $k' \sim x_0^{-1}$, where the spectrum levels off and the rms value of A is

$$A_k \sim G\rho_b(m_0/M_u)^{1/2}(kx_0)^2, \qquad kx_0 \lesssim 1. \tag{28.31}$$

This can be compared to the linear source term

$$4\pi G\rho_b|\delta_{\mathbf{k}}| = 4\pi G\rho_b(m_0/M_u)^{1/2}(kx_0)^{n/2}. \tag{28.32}$$

If $n < 4$, the linear term is larger than A, and so the long wavelength part $\lambda \gg x_0$ grows as expected under ordinary linear perturbation theory. If $n > 4$, the nonlinear term dominates. The coherence time for A is on the order of the expansion time t because we have suppressed the contribution to A from subclustering on scales less than x_0. Thus in an expansion time the spectrum grows a tail

$$|\delta_{\mathbf{k}}| \sim t^2 A_k \sim (m_0/M_u)^{1/2}(kx_0)^2, \tag{28.33}$$

where x_0 and $m_0 \sim \rho_b(t)(a(t)x_0)^3$ are the characteristic size and mass of the clusters that are forming at epoch t. This agrees with equation (28.10).

E. Summary

The results of this section show the consistency of the idea that if the primeval matter distribution approximates a random Gaussian process with power spectrum $\propto k^n$, $-3 < n < 4$, then the spectrum at fixed comoving wavelength k^{-1} grows as in linear perturbation theory until the density fluctuations $\delta\rho/\rho$ on this scale reach unity and fragment into bound clusters. This is the process discussed in Section 26. In an Einstein-de Sitter model the characteristic size of the clusters forming at epoch t is (eq. 26.9)

$$r_0(t) \propto t^{(10+2n)/(9+3n)}. \tag{28.34}$$

The corresponding coordinate size is

$$k_0^{-1} \sim x_0 \propto t^{4/(9+3n)}, \tag{28.35}$$

and the mass is

$$m_0(t) \propto x_0^{\,3} \propto t^{4/(3+n)}. \tag{28.36}$$

This applies at $n < 4$; one might guess that if n were greater than 4 to begin with, the nonlinear motions on small scale would force the development of a k^4 tail, and the linear gravity source term for $\delta_{\mathbf{k}}(t)$ would then force m_0 (t) to grow at the minimum rate $m_0 \propto t^{4/7}$ (Zel'dovich 1974, Peebles 1974c).

The consistency of this picture rests on the result that the nonlinear term A in the equation for $\delta_{\mathbf{k}}(t)$ is less than the linear term $4\pi G\rho_b\delta_{\mathbf{k}}$ (eqs. 28.31, 28.32) at $x > x_0(t)$ and only becomes comparable to the linear term as $(\delta\rho/\rho)_{k^{-1}}$ approaches unity. One might object that although the picture thus appears to be consistent, it has not been shown that it is necessary: one might imagine that the nonlinear process of clustering on the scale $x \sim x_0(t)$ forces the clustering to develop faster than in equation (28.36) (eg. Press and Schechter 1974). No completely clean test of this has yet been devised; a reasonable argument goes as follows. Let x be a fixed comoving length, $x \sim x_0(t)$ at some chosen epoch t. If a sphere of radius x is placed on a developing protoobject, the mass excess is $\delta M/M \sim 1$, and numerical experiments verify that in an Einstein-de Sitter model the excess grows as $t^{2/3}$ at fixed x even when $\delta M/M$ is close to unity (Peebles and Groth 1976; § 70 below). This shows $|\delta_{\mathbf{k}}|^2 \propto t^{4/3}$ (eq. 28.1), consistent with the picture.

One can imagine that at very high redshift nongravitational processes forced the mass to pile up into nonlinear lumps, this in turn generating the k^4 tail and the subsequent gravitational development of clustering, $m_0 \propto t^{4/7}$. However, one would expect from causality that the lump size is smaller than the horizon. (This is demanded also if the density fluctuations are not to generate strong curvature fluctuations; compare § 95.) The number $r_0(t)/ct$, representing the ratio of the largest scale of appreciable clustering to the distance to the horizon, decreases with increasing time if $n = 4$ and nongravitational forces may be neglected. Since this ratio now is not much less than 0.01, it seems difficult to see how causal quantum processes operating at very high redshift could have played any role. It is well to bear in mind, however, that the classical field theory aspects of this line of speculation have not yet been analyzed in full detail, and the quantum aspects are still only a subject of speculation.

III. n-POINT CORRELATION FUNCTIONS: DESCRIPTIVE STATISTICS

29. STATISTICAL MEASURES OF THE GALAXY DISTRIBUTION

Two general approaches to the empirical study of the large-scale matter distribution might be called the botanical and the statistical. Reduction of the phenomena to specific sorts of objects like galaxies and Abell clusters of galaxies is direct and certainly has proved profitable. On the other hand, one can see since the 1930s a tendency to think that the general distribution is so complicated, and the data we can hope to have so schematic that a full reduction to genera and species of clustering might not be profitable or even possible. The alternative is to resort to statistical measures. For example, Hubble (1934) studied the frequency distribution of the count N of galaxies found in a telescope field. The distribution of N is strongly skew, but he found that the distribution of log N is quite close to a Gaussian. He suggested that such a simple and curious property must be an important clue to the nature of the general irregular distribution of galaxies on small scales. Bok (1934) and Mowbray (1938) compared the variance of N with what would be expected for a statistically uniform random distribution taking account of reasonable estimates of the variation of limiting magnitude from plate to plate and the variations in obscuration in the galaxy (which one could estimate from fluctuations of star counts). They both found that the variance of N is considerably larger than expected for a random galaxy distribution, again showing that galaxies generally cluster. Shapley and his associates made maps of angular positions of galaxies (detected at some limiting apparent magnitude) in sample areas of the sky. An example is reproduced in Figure 29.1 (Shapley 1935). Part of the large-scale irregularity apparent in the map is due to the decrease of sensitivity toward the edges of the several plates that cover the sample area (and of course in the large blank areas in the four corners no galaxies are marked because these areas are outside the sample). The map gives a striking impression of the complexity of the distribution. If the true spatial distribution were known, it would certainly help clarify the picture; whether the distribution would resolve into distinct clusters of diverse sorts is not known. (There is a further practical complication: the spatial distribution could be estimated by using redshift as a measure of distance, but the peculiar motions within clusters are

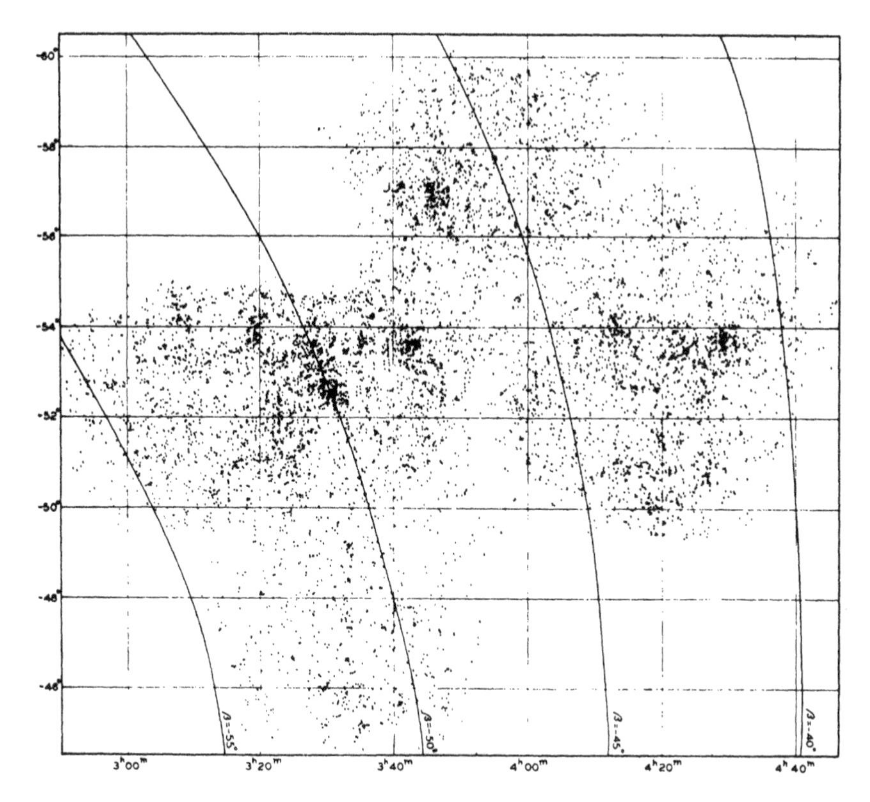

FIG. 29.1. One of Shapley's maps of the angular positions of galaxies brighter than $m \sim 18$ (Shapley 1935).

expected to cause errors in distance comparable to the clustering length. This is discussed in Section 76.)

For the choice of statistics to measure galaxy clustering the clear historical favorites have been variants of the autocorrelation function. Bok's statistic, the dispersion or variance of the counts N in cells, is an integral over the two-point correlation function (§ 36). Zwicky's (1953) index of clumpiness is the ratio of the variance of N to what would be expected for a uniform random distribution. Rubin (1954) used estimates of the variance of N based on the Harvard counts to measure the spatial galaxy autocorrelation function. Limber (1953, 1954) estimated the autocorrelation function of galaxy counts in the preliminary results from the Lick survey then in progress. He showed that there is a linear integral equation relating this angular correlation function to the corresponding spatial correlation function $\xi(r)$, and he used this to place some limits on $\xi(r)$.

Limber followed Chandrasekhar and Münch (1952), who had discussed the relation between the angular fluctuations in the brightness of the Milky Way and the clumpy distribution of the absorbing interstellar dust. In the problem as originally formulated by Ambartsumian (1944) it is assumed that the stars are uniformly distributed through the Milky Way and the dust is in randomly placed clouds. After analyzing this problem in some detail, Chandrasekhar and Münch turned to another approach where the dust distribution is characterized by its spatial autocorrelation function. Their equation relating this function to the autocorrelation of the brightness of the Milky Way (Chandrasekhar and Münch 1952, eq. 77) was a fairly close progenitor of Limber's equation relating the angular and spatial correlation functions (eq. 51.7 below). A special version of Limber's equation, relating the variance of counts of galaxies in cells in the sky to the spatial galaxy correlation function, was independently derived by Rubin (1954).[1]

During the 1950s the most extensive statistical program was that of Neyman and Scott (Neyman 1962, Scott 1962, and references therein). An important motivation for their work was the Lick survey: Shane emphasized that the large amount of data coming from this survey was a considerable challenge and opportunity for the statistical approach. Neyman and Scott devised *a priori* statistical models of clustering and then adjusted the parameters to fit model statistics to estimates from the data. Their mathematical methods were remarkable and perhaps have not yet been fully appreciated and exploited. The main empirical statistics they used were the angular autocorrelation function of the galaxy counts (for example, Neyman, Scott, and Shane 1956) and Zwicky's index of clumpiness (Neyman, Scott, and Shane 1954).

The Fourier or spherical harmonic transform of the autocorrelation function is the power spectrum (§§ 41, 46). This statistic was used by Yu and Peebles (1969) to test for clustering among positions of Abell clusters, by Peebles and Hauser (1974) to measure galaxy clustering in the Lick sample, by Shectman (1974) to measure the angular fluctuations in the integrated background light from the galaxies (§ 58), and by Webster (1976b) to test for clustering among 4C radio sources.

[1]The Limber-Rubin approach was criticized by Neyman and Scott (1955) because the galaxy distribution was treated as a continuous density function $\rho(\mathbf{r})$. As Neyman and Scott showed, the resulting expression for the variance of counts of galaxies in cells can contradict the condition that the N are nonnegative integers (because the term $\langle N\rangle$ is missing from the right side of eq. 36.7 below) when $\langle N\rangle$ is small. Layzer (1956) and Limber (1957) showed that this can be resolved by letting $\rho(\mathbf{r})$ be a probability density: the chance of finding a galaxy in δV at $\mathbf{r}$ is $\rho(\mathbf{r})\,\delta V$ (§§ 33, 38 below). Another approach is to replace the autocorrelation function of $\rho(\mathbf{r})$ with the two-point correlation function for a distribution of point-like objects (§ 31).

One can make out several reasons for the popularity of the autocorrelation function and its variants. Most directly, the approach has proved useful in many other applications, so it is natural that one should think of it here. Of considerable practical importance has been the fact that there is a simple linear equation relating the directly observable angular correlation function to the wanted spatial function. This means the translation from one to the other is fairly easy, and equally important it makes it easy to say how the statistical estimates ought to scale with the depth of the survey and hence to test for possible contamination of the estimates by systematic errors (Hauser and Peebles 1973, Peebles and Hauser 1974). A third useful result is that the dynamics of the galaxy distribution can be treated in terms of the mass correlation functions: the statistic that proves useful for the reduction of the data may also be useful for the analysis of the theory. This is the subject of Chapter IV.

The autocorrelation function $\xi(r)$ certainly is not the only useful statistic. It is not what one would use to characterize the abundance of rare extreme fluctuations like the Abell clusters because it is not very sensitive to them: one would invent a more powerful measure. Also, no matter how accurately ξ is known it contains definitely limited and restricted information. For example, to test for clustering among positions of rich clusters Abell (1958) considered the frequency distribution of the counts N of clusters found in cells of fixed angular size. If the clusters were placed at random, N would have a Poisson distribution. Abell used the deviation of the observed distribution from Poisson as a measure of clustering of the objects. As mentioned above the two-point correlation function $\xi(r)$ determines the second moment of this distribution of N. To predict the third moment, one would need the three-point correlation function, and to predict the full expected shape of the distribution in N, one would need to know all orders of the correlation functions (§ 39). Thus Abell's statistic in general can be sensitive to different aspects of the clustering than are revealed by $\xi(r)$.

Another popular statistic has been the nearest neighbor distance. This was used by Bogart and Wagoner (1973) to demonstrate the clustering of Abell clusters and by Turner and Gott (1975) in an attempt to isolate a uniformly distributed field population of galaxies. Gott and Turner (1977) introduced a multiplicity function as an analog of a luminosity function for the distribution of richnesses of groups and clusters of galaxies, and Bhavsar (1978) applied this approach in some detail. A problem with such statistics is that it is not easy to see how the angular data translates to the wanted statistics of the spatial distribution. Some aspects of this are discussed by White (1979) and in Sections 36 to 39 and 59 to 62 below.

The response to the limited information in any statistic must be tempered by the observational situation. If the data are not very extensive,

ξ might be as detailed a measure as it is reasonable to use. This was the situation in the discussions in the 1950s, and it certainly meant the statistical models of Neyman and Scott had a good deal more detail than what was available from the data. With the completion of the Zwicky and Lick catalogs and the development of high speed computers that made it easy to handle all the data, the situation changed. It became clear that ξ could be estimated with some precision and reproducibility and that a good deal more statistical information could be drawn from the data (Peebles 1974a). One systematic way to do this is to use higher order correlation functions. The approach was introduced by Peebles and Groth (1975). The estimation proves to work very well in the third order and marginally in the fourth order (Fry and Peebles 1978). The result is a substantial increase in the information and constraints on clustering models, though there is by no means a full statistical characterization.

The discussion in this chapter is completely devoted to n-point correlation functions and related statistics; one should of course bear in mind that while these statistics have proved their worth there may well emerge other equally useful measures.

30. Fair sample hypothesis

A central tenet of the standard cosmology is that, as Hubble put it, the part of the universe accessible to observation is a fair sample of the whole. Since people have considered cosmological models that are homogeneous but anisotropic, the fair sample hypothesis will be taken to mean that the universe is statistically homogeneous and isotropic. As has been discussed in Section 2 this hypothesis has successfully met the available tests, some quite precise, and it will be adopted in all the following discussion.

Matter certainly is distributed in a strongly clumpy fashion, in stars, galaxies, and clusters of galaxies, so the homogeneity and isotropy assumption must be supposed to apply in a statistical sense, in the average over large enough regions: the universe is assumed to be a homogeneous and isotropic random process (Neyman 1962). One imagines that the matter distribution has been determined by some physical process involving a lengthy and complex sequence of events, so that what is found at any particular place is the result of many slight variations of parameters within the sequence. Samples from well separated spots are uncorrelated, and the collection of such samples is a statistical ensemble generated by many independent applications of the process. Statistics such as the n-point correlation functions can be considered to be averages across the ensemble. The fair sample hypothesis states first that it makes sense to think of well separated parts of the universe as being independent realizations of the same physical process, second that within the visible part of the universe there are many independent samples that can be lumped to approximate a statistical ensemble, and third that averages across the ensemble are unaffected by a rotation.

Of particular importance in the data analysis is the test of reproducibility of the statistics derived from different parts of space: from different parts of the sky or from surveys at different depth. This is an essential test of whether the data are seriously affected by systematic errors, and of course it also is a positive test of the fair sample hypothesis. The results to date agree fairly well with the hypothesis (§§ 46, 57).

In this chapter space curvature and expansion are ignored for the most part; usual Euclidian geometry is used, and the fact that galaxies are observed along the back light cone, not at a fixed instant of cosmic time, is ignored. This is a good approximation because most of the available data samples only a small fraction of the Hubble distance cH^{-1}. For deeper surveys one must of course take account of the fact that the process is a function of time (§ 56).

31. Two-point spatial correlation function $\xi(r)$

Since many catalogs list positions of galaxies, or clusters of galaxies, or radio sources, it is often useful to think of the matter distribution as a distribution of point-like objects that could be nucleons, or galaxies, or clusters of galaxies. If differences among objects are ignored, it is only a question of the nature of the distribution of positions $\mathbf{r}_j$, and this can be described by the n-point correlation functions. These statistics as defined here and the following sections are almost identical to what is standard in the theory of nonideal gases, but, as described in Section 32, a slight formal difference has been introduced.

The probability that an object is found in the infinitesimal volume δV is

$$\delta P = n \delta V, \tag{31.1}$$

where the mean number density n is independent of position. This can be understood as an average across the ensemble: if M realizations are examined, an object is found in δV in

$$N = Mn\delta V \tag{31.2}$$

cases. The probability is proportional to the size of the volume element because doubling δV doubles the chance of finding an object, while the chance of finding more than one object is an infinitesimal of higher order (if one assumes, as is the case of practical interest, the objects are not in arbitrarily tight groups).

The mean number of objects found within the finite volume V is the integral of equation (31.1),

$$\langle N \rangle = nV. \quad (31.3)$$

One direct way to see this is to imagine adding the count in equation (31.2) over all volume elements δV in V.

The two-point correlation function ξ is defined by the joint probability of finding an object in both of the volume elements δV_1 and δV_2 at separation r_{12},

$$\delta P = n^2 \delta V_1 \delta V_2 [1 + \xi(r_{12})]. \quad (31.4)$$

Consistent with homogeneity and isotropy ξ has been written as a function of the separation alone. The factor n^2 makes the correlation function dimensionless. As in equation (31.1) the probability is proportional to $\delta V_1 \delta V_2$ because doubling either infinitesimal element doubles the chance of finding an object. In a uniform random Poisson point process the probabilities of finding objects in δV_1 and δV_2 are independent so the joint probability is the product of the single point probabilities in equation (31.1),

$$\delta P = n^2 \delta V_1 \delta V_2 \qquad \text{(uniform Poisson).} \quad (31.5)$$

In this case $\xi \equiv 0$; if the object positions are correlated, $\xi > 0$, if the positions are anticorrelated, $-1 \leq \xi < 0$.

Since the chance of finding an object in δV_1 is $n\delta V_1$, the conditional probability of finding an object in δV_2 given that there is an object in δV_1 is

$$\delta P(2 \mid 1) = n\delta V_2 [1 + \xi(r_{12})]. \quad (31.6)$$

Another way to put this is that if an object is chosen at random from the ensemble, the probability of finding that it has a neighbor at distance r in δV (and in the same realization) is

$$\delta P = n\delta V [1 + \xi(r)]. \quad (31.7)$$

The mean number of neighbors within distance r of a randomly chosen object is the integral of equation (31.7),

$$\langle N \rangle_p = \frac{4}{3}\pi r^3 n + n\int_0^r \xi(r)\,dV. \tag{31.8}$$

The subscript distinguishes this mean from equation (31.3) for randomly placed V. For a Poisson distribution $\xi = 0$, and the two means agree because the fact that V has been centered on an object does not affect the chance of finding objects anywhere else in V. If the positions are correlated, it does bias the mean: the integral of $n\xi$ is the mean number of neighbors in excess of what is expected for a uniform random distribution.

The integral in equation (31.8) extended to all r gives the total number of neighbors in excess of random,

$$n_c - 1 = n\int \xi\,dV. \tag{31.9}$$

If the integral converges, n_c is one measure of the mean number of objects per cluster. As an example of this suppose all objects are in clusters each with diameter D and each containing n_c members, and suppose the cluster centers are distributed like a uniform random Poisson process. Having chosen an object at random, one has also chosen the cluster to which it belongs. Since the clusters are supposed to be randomly distributed, this does not bias the distribution of other clusters, so they contribute nV to $\langle N \rangle_p$, the same as for a randomly placed volume. There are in addition $n_c - 1$ neighbors from the chosen cluster, so the integral of $n\xi$ is $n_c - 1$. If the number of objects per cluster is a random variable, the probability of choosing an object from a cluster with n_α members is proportional to n_α, so

$$n\int \xi dV = \langle n_\alpha(n_\alpha - 1)\rangle / \langle n_\alpha \rangle, \tag{31.10}$$

where the average is over the abundance of clusters in space. Some further aspects of such cluster models are discussed in Sections 40 and 61 below.

32. Two-point correlation function: another definition

There is a formally different way to define the n-point correlation functions (for example, Montgomery and Tidman 1964) that makes an interesting difference in equation (31.9). Suppose there are m objects confined to a fixed region R, volume V_u. Each object has a label, so $\mathbf{x}_1$ means the position of object (1) and so on. The probability that object (1) is found in the volume element δV_1, object (2) is in δV_2, and so on, is

$$\delta P = f(\mathbf{x}_1, \mathbf{x}_2, \ldots \mathbf{x}_m)\,\delta V_1 \delta V_2 \ldots \delta V_m, \tag{32.1}$$

where the normalizing condition is

$$\int_R f\, dV_1 \ldots dV_m = 1. \tag{32.2}$$

It will be assumed that the objects are identical (aside from the labels), so f is symmetric in its arguments. Then the probability of finding any object in δV_1 and any object in δV_2 is

$$\begin{aligned} \delta P &= m(m-1)\delta V_1 \delta V_2 \int_R f(\mathbf{r}_1, \mathbf{r}_2, \mathbf{r}_3 \ldots)\, dV_3 \ldots dV_m \\ &= n^2[1 + \xi(r_{12})]\delta V_1 \delta V_2. \end{aligned} \tag{32.3}$$

The integral gives the probability that object (1) is in δV_1 and object (2) is in δV_2, and the factor $m(m-1)$ converts this to the probability that any of the m objects is in δV_1, any of the $m-1$ others is in δV_2. The second line defines the two-point correlation function, with

$$n = m/V_u. \tag{32.4}$$

The normalization in equation (32.2) implies the constraint

$$n^2 \int_R dV_1 dV_2 \xi(\mathbf{r}_{12}) = -m, \tag{32.5}$$

or, since it will be assumed that the correlation length is much less than the size of R,

$$n \int d^3r \xi(r) = -1. \tag{32.6}$$

Here $n_c = 0$ regardless of how the objects are clustered. This is so because each object has $m-1$ neighbors in R, just one fewer than the number expected for a uniform random distribution. Equation (32.6) differs from equation (31.9) because there the number of objects in each realization is a random variable, and in choosing an object at random from the ensemble, one has biased the number of objects in the realization by the amount n_c. If the objects are arranged in randomly placed clusters of size D with n_c members each, then with the present definition the correlation function at large separation is

$$\xi(r) \approx -n_c/m, \qquad r > D. \tag{32.7}$$

It is biased negative because the background outside a chosen cluster contains $m - n_c$ members, for a mean density

$$\langle n \rangle \approx (m - n_c)/V_u = n(1 - n_c/m), \tag{32.8}$$

lower than the overall density by the fractional amount n_c/m.

The estimates of the correlation functions from catalogs of objects roughly agree with equations (32.6) and (32.7) (apart from boundary effects) because one must estimate the density n from the observed count (§ 47 below). In the discussion of the theory it will be more convenient to use the definitions given in equations (31.4) and (31.7) where these constraints do not apply. Of course, the practical difference is negligible because one must assume any useful sample contains many more objects than does a single cluster, $m \gg n_c$.

33. Two-point correlation function: Poisson model

The distribution of objects might be approximated by a continuous density function $\rho(\mathbf{r})$. The mean is

$$\langle \rho(\mathbf{r}) \rangle = n, \tag{33.1}$$

and the dimensionless autocorrelation function is

$$\xi(r) = \langle [\rho(\mathbf{x} + \mathbf{r}) - \langle \rho \rangle][\rho(\mathbf{x}) - \langle \rho \rangle] \rangle / \langle \rho \rangle^2, \tag{33.2}$$

which can be rewritten as

$$\langle \rho(\mathbf{x} + \mathbf{r}) \rho(\mathbf{x}) \rangle = n^2[1 + \xi(r)]. \tag{33.3}$$

Equations (33.3) and (31.4) are very similar, and the correspondence can be made more formal (Layzer 1956, Limber 1957). To construct a distribution of objects first select $\rho(\mathbf{r})$ from an ensemble, and then at each volume element δV in the sample region place an object with probability

$$\delta P = \rho(\mathbf{r}) \delta V. \tag{33.4}$$

This is a Poisson process in which the probability density is a function of position, $\rho(\mathbf{r})$. For given ρ the joint probability that objects are assigned to the elements δV_1 and δV_2 is the product

$$\delta P = \rho(\mathbf{r}_1) \delta V_1 \rho(\mathbf{r}_2) \delta V_2. \tag{33.5}$$

On averaging this over the ensemble of functions and using equation (33.3), one finds as before

$$\delta P = n^2[1 + \xi(r)]\delta V_1 \delta V_2, \tag{33.6}$$

where now ξ is the autocorrelation function of $\rho(\mathbf{r})$.

One should bear in mind that ξ as defined here is a model: it is easy to think of point processes it cannot describe such as that of hard spheres where $\xi = -1$ for $0 \leq r <$ sphere diameter. The definition in Section 31 makes ξ a descriptive statistic, for the definition is at the same time a prescription for how to estimate it. The Poisson model does have some convenient properties (§ 38), and, so far as is known, it does not conflict with the empirical galaxy distribution.

34. Three-point correlation function

Following equation (31.4), one can define the three-point function by the joint probability of finding objects in each of the three elements δV_1, $\delta V_2, \delta V_3$,

$$\delta P = n^3 \delta V_1 \delta V_2 \delta V_3 [1 + \xi(r_a) + \xi(r_b) + \xi(r_c) + \zeta(r_a, r_b, r_c)], \tag{34.1}$$

where r_a, r_b, and r_c are the sides of the triangle defined by the three points. The assumption of homogeneity and isotropy means that ζ is a symmetric function of these three lengths. The quantity in square brackets is the full three-point correlation function, and ζ is the reduced part.

There are several advantages to using the reduced function ζ. If δV_1 and δV_2 are close together and δV_3 is so far away that the chance of finding an object in it is unaffected by what happens in δV_1 and δV_2, then equations (31.1) and (31.4) say

$$\delta P = n^2 \delta V_1 \delta V_2 [1 + \xi(r_{12})] \cdot n \delta V_3. \tag{34.2}$$

Comparing this with equation (34.1), one sees that ζ vanishes. In a dilute nonideal gas typically $1 \gg \xi \gg \zeta$, so one can conveniently treat the reduced function ζ as a perturbation. For the galaxy distribution the dominant term at small r is $\zeta \gg \xi \gg 1$, but here there is another important practical advantage of the reduced function. The most extensive data available on the galaxy distribution are catalogs of angular positions. The two- and reduced three-point angular correlation functions are defined by equations like (31.4) and (34.1), and these two angular functions are integrals over ξ and ζ respectively (§ 54). In the analog of equation (34.1) for the angular distribution the dominant term in the square brackets may be the first, which represents triplets of galaxies at three very different distances accidently seen close together in the sky. The next largest terms are the three two-point functions that represent a pair of galaxies close together in space with the triplet accidently completed by a third galaxy at a very different distance. The form of equation (34.1) thus makes convenient the separation of accidental and physical triplets in the angular data.

The conditional probability of finding two objects to complete the triangle r_a, r_b, r_c with a randomly chosen object is (compare eq. 31.6)

$$\delta P = n^2 \delta V_2 \delta V_3 (1 + \xi_a + \xi_b + \xi_c + \zeta_{abc}), \tag{34.3}$$

where $\xi_a = \xi(r_a)$ and so on. The conditional probability of finding an object to complete the triangle having found a pair of objects at separation r_a is (by eq. 31.4)

$$\delta P = n \delta V_3 (1 + \xi_a + \xi_b + \xi_c + \zeta_{abc})/(1 + \xi_a). \tag{34.4}$$

If the distribution of objects is approximated as a continuous function $\rho(r)$ (§ 33), the third central moment for the distribution is

$$\zeta(r, s, |\mathbf{r} - \mathbf{s}|) = \langle [\rho(\mathbf{x} + \mathbf{r}) - \langle \rho \rangle][\rho(\mathbf{x} + \mathbf{s}) - \langle \rho \rangle][\rho(\mathbf{x}) - \langle \rho \rangle] \rangle / \langle \rho \rangle^3. \tag{34.5}$$

On multiplying out the right hand side and using equation (33.3), one finds

$$\langle \rho(\mathbf{x} + \mathbf{r}) \rho(\mathbf{x} + \mathbf{s}) \rho(\mathbf{x}) \rangle = n^3 [1 + \xi(r) + \xi(s) + \xi(|\mathbf{r} - \mathbf{s}|) + \zeta], \tag{34.6}$$

with $\langle \rho \rangle = n$, which agrees with equation (34.1). As before one can make the correspondence more formal by using the Poisson model. For given density function $\rho(\mathbf{r})$ the chance that objects are placed at $\mathbf{x}$, $\mathbf{x} + \mathbf{r}$ and $\mathbf{x} + \mathbf{s}$ is (eqs. 33.4 and 33.5)

$$\delta P = \rho(\mathbf{x}) \rho(\mathbf{x} + \mathbf{r}) \rho(\mathbf{x} + \mathbf{s}) \delta V_1 \delta V_2 \delta V_3 . \tag{34.7}$$

On averaging this and using equation (34.6), one arrives at equation (34.1). In this sense the extra terms appear in the square brackets in equation (34.1) because ζ is the central moment.

For some purposes it has proved convenient to use new independent variables. Since ζ is a symmetric function of the three sides of the triangle, we can set

$$r_a < r_b < r_c, \tag{34.8}$$

and then let (Peebles and Groth 1975)

$$\begin{gathered} r = r_a, \qquad u = r_b/r_a, \qquad v = (r_c - r_b)/r_a, \\ 1 < u, \qquad 0 < v < 1, \end{gathered} \tag{34.9}$$

so r is a measure of the size of the triangle and the dimensionless parameters u, v measure the shape. If the points at either end of the side $r_a = r$ are held fixed, the range δu, δv in the shape parameters defines a single ring of volume

$$\delta V_3 = 2\pi r^3 u(u + v)\,\delta u\,\delta v. \tag{34.10}$$

Thus if the vertex joining r_a and r_c is held fixed, the six dimensional element defined by triangles with size and shape r, u, v in the range δr, δu, δv is

$$(dV)^2 = 8\pi^2 r^5 u(u + v)\,\delta r\,\delta u\,\delta v. \tag{34.11}$$

35. Four-point correlation function

This is defined through the joint probability of finding objects in each of the four volume elements δV_1, δV_2, δV_3 and δV_4,

$$\begin{aligned} \delta P = n^4 \delta V_1 \delta V_2 \delta V_3 \delta V_4 [1 & \\ + \xi(r_{12}) + \cdots & \quad \text{(6 terms)} \\ + \zeta(r_{12}, r_{23}, r_{31}) + \cdots & \quad \text{(4 terms)} \\ + \xi(r_{12})\xi(r_{34}) + \cdots & \quad \text{(3 terms)} \\ + \eta]. & \end{aligned} \tag{35.1}$$

The reduced function η is a function of the six variables needed to fix the relative positions of the four points (4×3 coordinates minus 3 translations minus 3 rotations).

As for the three-point function, there are several ways to see why all the extra terms are added to η in equation (35.1). In the corresponding equation for the angular distribution of galaxies, the dominant term in the square brackets proves to be the first, which arises from four galaxies at four quite different distances that accidently appear close together in the sky. The second line represents two galaxies close together in space and the other two accidently close in projection, the third line close triplets plus a fourth accidently close in projection, the fourth line two close pairs accidently seen close together in projection. With these projection effects removed, the reduced angular correlation function u is an integral over η (§ 55).

If δV_1 is close to δV_2 and δV_3 close to δV_4 but well removed from the other pair, the probability of equation (35.1) reduces to the product of two-point probabilities (31.4)

$$\begin{aligned}\delta P &= n^2 \delta V_1 \delta V_2 [1 + \xi_{12}] \cdot n^2 \delta V_3 \delta V_4 [1 + \xi_{34}] \\ &= n^4 [1 + \xi_{12} + \xi_{34} + \xi_{12}\xi_{34}] \delta V_1 \delta V_2 \delta V_3 \delta V_4 .\end{aligned} \tag{35.2}$$

Comparing this with equation (35.1), one sees that $\eta \longrightarrow 0$ when the pairs are well separated, as desired. When three points are close together and the fourth well removed, equation (35.1) reduces to the product of a three-point distribution with a one-point distribution, and again $\eta \longrightarrow 0$.

For a continuous density function, the fourth central moment is defined by the equation

$$\begin{aligned}\langle \rho \rangle^4 \eta = {} & \langle (\rho_1 - \langle \rho \rangle)(\rho_2 - \langle \rho \rangle)(\rho_3 - \langle \rho \rangle)(\rho_4 - \langle \rho \rangle) \rangle \\ & - \langle (\rho_1 - \langle \rho \rangle)(\rho_2 - \langle \rho \rangle) \rangle \langle (\rho_3 - \langle \rho \rangle)(\rho_4 - \langle \rho \rangle) \rangle \\ & - \langle (\rho_1 - \langle \rho \rangle)(\rho_3 - \langle \rho \rangle) \rangle \langle (\rho_2 - \langle \rho \rangle)(\rho_4 - \langle \rho \rangle) \rangle \\ & - \langle (\rho_1 - \langle \rho \rangle)(\rho_4 - \langle \rho \rangle) \rangle \langle (\rho_2 - \langle \rho \rangle)(\rho_3 - \langle \rho \rangle) \rangle .\end{aligned} \tag{35.3}$$

The last three terms make $\eta \longrightarrow 0$ when one pair of points is well away from the other pair. This moment is the analog of the excess kurtosis of a distribution $f(x)$,

$$u = \langle (x - \langle x \rangle)^4 \rangle - 3 \langle (x - \langle x \rangle)^2 \rangle^2 . \tag{35.4}$$

The last term here makes u vanish if f is a Gaussian.

In the Poisson model the chance of finding objects in the four volume elements is

$$\delta P = \langle \rho_1 \rho_2 \rho_3 \rho_4 \rangle \delta V_1 \delta V_2 \delta V_3 \delta V_4 . \tag{35.5}$$

On multiplying out the top line of equation (35.3) and using the definitions of ξ and ζ in equations (33.3) and (34.5), one finds that equations (35.3) and (35.5) can be reduced to equation (35.1).

36. Moments of counts of objects

A useful illustration of the meaning of the n-point correlation functions is their relation to the moments of counts of objects. There are two interesting cases: where the objects are counted in a randomly placed cell and where the cell is centered on a randomly chosen object. The general relations are derived here. Some numerical results for the observed galaxy distribution are given in Section 59.

A. Randomly placed cells

A convenient way to compute the moments of the count N of objects in a cell is to imagine the cell is divided into infinitesimal elements with n_1 objects in the element δV_1. The probability that $n_1 = 1$ is $n\delta V_1$ (eq. 31.2), and the probability that $n_1 > 1$ is an infinitesimal of higher order. Thus

$$n\delta V_1 = \langle n_1 \rangle = \langle n_1^2 \rangle = \langle n_1^3 \rangle = \cdots \tag{36.1}$$

to order δV because $n_1^m = n_1$ if $n_1 = 0, 1$. The product $n_1 n_2$ for the counts in the disjoint elements δV_1, δV_2 is equal to unity if there are objects in both elements, and the probability for this is given by equation (31.4), so

$$\begin{aligned} \langle n_1 n_2 \rangle &= n^2 \delta V_1 \delta V_2 (1 + \xi_{12}), \\ \langle (n_1 - \langle n_1 \rangle)(n_2 - \langle n_2 \rangle) \rangle &= n^2 \delta V_1 \delta V_2 \xi_{12}. \end{aligned} \tag{36.2}$$

The count in the cell V is

$$N = \Sigma n_1. \tag{36.3}$$

By equation (36.1), the mean count is

$$\langle N \rangle = \Sigma \langle n_1 \rangle = \int_V n dV = nV, \tag{36.4}$$

as before (eq. 31.3). The second moment is

$$\langle N^2 \rangle = \Sigma \langle n_1^2 \rangle + \Sigma \langle n_1 n_2 \rangle. \tag{36.5}$$

The squared terms are given by equation (36.1), the cross terms by equation (36.2):

$$\langle N^2 \rangle = nV + (nV)^2 + I_2, \qquad I_2 = n^2 \int_V dV_1 dV_2 \xi_{12},$$
$$\mu_2 = \langle (N - nV)^2 \rangle = nV + I_2. \tag{36.6}$$

If $\xi = 0$, this reduces to $\mu_2 = nV$ as for a Poisson distribution. If the objects are correlated, $\xi > 0$, the dispersion in N is increased. If V is large compared to the clustering length, equation (36.6) is

$$\mu_2 = nV + n^2 V \int \xi(r) d^3 r = n_c nV, \tag{36.7}$$

where n_c is the number of objects per cluster (eq. 31.9), or

$$\langle (N/n_c)^2 \rangle = \langle N/n_c \rangle^2 + \langle N/n_c \rangle. \tag{36.8}$$

In effect there are n/n_c clusters per unit volume randomly placed.

The mean of N^3 depends on the three-point correlation function. As in equation (36.2), the probability that $n_1 n_2 n_3 = 1$ for disjoint cells is given by equation (34.1), and

$$\langle n_1 n_2 n_3 \rangle = n^3 \delta V_1 \delta V_2 \delta V_3 (1 + \xi_{12} + \xi_{23} + \xi_{31} + \zeta),$$
$$\langle (n_1 - \langle n_1 \rangle)(n_2 - \langle n_2 \rangle)(n_3 - \langle n_3 \rangle) \rangle = n^3 \delta V_1 \delta V_2 \delta V_3 \zeta. \tag{36.9}$$

The second line uses equations (36.1) and (36.2). It is convenient to evaluate the central moment,

$$\begin{aligned} \mu_3 &= \langle (N - nV)^3 \rangle = \langle (\Sigma n_i - \langle n_i \rangle)^3 \rangle \\ &= \Sigma \langle (n_1 - \langle n_1 \rangle)(n_2 - \langle n_2 \rangle)(n_3 - \langle n_3 \rangle) \rangle \\ &\quad + 3\Sigma \langle (n_1 - \langle n_1 \rangle)^2 (n_2 - \langle n_2 \rangle) \rangle \\ &\quad + \Sigma \langle (n_1 - \langle n_1 \rangle)^3 \rangle. \end{aligned} \tag{36.10}$$

The first term, where the three cells all are different, is given by equation (36.9). The second term is

$$\Sigma \langle (n_1 - \langle n_1 \rangle)^2 (n_2 - \langle n_2 \rangle) \rangle = n^2 \int \xi_{12} dV_1 dV_2 = \mu_2 - nV, \tag{36.11}$$

and the third term is

$$\Sigma\langle(n_1 - \langle n_1\rangle)^3\rangle = \Sigma\langle n_1{}^3\rangle = nV, \tag{36.12}$$

so the third moment is (Peebles 1975)

$$\begin{gathered}\mu_3 = 3\mu_2 - 2\,nV + I_3, \qquad I_3 = n^3 \int_V dV_1 dV_2 dV_3 \zeta, \\ \langle N^3\rangle = nV + 3\,(nV)^2 + (nV)^3 + 3\,(nV+1)I_2 + I_3.\end{gathered} \tag{36.13}$$

If $\xi = \zeta = 0$, then $\mu_3 = \mu_2 = nV$, the result for a Poisson distribution.

The probability that $n_1 n_2 n_3 n_4 = 1$ for four disjoint cells is given by equation (35.1). With equations (36.2) and (36.9) one finds

$$\begin{gathered}\langle(n_1 - \langle n_1\rangle)(n_2 - \langle n_2\rangle)(n_3 - \langle n_3\rangle)(n_4 - \langle n_4\rangle)\rangle \\ = n^4 \delta V_1 \delta V_2 \delta V_3 \delta V_4(\eta + \xi_{12}\xi_{34} + \xi_{13}\xi_{24} + \xi_{14}\xi_{23}).\end{gathered} \tag{36.14}$$

The fourth central moment is

$$\begin{aligned}\mu_4 &= \langle(\Sigma n_i - \langle n_i\rangle)^4\rangle \\ &= \Sigma\langle(n_1 - \langle n_1\rangle)(n_2 - \langle n_2\rangle)(n_3 - \langle n_3\rangle)(n_4 - \langle n_4\rangle)\rangle \\ &\quad + 6\,\Sigma\langle(n_1 - \langle n_1\rangle)^2(n_2 - \langle n_2\rangle)(n_3 - \langle n_3\rangle)\rangle \\ &\quad + 3\,\Sigma\langle(n_1 - \langle n_1\rangle)^2(n_2 - \langle n_2\rangle)^2\rangle \\ &\quad + 4\,\Sigma\langle(n_1 - \langle n_1\rangle)^3(n_2 - \langle n_2\rangle)\rangle + \Sigma\langle(n_1 - \langle n_1\rangle)^4\rangle.\end{aligned} \tag{36.15}$$

By equation (36.1) the last term is nV. The second last term is

$$4\,\Sigma(\langle n_1{}^3 n_2\rangle - \langle n_1{}^3\rangle\langle n_2\rangle) = 4\,I_2; \tag{36.16}$$

the third term is

$$3\,\Sigma\langle n_1{}^2 n_2{}^2\rangle = 3\,[I_2 + (nV)^2]; \tag{36.17}$$

the second term is

$$6\,I_3 + 6\,nVI_2, \tag{36.18}$$

and the first term is

$$I_4 + 3\,I_2{}^2, \qquad I_4 = n^4 \int_V \eta\, dV_1 dV_2 dV_3 dV_4. \tag{36.19}$$

The sum is (Fry and Peebles 1978)

$$\begin{aligned}\mu_4 &= I_4 + 6\, I_3 + 3\, I_2^{\,2} + (7 + 6\, nV)I_2 + 3(nV)^2 + nV \\ &= I_4 + 6\, \mu_3 + 3\, \mu_2^{\,2} - 11\, \mu_2 + 6\, nV.\end{aligned} \tag{36.20}$$

This is the fourth central moment in terms of integrals over the correlation functions.

B. Moments of the counts of neighbors

Here the cell is centered on an object and N is the count of objects in V excluding the one on which V was centered. The subscript p on the averages will distinguish them from the averages for randomly placed cells.

The count of neighbors is written as before as

$$N = \Sigma n_1, \tag{36.21}$$

where the cell has been divided into infinitesimal elements and $n_1 = 0, 1$ is the count in the element δV_1. The mean of n_1 is the conditional probability (eq. 31.6),

$$\langle n_1 \rangle = n\delta V_1 [1 + \xi(r_1)], \tag{36.22}$$

where r_1 is the distance from the chosen object to δV_1. The mean count is then

$$\langle N \rangle_p = nV + n \int_V \xi\, dV, \tag{36.23}$$

which agrees with equation (31.8).

The mean of $n_1 n_2$ for disjoint elements $\delta V_1 \delta V_2$ is the conditional probability given by equation (34.3):

$$\begin{aligned}\langle n_1 n_2 \rangle = n^2 \delta V_1 \delta V_2 [1 &+ \xi(r_1) + \xi(r_2) \\ &+ \xi(|\mathbf{r}_1 - \mathbf{r}_2|) + \zeta(\mathbf{r}_1, \mathbf{r}_2)],\end{aligned} \tag{36.24}$$

where the two elements are at positions $\mathbf{r}_1$, $\mathbf{r}_2$ relative to the chosen object. The second moment of N is then

$$\begin{aligned}\langle N^2 \rangle_p &= \Sigma \langle n_1^{\,2} \rangle + \Sigma \langle n_1 n_2 \rangle \\ &= \langle N \rangle_p (1 + 2nV) - (nV)^2 + n^2 \int dV_1 dV_2 [\zeta + \xi(r_{12})],\end{aligned} \tag{36.25}$$

and the central moment is

$$\langle (N - \langle N \rangle_p)^2 \rangle_p = \langle N \rangle_p + n^2 \int dV_1 dV_2 (\zeta - \xi_1 \xi_2 + \xi_{12}). \quad (36.26)$$

By a similar computation one finds that the third central moment of the count of neighbors is

$$\begin{aligned} \langle (N - \langle N \rangle_p)^3 \rangle_p = {} & 3 \langle (N - \langle N \rangle_p)^2 \rangle_p - 2 \langle N \rangle_p \\ & + n^3 \int dV_1 dV_2 dV_3 [\eta - 3\xi_1 \zeta(r_2, r_3, r_{23}) \\ & + \zeta(r_{12}, r_{23}, r_{31}) + 2\xi_1 \xi_2 \xi_3], \end{aligned} \quad (36.27)$$

and the mean of the cube of the number of neighbors is

$$\begin{aligned} \langle N^3 \rangle_p = {} & (nV)^3 - \langle N \rangle_p (2 + 3nV + 3(nV)^2) \\ & + 3 \langle N^2 \rangle_p (1 + nV) + n^3 \int dV_1 dV_2 dV_3 \zeta_{123} \\ & + 3n^3 \int \xi \, dV \int \xi_{12} \, dV_1 dV_2 + n^3 \int \eta \, dV_1 dV_2 dV_3. \end{aligned} \quad (36.28)$$

37. Constraints on ξ and ζ

Not all functions ξ, ζ correspond to realizable distributions of objects. Two constraints on these functions follow from the condition that the second central moments cannot be negative. Equation (36.6) gives

$$\mu_2 = nV + n^2 \int dV_1 dV_2 \xi_{12} > 0. \quad (37.1)$$

If the objects are distributed like hard spheres, with no correlation at separations greater than the sphere diameter,

$$\xi = -1, \qquad r \le r_0; \qquad \xi = 0, \qquad r > r_0; \quad (37.2)$$

equation (37.1) with the size of the cell $\gg r_0$ says

$$n < \frac{3}{4\pi} r_0^{-3}. \quad (37.3)$$

If n were larger than this, the positions would have to be correlated at $r \sim r_0$. If the spheres were in a cubic close packed lattice, the density would be

$$n = 2^{1/2} r_0^{-3}, \tag{37.4}$$

5.9 times the density allowed by equations (37.2) and (37.3).

The second central moment of the count of neighbors (eq. 36.26) is

$$n \int (1 + \xi)\, dV + n^2 \int dV_1 dV_2 (\zeta - \xi_1 \xi_2 + \xi_{12}) > 0. \tag{37.5}$$

This simplifies if n is large, so the first term can be neglected, and if $\xi \gg 1$, to

$$\int_V dV_1 dV_2 [\zeta - \xi(r_1)\xi(r_2)] > 0. \tag{37.6}$$

The galaxy two- and three-point correlation functions are found to be given to good accuracy by the power law model

$$\begin{gathered} \xi = Br^{-\gamma}, \qquad \gamma \simeq 1.77, \\ \zeta(\mathbf{r}_1, \mathbf{r}_2) = Q[\xi(r_1)\xi(r_2) + \xi(r_1)\xi(r_{12}) + \xi(r_2)\xi(r_{12})], \\ Q \approx 1.3. \end{gathered} \tag{37.7}$$

This model in equation (37.6) gives

$$(1 - Q)\left(\int dV\xi\right)^2 < 2Q \int dV_1 dV_2 \xi(r_1)\xi(r_{12}). \tag{37.8}$$

If V is a sphere of radius r centered on the chosen object, then

$$\begin{gathered} \int dV\xi = 4\pi B r^{(3-\gamma)}/(3 - \gamma), \\ \int dV_1 dV_2 \xi(r_1)\xi(r_{12}) = 88 B^2 r^{6-2\gamma}, \qquad \gamma = 1.8, \end{gathered} \tag{37.9}$$

where the numerical coefficient in the second equation is obtained by the integrations summarized in Section 59. Equations (37.8) and (37.9) yield the inequality

$$Q > 0.372. \tag{37.10}$$

This limit is about one third the observed value. Thus it is interesting that ζ for galaxies at large n is a fixed and modest multiple of the minimum allowed by the constraint.

38. Probability generating function

The labor of finding the expressions for the moments of the counts of objects in terms of integrals over the n-point functions is reduced by using a generating function. The general case is considered by White (1979). The Poisson model from Section 33 is discussed here.

The probability generating function for the cell V is

$$\psi(t) = \Sigma P_N e^{Nt}, \tag{38.1}$$

where P_N is the probability that V is found to contain N objects. A convenient property of ψ is that if the counts in separate regions are statistically independent, the generating functions multiply: the probability of finding the total count N in regions V_1 plus V_2 is

$$P_N = \Sigma P_{N-N'}(1) P_{N'}(2), \tag{38.2}$$

so

$$\psi_{12} = \psi_1 \psi_2, \tag{38.3}$$

where ψ_1 belongs to V_1, ψ_2 to V_2, and ψ_{12} to the combined region.

In the Poisson model of Section 33 the probability of finding an object in the volume element δV at $\mathbf{r}$ is

$$P_1 = \rho(\mathbf{r}) \delta V; \tag{38.4}$$

the chance of finding no object is

$$P_0 = 1 - \rho \delta V, \tag{38.5}$$

and the chance of finding more than one object is an infinitesimal of higher order than δV, so the ψ belonging to δV is

$$\psi = 1 - \rho(\mathbf{r}) \delta V + \rho(\mathbf{r}) \delta V e^t = \exp\left[(e^t - 1) \rho \delta V\right]. \tag{38.6}$$

The ψ belonging to the finite cell V is then (Layzer 1956)

$$\psi(t) = \langle \exp (e^t - 1) \textstyle\int \rho \, dV \rangle, \tag{38.7}$$

where the integral is over V and the brackets represent the average across the ensemble of density functions ρ.

The mth derivative of ψ at $t = 0$ is the mth moment

$$\psi^{(m)}(t = 0) = \Sigma N^m P_N = \langle N^m \rangle. \tag{38.8}$$

The moments about the mean nV are obtained from the generating function

$$\begin{aligned} \chi(t) &= \psi(t) e^{-nVt} = \langle e^f \rangle, \\ f &= (e^t - t - 1) nV + (e^t - 1) \int \delta\rho \, dV, \end{aligned} \tag{38.9}$$

where the density has been written as the sum of the mean and the fluctuating part

$$\rho(\mathbf{r}) = n + \delta\rho(\mathbf{r}). \tag{38.10}$$

The mth derivative of χ at $t = 0$ is the central moment

$$\chi^{(m)}(0) = \langle (N - nV)^m \rangle = \mu_m. \tag{38.11}$$

The first four derivatives are

$$\begin{aligned} \chi'(0) &= \langle f'(0) \rangle, \\ \chi''(0) &= \langle f'' + (f')^2 \rangle, \\ \chi'''(0) &= \langle f''' + 3f'f'' + (f')^3 \rangle, \\ \chi''''(0) &= \langle f'''' + 4f'f''' + 3(f'')^2 + 6(f')^2 f'' + (f')^4 \rangle, \end{aligned} \tag{38.12}$$

where

$$\begin{aligned} f'(0) &= \int \delta\rho \, dV, \\ f^{(m)}(0) &= nV + \int \delta\rho \, dV, \qquad m > 1. \end{aligned} \tag{38.13}$$

By using the definitions of the n-point correlation functions in terms of the $\delta\rho$ (eqs. 33.2, 34.5, 35.3), one can verify that equations (38.12) yield the central moments obtained in Section 36A.

The moments of the number of neighbors in the cell V centered on a randomly chosen object are given by the generating function

$$\begin{aligned} \psi(t) &= \Sigma P_N e^{Nt} = \langle \rho(0) \exp{(e^t - 1)} \int \rho \, dV \rangle / n, \\ \psi^{(m)}(t = 0) &= \langle N^m \rangle_p. \end{aligned} \tag{38.14}$$

The factor $\rho(0)$ in the expectation value takes account of the fact that the cell is centered on an object at $r = 0$, and N is the number of objects in V not counting this chosen one. By following the above calculation, one can verify that equations (38.14) reproduce the moments obtained in Section 36B.

39. Estimates of P_N

If all n-point correlation functions are known, then in principle one can compute such statistics as the distribution P_N of counts in a cell and the distribution of nearest neighbor distances. There are two interesting limiting cases where P_N can be estimated from the low order correlation functions. If V is larger than the maximum clustering length, P_N approaches a Gaussian with variance fixed by the integral of ξ. If V is small, P_N rapidly decreases at large N, and we can use an approximation scheme as follows.

The P_N for randomly placed V will be considered first. Suppose V is so small there is negligible chance it contains four or more objects. Then the integral over V of the joint probability in equation (34.1) of finding objects in δV_1, δV_2, δV_3 is 6 P_3 because P_3 is the probability there is a triplet somewhere in V and the integral counts the triplet 6 times. Therefore

$$P_3 = \tfrac{1}{6}(nV)^3 + \tfrac{1}{2}nVI_2 + \tfrac{1}{6}I_3 \tag{39.1}$$

where the integrals I_2 and I_3 over ξ and ζ are defined in equations (36.6) and (36.13). The integral over V of the joint probability in equation (31.4) of finding objects in δV_1 and δV_2 is

$$(nV)^2 + I_2 = 2P_2 + 6P_3 \tag{39.2}$$

because if there are two objects in V the integral counts the pair twice and if there are three objects the integral picks up a pair 6 times. With equation (39.1) this gives

$$P_2 = \tfrac{1}{2}(nV)^2(1 - nV) + (\tfrac{1}{2} - \tfrac{3}{2}nV)I_2 - \tfrac{1}{2}I_3. \tag{39.3}$$

The mean number in V is

$$nV = P_1 + 2P_2 + 3P_3, \tag{39.4}$$

and this with equations (39.1) and (39.3) yields

$$P_1 = nV(1 - nV + \tfrac{1}{2}(nV)^2) + (\tfrac{3}{2}nV - 1)I_2 + \tfrac{1}{2}I_3. \tag{39.5}$$

Finally, P_0 follows from the condition that the P_N add up to unity:

$$P_0 = 1 - nV + \tfrac{1}{2}(nV)^2 - \tfrac{1}{6}(nV)^3 + \tfrac{1}{2}(1 - nV)I_2 - \tfrac{1}{6}I_3. \qquad (39.6)$$

These are the wanted approximations to P_0 through P_3 in terms of integrals over ξ and ζ. Of course, the accuracy would be improved if we started with a larger N and higher order correlation function.

Under equations (37.7) the dominant term in P_3 at small r is the last one,

$$P_3 \approx \tfrac{1}{6}I_3 = \tfrac{1}{6}n^3 \int dV_1 dV_2 dV_3 \zeta \sim QB^2 n^3 r^{9-2\gamma}, \qquad (39.7)$$

for a cell of radius r. The dominant term in P_2 is

$$P_2 \approx \tfrac{1}{2}I_2 = \tfrac{1}{2}n^2 \int dV_1 dV_2 \xi \sim Bn^2 r^{6-\gamma}, \qquad (39.8)$$

while the dominant term in P_1 in the limit of small r is nV.

An approximation scheme nearly equivalent to the above is based on the probability generating function. It is convenient here to change the variable from e^t to t in equation (38.1), which changes equation (38.7) to

$$\phi(t) = \Sigma P_N t^N = \langle \exp\,(t - 1) \int \rho\, dV \rangle. \qquad (39.9)$$

With

$$\rho = n + \delta\rho(\mathbf{r}), \qquad (39.10)$$

this is

$$\begin{aligned} \phi(t) &= e^{-nV} \langle e^{-\int \delta\rho dV} \exp t(nV + \int \delta\rho\, dV) \rangle \\ &= e^{-nV} \langle e^{-\int \delta\rho dV} [1 + t(nV + \int \delta\rho\, dV) + \cdots] \rangle. \end{aligned} \qquad (39.11)$$

By identifying powers of t in equations (39.9) and (39.11), one sees

$$\begin{aligned} P_0 &= e^{-nV} \langle e^{-\int \delta\rho dV} \rangle, \\ P_m &= \frac{1}{m!} e^{-nV} \langle (nV + \int \delta\rho\, dV)^m e^{-\int \delta\rho dV} \rangle. \end{aligned} \qquad (39.12)$$

By expanding the exponential as a series in $\delta\rho$, one can now write each P_m

as a sum over integrals of all the n-point correlation functions. For example,

$$P_0 = e^{-nV}(1 + \tfrac{1}{2}I_2 - \tfrac{1}{6}I_3 + \cdots), \tag{39.13}$$

and so on. This agrees with equation (39.6) through terms of third order in V. The same applies to the expressions for P_1, P_2 and P_3.

The distribution P_N in the count of neighbors of an object may be computed using the iteration scheme or from the probability generating function (eq. 38.14). For example, the probability that there is no neighbor within distance r is

$$P_0 = \left\langle \rho(0) \exp - \int_0^r \rho\, dV \right\rangle \Big/ n = ge^{-nV},$$

$$\begin{aligned} g &= n^{-1}\langle (n + \delta\rho(0))e^{-\int \delta\rho dV} \rangle \\ &= n^{-1}\left\langle (n + \delta\rho(0))\left[1 - \int \delta\rho\, dV + \tfrac{1}{2}\left(\int \delta\rho\, dV\right)^2 + \cdots\right]\right\rangle \\ &= 1 - n\int_0^r \xi(r_1)\, dV_1 + \tfrac{1}{2}n^2 \int_0^r dV_1 dV_2(\zeta(\mathbf{r}_1, \mathbf{r}_2) + \xi_{12}) + \cdots . \end{aligned} \tag{39.14}$$

This result was obtained by Fall, Geller, Jones, and White (1976).

Another interesting example is the distribution in distance to the nearest neighbor. The chosen object is at $\mathbf{r} = 0$. If the density function ρ is given, the probability that there is no object within distance $r = |\mathbf{r}|$ and there is an object in δV at $\mathbf{r}$ is

$$\delta P = \rho(\mathbf{r})\delta V \exp - \int_0^r \rho\, dV. \tag{39.15}$$

Therefore the probability that the nearest neighbor is at distance r in the range δr is

$$\delta P = 4\pi r^2 \delta r \left\langle \rho(0)\rho(\mathbf{r}) \exp - \int_0^r \rho\, dV \right\rangle \Big/ n, \tag{39.16}$$

where the factor $\rho(0)/n$ takes account of the fact that there is an object at $\mathbf{r} = 0$. The result of expanding this in $\delta\rho$ is

$$\delta P = 4\pi r^2 \delta r n f e^{-nV},$$

$$\begin{aligned} f &= \langle (n + \delta\rho(0))(n + \delta\rho(\mathbf{r})) \exp - \int \delta\rho\, dV \rangle / n^2. \\ &= 1 + \xi(r) - n\int d^3 r_1 [\zeta(\mathbf{r}_1, \mathbf{r}) + \xi(r_1) + \xi(|\mathbf{r} - \mathbf{r}_1|)] + \cdots , \end{aligned} \tag{39.17}$$

to order nV. When ξ and ζ are given by equations (37.7), the distribution of nearest neighbor distances at small r is

$$\delta P \propto r^{0.2}\delta r, \tag{39.18}$$

and this power law variation is cut off when $nV(1 + \xi(r))$ reaches unity.

The mean distance to the nearest neighbor is found by multiplying equation (39.16) by r and integrating over r. The expression is simplified by integrating by parts:

$$\langle r \rangle = \int dr \Big\langle \rho(0) \exp - \int_0^r \rho \, dV \Big\rangle \Big/ n, \tag{39.19}$$

which again can be expanded as a sum of integrals over the correlation functions.

These results of course apply equally well to the two-dimensional distribution in a catalog of galaxy angular positions. Here there is the advantage that the reduced correlation functions are small because many of the close groupings are accidental projection effects so expansions like equation (39.14) may converge rapidly. This was used by Fall, Geller, Jones, and White (1976) in a discussion of the nearest neighbor distribution in the Zwicky catalog. For further discussion see White (1979).

40. CLUSTER MODEL

In the clustering model approach pioneered by Neyman and Scott (1952) galaxies are placed in clumps—structures that might contain a single galaxy, or a cluster of galaxies, or a nested clustering hierarchy—and the clumps are distributed like a uniform random Poisson point process. This could not be a literally true representation of the physical process by which the galaxies were placed because each clump must be perturbed by the neighboring clumps. It is a convenient computational device, and it has the great advantage that one can visualize the nature of the clustering that is being modeled. The general relations between the clump structure and the n-point correlation functions are derived here. A somewhat different and still more general treatment is given by McClelland and Silk (1977). The results are applied to specific clump models in Sections 61 and 62.

The probability of finding an object at distance r to $r + \delta r$ from a randomly chosen object is (eq. 31.7)

$$\delta P = n[1 + \xi(r)]\delta V, \qquad \delta V = 4\pi r^2 \delta r. \tag{40.1}$$

The probability of finding a neighbor from another clump is

$$\delta P = n\delta V, \tag{40.2}$$

the same as for a randomly placed δV because the presence of the clump to which the chosen object belongs does not affect the mean abundance of other clumps. The probability of finding a neighbor from the same clump is thus

$$\delta P = n\xi(r)\delta V. \tag{40.3}$$

Suppose that in some large region V_u there are N_α clumps of type α, such a clump having n_α members and containing on the average $\delta N_P(\alpha)$ pairs of objects at separation r to $r + \delta r$. Then equation (40.3) can be written as

$$n\xi(r)\delta V = \sum_\alpha (N_\alpha n_\alpha)(2\delta N_P(\alpha)/n_\alpha)\Big/\sum_\beta N_\beta n_\beta. \tag{40.4}$$

The first factor on the right side gives the probability that the randomly chosen object is in a clump of type α. The second factor is the probability that it is at one end or the other of one of the pairs at separation r to $r + \delta r$. The correlation function is then

$$n\xi(r) = 2\langle\delta N_P(\alpha)\rangle/(\delta V\langle n_\alpha\rangle), \tag{40.5}$$

where the averages are weighted by the spatial abundances of the clump types. The right-hand side of this equation is determined by the clump structures and their relative abundances. When these are fixed, ξ varies inversely as n, the mean space density of objects.

Since the total number of pairs in a clump containing n_α objects is

$$\int dN_P(\alpha) = n_\alpha(n_\alpha - 1)/2, \tag{40.6}$$

equation (40.5) says the integral of the two-point correlation function is

$$n\int \xi\, dV = \langle n_\alpha(n_\alpha - 1)\rangle/\langle n_\alpha\rangle, \tag{40.7}$$

which agrees with equation (31.10).

The probability of finding neighboring objects in δV_2 and δV_3 is

$$\delta P = n^2\delta V_2\delta V_3(1 + \xi_{12} + \xi_{13} + \xi_{23} + \zeta). \tag{40.8}$$

For definiteness it will be supposed that r_{12} is the short side of the triangle, r_{13} the long side, the chosen object being at the vertex of these two sides. The probability of finding neighbors each of which belongs to a different clump is (eq. 40.2)

$$\delta P = n^2 \delta V_2 \delta V_3. \tag{40.9}$$

The probability of finding one neighbor from the clump to which the chosen object belongs and the other neighbor from another clump is (eq. 40.3)

$$\delta P = n^2 \delta V_2 \delta V_3 (\xi_{12} + \xi_{13}), \tag{40.10}$$

and the probability of finding neighbors from a common different clump is

$$\delta P = n^2 \delta V_2 \delta V_3 \xi_{23}. \tag{40.11}$$

The probability of finding all objects from a common clump is then

$$\delta P = n^2 \zeta dV_2 dV_3. \tag{40.12}$$

If $\delta N_t(\alpha)$ is the mean number of triplets in a clump of type α that define triangles with sizes and shapes r, u, v in the ranges δr, δu, δv, then this probability can be written as

$$n^2 \zeta \delta V_2 \delta V_3 = \Sigma (N_\alpha n_\alpha)(\delta N_t(\alpha)/n_\alpha)/\Sigma N_\beta n_\beta, \tag{40.13}$$

so the three-point function is

$$n^2 \zeta = \langle \delta N_t(\alpha) \rangle / [(\delta V)^2 \langle n_\alpha \rangle], \tag{40.14}$$

where $(\delta V)^2$ is given by equation (34.11). As in equation (40.5), the right hand side is determined by the structures and relative abundances of the clumps, and when these are fixed, ζ is fixed up to the factor n^2. Since the total number of triplets in each clump is known, equation (40.14) says

$$n^2 \int \zeta \, dV_2 dV_3 = \langle n_\alpha (n_\alpha - 1)(n_\alpha - 2) \rangle / \langle n_\alpha \rangle. \tag{40.15}$$

For the four-point function the probability of finding three neighbors is given by equation (35.1). As above one finds that all the terms added to the reduced function η in this expression describe all the different ways of making up a quadruplet of objects from two or more clumps, so the probability of finding a quadruplet all from the same clump is

$$\delta P = n^3 \eta \delta V_2 \delta V_3 \delta V_4 = \langle \delta N_q(\alpha) \rangle / \langle n_\alpha \rangle. \tag{40.16}$$

Thus $n^3\eta$ is determined by the mean number of quadruplets $\delta N_q(\alpha)$ per clump.

41. Power Spectrum

For a continuous function the Fourier transform of the autocorrelation function is the power spectrum. The analog for a random point process is obtained as follows. As in Section 27 it will be supposed that the universe is periodic in some large rectangular volume V_u. The Fourier transform of the distribution of objects is (eq. 27.8)

$$\delta_{\mathbf{k}} = \Sigma e^{i\mathbf{k}\cdot\mathbf{r}_j} / nV_u, \tag{41.1}$$

where the jth object is at r_j and exp $i\mathbf{k} \cdot \mathbf{r}$ is periodic in V_u. Following Section 36, one can divide V_u into infinitesimal cells with $n_1 = 0$ or 1 objects in the element δV_1, so equation (41.1) becomes

$$\delta_{\mathbf{k}} = \Sigma n_1 e^{i\mathbf{k}\cdot\mathbf{r}_1} / nV_u. \tag{41.2}$$

Since $\langle n_1 \rangle = n\delta V_1$ (eq. 36.1), the expectation value of this is

$$\begin{aligned} \langle \delta_{\mathbf{k}} \rangle &= \int dV e^{i\mathbf{k}\cdot\mathbf{r}} / V_u = 0, \qquad k \neq 0, \\ \langle \delta_0 \rangle &= 1. \end{aligned} \tag{41.3}$$

The mean of the product of two components is

$$\begin{aligned} (nV_u)^2 \langle \delta_{\mathbf{k}} \delta_{-\mathbf{k}'} \rangle &= \Sigma \langle n_1{}^2 \rangle e^{i(\mathbf{k}-\mathbf{k}')\cdot\mathbf{r}_1} + \Sigma \langle n_1 n_2 \rangle e^{i(\mathbf{k}\cdot\mathbf{r}_1 - \mathbf{k}'\cdot\mathbf{r}_2)} \\ &= n \int dV e^{i(\mathbf{k}-\mathbf{k}')\cdot\mathbf{r}} \\ &\quad + n^2 \int dV_1 dV_2 [1 + \xi_{12}] e^{i(\mathbf{k}\cdot\mathbf{r}_{12} + (\mathbf{k}-\mathbf{k}')\cdot\mathbf{r}_2)}. \end{aligned} \tag{41.4}$$

The last line uses equation (36.2). If $\mathbf{k} \neq \mathbf{k}'$ both integrals vanish: the Fourier components belonging to different $\mathbf{k}$ are statistically independent.

If $\mathbf{k} = \mathbf{k}'$, equation (41.4) reduces to the power spectrum,

$$\begin{aligned} \langle |\delta_{\mathbf{k}}|^2 \rangle &= \int \frac{d^3 r}{V_u} \xi(r) e^{i\mathbf{k}\cdot\mathbf{r}} + \frac{1}{nV_u}, \qquad \mathbf{k} \neq 0; \\ \langle |\delta_0|^2 \rangle &= \int \frac{d^3 r}{V_u} \xi(r) + 1 + \frac{1}{nV_u}. \end{aligned} \tag{41.5}$$

The last line agrees with equation (36.6).

By using

$$\sum_{\mathbf{k}} e^{-i\mathbf{k}\cdot\mathbf{r}} = V_u \delta(\mathbf{r}) \tag{41.6}$$

with equation (41.5), one finds the reciprocal relation,

$$\sum_{k\neq 0} (\langle |\delta_{\mathbf{k}}|^2 \rangle - 1/nV_u) e^{-i\mathbf{k}\cdot\mathbf{r}} = \xi(r) - V_u^{-1} \int d^3 r \xi. \tag{41.7}$$

In the limit of large V_u with the sum changed to an integral this becomes

$$\frac{V_u}{(2\pi)^3} \int d^3 k (\langle |\delta_{\mathbf{k}}|^2 \rangle - 1/nV_u) e^{-i\mathbf{k}\cdot\mathbf{r}} = \xi(r). \tag{41.8}$$

For another way to derive this equation see Peebles (1973b).

The Fourier transform of a continuous function $\rho(\mathbf{r})$, in the same normalization as equation (41.1), is

$$\delta_{\mathbf{k}} = \int \rho(\mathbf{r})\, d^3 r e^{i\mathbf{k}\cdot\mathbf{r}} / \langle \rho \rangle V_u. \tag{41.9}$$

The result of squaring this, averaging, and using equation (33.3) for the autocorrelation function of ρ is the usual expression

$$\begin{aligned} \langle |\delta_{\mathbf{k}}|^2 \rangle &= V_u^{-1} \int d^3 r \xi(r) e^{i\mathbf{k}\cdot\mathbf{r}}, \qquad \mathbf{k} \neq 0; \\ \xi(r) &= \frac{V_u}{(2\pi)^3} \int d^3 k \langle |\delta_{\mathbf{k}}|^2 \rangle e^{-i\mathbf{k}\cdot\mathbf{r}}. \end{aligned} \tag{41.10}$$

The particle relations in equation (41.5) and (41.8) are obtained by adding the Dirac delta function $n^{-1}\delta(\mathbf{r})$ to ξ in equation (41.10). This reflects the fact that there certainly is a mass concentration at the position of an object.

With

$$\int d\cos\theta d\phi e^{i\mathbf{k}\cdot\mathbf{r}} = 4\pi \sin(kr)/kr, \tag{41.11}$$

equation (41.5) can be written as

$$nV_u\langle|\delta_{\mathbf{k}}|^2\rangle - 1 = n\int d^3r\xi(r)\sin(kr)/kr. \tag{41.12}$$

The power spectrum thus measures the mean number of neighbors in excess of random within distance $\sim k^{-1}$ of a randomly chosen object (eq. 31.8). If k^{-1} is larger than the maximum correlation length, equation (41.12) becomes

$$nV_u\langle|\delta_{\mathbf{k}}|^2\rangle = n_c, \tag{41.13}$$

where n_c is the mean number of objects per cluster as defined in equation (31.9).

One can understand equations (41.12) and (41.13) as follows (Yu and Peebles 1973). If the objects were distributed like a uniform random Poisson process, the transform $\delta_{\mathbf{k}}$ (eq. 41.1) would be a random walk of nV_u steps of length $(nV_u)^{-1}$, so the power spectrum would be

$$\langle|\delta_{\mathbf{k}}|^2\rangle = (nV_u)^{-1}, \tag{41.14}$$

consistent with equation (41.13) with $n_c = 1$ object per cluster. Suppose now the objects are in clusters of $n_c > 1$ objects each. If the objects are placed more or less at random within the clusters and the wavelength $2\pi/k$ is much smaller than the cluster size, the phases $\mathbf{k}\cdot\mathbf{x}_j$ still are in effect random (mod. 2π), so equation (41.14) applies, consistent with equation (41.12) when k^{-1} is much less than the coherence length of $\xi(r)$. In the other limit, where the wavelength is much greater than the maximum clustering length, each cluster acts like a single object of mass n_c so $\delta_{\mathbf{k}}$ is a random walk of nV_u/n_c steps each of length n_c/nV_u, giving

$$\langle|\delta_{\mathbf{k}}|^2\rangle = n_c/(nV_u), \tag{41.15}$$

as in equation (41.13).

42. Power law model for the spectrum

The galaxy two-point correlation function is a close approximation to a power law,

$$\xi \propto r^{-\gamma}, \qquad \gamma \simeq 1.8, \tag{42.1}$$

over a substantial range in r. A fairly common speculation is that the power spectrum of density fluctuations at large redshift was a power law (§§ 72, 95).

$$\langle |\delta_{\mathbf{k}}|^2 \rangle \propto k^n. \tag{42.2}$$

In discussions of such models it is useful to have the relation between ξ and the power spectrum. The relation is derived here for equation (41.10) for a continuous function $\rho(\mathbf{r})$.

The power spectrum will be taken to be

$$\langle |\delta_k|^2 \rangle = Ak^n e^{-\lambda_0 k}, \qquad n > -3, \tag{42.3}$$

where A, n and λ_0 are constants, the last a short wavelength cutoff needed if $n \geq 0$. The lower bound on n assures convergence of the integral over the spectrum at long wavelength, $k \rightarrow 0$. The autocorrelation function is

$$\begin{aligned} \xi(r) &= \frac{V_u A}{(2\pi)^3} \int d^3k k^n e^{-\lambda_0 k - i\mathbf{k}\cdot\mathbf{r}} \\ &= \frac{V_u A}{2\pi^2 r} \int_0^\infty dk k^{1+n} \sin kr\, e^{-k\lambda_0}, \end{aligned} \tag{42.4}$$

where the second line is the result of integrating over angles (eq. 41.11). It is convenient to integrate once by parts:

$$\begin{aligned} \xi(r) &= \frac{V_u A}{2\pi^2 (2+n) r} \int_0^\infty dk k^{2+n} (\lambda_0 \sin kr - r \cos kr) e^{-k\lambda_0} \\ &= -\frac{V_u A}{2\pi^2 (2+n) r} (\lambda_0 \mathrm{Im} + r\mathrm{Re}) \int_0^\infty dk k^{2+n} e^{-k(\lambda_0 + ir)}. \end{aligned} \tag{42.5}$$

With the variable changed to $z = k(\lambda_0 + ir)$ and the contour shifted back to the real axis this becomes

$$\xi(r) = \frac{V_u A\Gamma(3+n)}{2\pi^2(2+n)r}(\lambda_0 \mathrm{Im} + r\mathrm{Re})\frac{1}{(\lambda_0 + ir)^{3+n}}$$

$$= \frac{V_u A\Gamma(3+n)}{2\pi^2(2+n)r}\frac{\sin(2+n)\phi}{(\lambda_0^2 + r^2)^{1+n/2}}, \tag{42.6}$$

$$\phi = \tan^{-1} r/\lambda_0, \qquad \Gamma(3+n) = \int_0^\infty dz z^{2+n} e^{-z}.$$

The limiting values are

$$r \ll \lambda_0, \qquad \xi = \frac{V_u A\Gamma(3+n)}{2\pi^2\lambda_0^{3+n}},$$

$$r \gg \lambda_0, \qquad \xi = \frac{V_u A\Gamma(3+n)}{2\pi^2}\frac{\sin(2+n)\pi/2}{(2+n)}\frac{1}{r^{3+n}}. \tag{42.7}$$

If $n = 0$, the second of equations (42.7) yields $\xi = 0$ at $r \gg \lambda_0$ as expected because the autocorrelation function vanishes for white noise. If $-3 < n < 0$, ξ is positive and varies as $r^{-(3+n)}$ longward of the cutoff λ_0. If $0 < n < 2$, ξ is negative at large r and approaches zero as $r^{-(3+n)}$, and it is positive and nearly constant at $r \ll \lambda_0$. One sees from equation (42.6) that the zero of ξ is at

$$r_0 = \lambda_0 \tan \pi/(2+n). \tag{42.8}$$

It is apparent that $\xi(r)$ must pass through zero because the Fourier transform of ξ is the power spectrum, and the limit $k \longrightarrow 0$ gives

$$\int_0^\infty d^3 r \xi(r) = 0, \qquad n > 0, \tag{42.9}$$

because the spectrum given by equation (42.3) vanishes at $k = 0$. Since $\xi(0)$ must be positive (because ξ is the autocorrelation function of a continuous function), $\xi(r)$ must pass through zero: the objects must be anticorrelated at some r. If $2 < n < 4$, the second of equations (42.7) says that at large r, $\xi(r) > 0$ and approaches zero as $r^{-(3+n)}$. Here one sees from equation (42.6) that as r increases from 0 to $r \gg \lambda_0$ and ϕ increases from 0 to $\pi/2$, $\xi(r)$ has two zeros both at r on the order of λ_0. If $4 < n < 6$, ξ has three zeros, and then approaches zero from negative values.

The details of these oscillations of $\xi(r)$ of course depend on the detailed shape of the adopted short wavelength cutoff of the spectrum. Another derivation of the way ξ approaches zero at large r is given by Peebles and Groth (1976). For another discussion see Bonometto and Lucchin (1978).

43. Bispectrum

Just as the Fourier transform of ξ is the power spectrum $|\delta_{\mathbf{k}}|^2$, the Fourier transform of the three-point function ζ is the product of three $\delta_{\mathbf{k}}$, the bispectrum.

The analysis is simplified by subtracting the mean from the distribution, so equation (41.2) for the Fourier transform is modified to

$$\delta_{\mathbf{k}} = \Sigma(n_1 - \langle n_1 \rangle)e^{i\mathbf{k}\cdot\mathbf{r}_1}/nV_u. \tag{43.1}$$

This differs from the original $\delta_{\mathbf{k}}$ only for δ_0, which now has zero mean. The mean of the product of three components is

$$\begin{aligned}
&(nV_u)^3\langle \delta_{\mathbf{k}_1}\delta_{\mathbf{k}_2}\delta_{\mathbf{k}_3}\rangle \\
&\quad = \Sigma\langle (n_i - \langle n_i\rangle)(n_j - \langle n_j\rangle)(n_k - \langle n_k\rangle)\rangle \\
&\qquad\qquad \cdot \exp i(\mathbf{k}_1\cdot\mathbf{r}_i + \mathbf{k}_2\cdot\mathbf{r}_j + \mathbf{k}_3\cdot\mathbf{r}_k) \\
&\quad = n^3 \int dV_1 dV_2 dV_3 \zeta(\mathbf{r}_1 - \mathbf{r}_3, \mathbf{r}_2 - \mathbf{r}_3) \\
&\qquad\qquad \cdot \exp i(\mathbf{k}_1\cdot\mathbf{r}_1 + \mathbf{k}_2\cdot\mathbf{r}_2 + \mathbf{k}_3\cdot\mathbf{r}_3) \qquad\qquad (43.2)\\
&\qquad + n^2 \int dV_1 dV_2 \xi(|\mathbf{r}_1 - \mathbf{r}_2|)\,[\exp i(\mathbf{k}_1\cdot\mathbf{r}_1 + (\mathbf{k}_2 + \mathbf{k}_3)\cdot\mathbf{r}_2) \\
&\qquad\qquad + \exp i((\mathbf{k}_1 + \mathbf{k}_2)\cdot\mathbf{r}_1 + \mathbf{k}_3\cdot\mathbf{r}_2) \\
&\qquad\qquad + \exp i((\mathbf{k}_1 + \mathbf{k}_3)\cdot\mathbf{r}_1 + \mathbf{k}_2\cdot\mathbf{r}_2)] \\
&\qquad + n \int dV \exp i(\mathbf{k}_1 + \mathbf{k}_2 + \mathbf{k}_3)\cdot\mathbf{r}.
\end{aligned}$$

As in equation (36.10), the first integral comes from those terms in the sum where all three cells are different, the second integral from those terms where two cells are the same, and the third integral from the terms where all three cells are the same. On changing the variables in the first integral to

$$\mathbf{r} = \mathbf{r}_1 - \mathbf{r}_3, \qquad \mathbf{s} = \mathbf{r}_2 - \mathbf{r}_3, \qquad \mathbf{r}_3 \tag{43.3}$$

and in the second to

$$\mathbf{r} = \mathbf{r}_1 - \mathbf{r}_2, \qquad \mathbf{r}_2, \tag{43.4}$$

one sees that all three integrals vanish unless

$$\mathbf{k}_1 + \mathbf{k}_2 + \mathbf{k}_3 = 0, \tag{43.5}$$

and that

$$\langle \delta_{\mathbf{k}_1}\delta_{\mathbf{k}_2}\delta_{-\mathbf{k}_1-\mathbf{k}_2}\rangle = V_u^{-2}\int d^3r d^3s \zeta(\mathbf{r},\mathbf{s})e^{i(\mathbf{k}_1\cdot\mathbf{r}+\mathbf{k}_2\cdot\mathbf{s})} + (nV_u)^{-1}[\langle|\delta_{\mathbf{k}_1}|^2\rangle + \langle|\delta_{\mathbf{k}_2}|^2\rangle + \langle|\delta_{\mathbf{k}_1+\mathbf{k}_2}|^2\rangle] - 2(nV_u)^{-2}, \tag{43.6}$$

where the integrals over ξ have been replaced using equation (41.5).

44. Cross correlation function

The cross correlation function for the two continuous functions $\rho_a(r)$ and $\rho_b(r)$ is

$$\xi_{ab}(r) = \langle(\rho_a(\mathbf{x}+\mathbf{r}) - \langle\rho_a\rangle)(\rho_b(\mathbf{x}) - \langle\rho_b\rangle)\rangle/\langle\rho_a\rangle\langle\rho_b\rangle,$$
$$\langle\rho_a(\mathbf{x}+\mathbf{r})\rho_b(\mathbf{x})\rangle = \langle\rho_a\rangle\langle\rho_b\rangle[1+\xi_{ab}(r)]. \tag{44.1}$$

The analog for the correlation among positions of point-like objects of types a and b is defined by the joint probability of finding an object of type a in the element δV_a and an object of type b in δV_b at distance r,

$$\delta P = n_a n_b[1+\xi_{ab}(r)]\delta V_a \delta V_b, \tag{44.2}$$

where n_a and n_b are the mean number densities of the two types of objects. The probability of finding an object b in δV at distance r from a randomly chosen object a is then

$$\delta P = n_b[1+\xi_{ab}(r)]\delta V_b, \tag{44.3}$$

so in effect the mean density of b-type objects at distance r from an a-type is

$$n(r) = n_b[1+\xi_{ab}(r)]. \tag{44.4}$$

The calculation in Section 41 leading to the relation between the autocorrelation function and the power spectrum can be repeated here. The Fourier transform of the distribution a is (eq. 41.2)

$$\delta_{\mathbf{k}}(a) = \frac{1}{n_a V_u}\Sigma n_i(a)e^{i\mathbf{k}\cdot\mathbf{r}_i}, \tag{44.5}$$

where $n_i(a) = 0, 1$ is the number of a-type objects found in the volume element δV_1 at r_1. The mean of the product of this with the transform of the b-distribution is the cross spectrum

$$\begin{aligned}\langle \delta_{\mathbf{k}}(a)\delta_{-\mathbf{k}}(b)\rangle &= \frac{1}{n_a n_b V_u^2}\Sigma\langle n_1(a)n_2(b)\rangle e^{i\mathbf{k}\cdot(\mathbf{r}_1-\mathbf{r}_2)} \\ &= V_u^{-2}\int dV_1 dV_2[1+\xi_{ab}(r_{12})]e^{i\mathbf{k}\cdot\mathbf{r}_{12}} \\ &= V_u^{-1}\int d^3r\xi_{ab}(r)e^{i\mathbf{k}\cdot\mathbf{r}}, \qquad \mathbf{k}\neq 0.\end{aligned} \tag{44.6}$$

This is the same as equation (41.5) except for the term $(nV_u)^{-1}$.

The cross correlation function has been used to study the mean distribution of galaxies around Abell cluster positions—as indicated in equation (44.4), the function $n_g(1+\xi_{cg}(r))$ in effect is the mean density in the halo of galaxies around a cluster—and the distribution of galaxies around radio sources and quasars (Peebles 1974e, Seldner and Peebles 1977a, 1978, 1979).

The cross correlation function can be generalized to higher moments. For example, one would write the joint probability of finding galaxies in δV_1 and δV_2 and a cluster center in δV_3 as

$$\begin{aligned}\delta P = n_g^2 n_c[1 &+ \xi_{cg}(r_{13}) + \xi_{cg}(r_{23}) + \xi_{gg}(r_{12}) \\ &+ \zeta_{cgg}(r_{12}, r_{23}, r_{31})]\delta V_1 \delta V_2 \delta V_3,\end{aligned} \tag{44.7}$$

where cg means the galaxy-cluster cross correlation function and gg the galaxy two-point correlation function. The role of these added terms is to make ζ approach zero when any one of the points is well away from the other two.

By following the methods of Section 36, one finds that the first two moments of the galaxy count in a region V around a cluster are (Fry and Peebles 1980a)

$$\langle N\rangle = nV + n\int d^3r\xi_{cg}(r), \tag{44.8}$$

$$\langle N^2\rangle = \langle N\rangle + \langle N\rangle^2 + n^2\int d^3r_1 d^3r_2[\xi_{gg}(r_{12}) + \zeta_{cgg} - \xi_{cg}(r_1)\xi_{cg}(r_2)].$$

The integrals are over V. The second expression can be compared to equation (36.6); we see that the combination $\zeta_{cgg} - \xi_{cg}\xi_{cg}$ fixes the extra variance when V is centered on a cluster rather than placed at a randomly chosen spot.

45. Angular two-point correlation function

Catalogs of angular positions of objects—galaxies, clusters of galaxies, radio sources, and so on—are the main sources of information on the nature of the large-scale clustering of matter. Since the measurement of distances of individual objects in such a survey usually is highly uncertain if not impractical, one can proceed in two steps: first obtain the correlation functions for the angular distribution, then analyze the relation between these statistics and the spatial correlation functions under the assumption that we are observing a spatially homogeneous and isotropic random process. In this second step it is convenient to imagine a statistical ensemble of catalogs obtained from random positions in space. The fair sample hypothesis is that available catalogs are deep enough that statistical estimates from the catalogs are good approximations to averages across this ensemble.

The probability of finding an object in the element of solid angle $\delta\Omega$ is (eq. 31.1)

$$\delta P = \mathcal{N}\delta\Omega, \tag{45.1}$$

where $\mathcal{N}$ is the mean density of objects in the sky. The mean number of objects in the finite cell Ω is

$$\langle N \rangle = \mathcal{N}\Omega. \tag{45.2}$$

The two-point correlation function is defined by the joint probability of finding objects in both of the elements of solid angle $\delta\Omega_1$ and $\delta\Omega_2$ placed at separation θ_{12} (eq. 31.4),

$$\delta P = \mathcal{N}^2\delta\Omega_1\delta\Omega_2[1 + w(\theta_{12})]. \tag{45.3}$$

All the discussion of the spatial function ξ in Sections 31 to 33 applies here also. The conditional probability of finding an object in $\delta\Omega$ at distance θ from a randomly chosen object in the ensemble is

$$\delta P = \mathcal{N}\delta\Omega[1 + w(\theta)], \tag{45.4}$$

so the expected number of neighbors within distance θ of an object is (eq. 36.23)

$$\langle N \rangle_p = \mathcal{N}\int_0^\theta d\Omega[1 + w(\theta)]. \tag{45.5}$$

Following equation (36.6), one sees that the variance of the count of objects in a randomly placed cell Ω is

$$\langle (N - \mathcal{N}\Omega)^2 \rangle = \mathcal{N}\Omega + \mathcal{N}^2 \int d\Omega_1 d\Omega_2 w(\theta_{12}). \tag{45.6}$$

The cell can be the whole sky: in this case equation (45.6) is the expected variance in the number of objects found in a whole sky catalog.

46. Angular power spectrum

Corresponding to the relation between $\xi(r)$ and the power spectrum (§ 41) is the relation between $w(\theta)$ and the square of the spherical harmonic transform for the angular distribution,

$$a_l^m = \sum_j Y_l^m(j), \tag{46.1}$$

where the sum is over the spherical harmonic Y_l^m evaluated at the angular positions of the objects in the sample.

The following properties of the Y_l^m will be used. With the polar angles θ, ϕ,

$$Y_l^m(\theta, \phi) = c_l^m P_l^{|m|}(\cos\theta) e^{im\phi}, \tag{46.2}$$

where $P_l^{|m|}(x)$ is the associated Legendre function of degree l and order m, $|m| \leq l$, and c_l^m normalizes the spherical harmonics to

$$\int d\Omega Y_l^m Y_{l'}^{-m'} = \delta_{ll'}\delta_{mm'}. \tag{46.3}$$

For $m \neq 0$, $P_l^{|m|}$ is close to zero near the polar caps,

$$\begin{aligned} P_l^{|m|} \approx 0, \qquad \theta \lesssim \theta_0 \quad \text{or} \quad \theta \gtrsim \pi - \theta_0, \\ \theta_0 = \sin^{-1} m/l, \end{aligned} \tag{46.4}$$

and at $\theta_0 < \theta < \pi - \theta_0$, P_l^m oscillates roughly like a sine wave with zeros spaced at $\Delta\theta = \pi/l$. The zeros of the real and imaginary parts of exp $im\,\phi$ are at angular separation

$$\Delta\chi = (\pi/m)\sin\theta, \tag{46.5}$$

at polar angle θ. At $\theta = \theta_0$ this separation amounts to

$$\Delta\chi(\theta_0) = \pi/l. \tag{46.6}$$

The zeros of the real and the imaginary parts of Y_l^m divide the sky into roughly rectangular cells. At low latitudes, $\theta_0 \lesssim \theta \lesssim \pi - \theta_0$, the minimum dimension of each cell is close to π/l. Near the polar caps the zeros of $\sin m\phi$ and $\cos m\phi$ crowd together, but here P_l^m is close to zero. Thus each spherical harmonic has a rather well-defined resolution π/l.

The addition theorem for spherical harmonics is (for example, Edmonds 1957)

$$\sum_m Y_l^m(j)Y_l^{-m}(k) = \frac{2l+1}{4\pi} P_l(\cos\theta_{jk}), \tag{46.7}$$

where the sum is $-l \le m \le l$ and θ_{jk} is the angular separation of the directions j and k. The Legendre polynomial P_l has the conventional normalization,

$$P_l(1) = 1, \qquad \int_{-1}^{1} P_l(\mu)P_{l'}(\mu)\, d\mu = 2\delta_{ll'}/(2l+1). \tag{46.8}$$

A function like $w(\theta)$ can be expanded in the P_l,

$$\begin{aligned} w(\theta) &= \Sigma\, c_l P_l(\cos\theta), \\ c_l &= (l + \tfrac{1}{2}) \int_{-1}^{+1} d\cos\theta\, w(\theta) P_l(\cos\theta), \end{aligned} \tag{46.9}$$

the second equation following from equation (46.8). This can be expressed as the completeness relation

$$\sum_l (l + \tfrac{1}{2}) P_l(\mu_1) P_l(\mu_2) = \delta(\mu_1 - \mu_2). \tag{46.10}$$

The completeness relation for the Y_l^m is (eq. 46.3)

$$\sum_{l,m} Y_l^m(1) Y_l^{-m}(2) = \delta(1, 2), \tag{46.11}$$

where here δ is the two-dimensional Dirac delta function in angular position.

The relation between the correlation function $w(\theta)$ and the power spectrum can be computed following the method used in Section 41. It will be supposed for the moment that the catalog covers the full sphere. With the sphere divided into small cells and $n_1 = 0$ or 1 objects in the element δV_1, the transform of the distribution (eq. 46.1) is

$$a_l^m = \Sigma\, n_1 Y_l^m(\Omega_1). \tag{46.12}$$

The ensemble average value of this is

$$\begin{aligned} \langle a_l^m \rangle &= \mathcal{N} \int d\Omega Y_l^m = 0, \qquad l \neq 0 \\ \langle a_0^0 \rangle &= (4\pi)^{1/2} \mathcal{N}. \end{aligned} \tag{46.13}$$

The mean square value for $l \neq 0$ is

$$\begin{aligned} \langle | a_l^m |^2 \rangle &= \Sigma \langle n_1^2 \rangle | Y_l^m |^2 + \Sigma \langle n_1 n_2 \rangle Y_l^m(1) Y_l^{-m}(2) \\ &= \mathcal{N} \int d\Omega | Y_l^m |^2 + \mathcal{N}^2 \int d\Omega_1 d\Omega_2 (1 + w_{12}) Y_l^m(1) Y_l^{-m}(2) \\ &= \bar{\mathcal{N}} + \mathcal{N}^2 \int d\Omega_1 d\Omega_2 w(\theta_{12}) Y_l^m(1) Y_l^{-m}(2). \end{aligned} \tag{46.14}$$

The last line can be simplified by writing w as a sum over the P_l. The addition theorem (46.7) gives

$$\begin{aligned} &\int d\Omega_1 d\Omega_2 P_{l'}(\cos\theta_{12}) Y_l^m(1) Y_l^{-m}(2) \\ &= 4\pi/(2l' + 1) \sum_{m'} \int d\Omega_1 d\Omega_2 Y_{l'}^{-m'}(1) Y_{l'}^{m'}(2) Y_l^m(1) Y_l^{-m}(2) \\ &= 4\pi \delta_{ll'}/(2l + 1), \end{aligned} \tag{46.15}$$

so with equation (46.9) equation (46.14) becomes

$$\begin{aligned} \langle | a_l^m |^2 \rangle &= \mathcal{N} + 4\pi \mathcal{N}^2 c_l/(2l + 1) \\ &= \mathcal{N} + 2\pi \mathcal{N}^2 \int_{-1}^{+1} d\cos\theta P_l(\cos\theta) w(\theta), \end{aligned} \tag{46.16}$$

for $l \neq 0$, independent of m. This relation between the angular correlation function and power spectrum can be compared to equation (41.12).

By the same calculation, one finds

$$\langle a_l^m a_{l'}^{-m'} \rangle = 0, \qquad l \neq l' \quad \text{or} \quad m \neq m'. \tag{46.17}$$

The a_l^m thus are statistically independent.

The component a_0^0 was written down in equation (45.6); because $Y_0^0 = (4\pi)^{-1/2}$,

$$\langle |a_0^0|^2 \rangle = \mathcal{N} + 4\pi\mathcal{N}^2 + 2\pi\mathcal{N}^2 \int w d\cos\theta. \qquad (46.18)$$

This agrees with equations (46.16) and (46.13).

The result of multiplying equations (46.16) and (46.18) by $P_l(\cos\theta)$, summing over all l, m, and using equation (46.10) is the reciprocal equation,

$$1 + w(\theta) = \sum_{l,m} (\langle |a_l^m|^2 \rangle - \mathcal{N}) P_l(\cos\theta)/(4\pi\mathcal{N}^2). \qquad (46.19)$$

Another way to arrive at this relation is based on the expression (Peebles 1973b)

$$\delta n_p = \pi \sum_{l>0} \sum_{jk} Y_l^m(j) Y_l^{-m}(k) \int_{\theta_a}^{\theta_b} d\cos\theta\, P_l(\cos\theta). \qquad (46.20)$$

The sums are over all l, except $l = 0$, all m from $-l$ to l, and all N objects j, k in the whole sky in some realization. The integral is over a thin ring, θ_b slightly larger than θ_a. By using the addition theorem (46.7), we can write equation (46.20) as

$$\delta n_p = \frac{1}{2} \sum_{l>0} \sum_{jk} (l + 1/2) \int_{\theta_a}^{\theta_b} P_l(\cos\theta_{jk})\, P_l(\cos\theta)\, d\cos\theta, \qquad (46.21)$$

and the completeness relation in equation (46.10) for the P_l gives

$$\begin{aligned} \delta n_p &= \frac{1}{2} \sum_{jk} \int_{\theta_a}^{\theta_b} [\delta(\cos\theta_{jk} - \cos\theta) - 1/2]\, d\cos\theta \\ &= n_p - 1/4 N^2(\cos\theta_b - \cos\theta_a). \end{aligned} \qquad (46.22)$$

The first term n_p is the number of distinct pairs at separation θ_a to θ_b in the realization and the second term is the expected number for a random distribution at density $N/4\pi$, so δn_p is the number of pairs in excess of random. The estimate of w at $\theta \sim \theta_a$ is then

$$\begin{aligned} w_e &= \frac{4\delta n_p}{N^2(\cos\theta_b - \cos\theta_a)} \\ &= \frac{4\pi}{N^2} \sum_{l>0} |a_l^m|^2 \int_{\theta_a}^{\theta_b} d\cos\theta\, P_l(\cos\theta)/(\cos\theta_b - \cos\theta_a). \end{aligned} \qquad (46.23)$$

With $a_0^0 = N/(4\pi)^{1/2}$ and (eq. 46.10)

$$\Sigma\,(2l + 1)P_l(\cos\theta) = 0, \qquad \theta \neq 0, \tag{46.24}$$

this can be rewritten as

$$1 + w_e(\theta) = \frac{4\pi}{N^2}\sum_{l,m} P_l\,(\cos\theta)(|\,a_l^m\,|^2 - N/4\pi), \tag{46.25}$$

which can be compared to equation (46.19).

As for equation (41.12), equation (46.16) says that $|\,a_l^m\,|^2$ is a measure of the mean number of neighbors in excess of random. The Legendre polynomial starts at $P_l(1) = 1$, reaches its first zero at (eq. 46.39 below)

$$\theta_1 \simeq 2.4/(l + \tfrac{1}{2})\text{ radians}, \tag{46.26}$$

and then oscillates with amplitude $\sim l^{-1/2}$ (eq. 46.8) and distance between zeros $\Delta\theta \approx \pi/l$. Thus at large l, equation (46.16) may be approximated as

$$u_l = \langle|\,a_l^m\,|^2\rangle/\mathcal{N} - 1 \simeq \mathcal{N}\int_0^{2.4l^{-1}} 2\pi\theta d\theta w(\theta). \tag{46.27}$$

The right-hand side is the mean number of neighbors in excess of random around a randomly chosen object. One can understand this by the same argument that was used to arrive at equation (41.15).

In a fair sample a_l^m would be a sum over the positions of many independent clumps of objects, so for $m \neq 0$ the probability distributions of the real and imaginary parts would be close to Gaussian and the distribution of $|\,a_l^m\,|^2$ close to exponential with mean given by the integral over $w(\theta)$.

In the case of practical interest where the catalog covers only part of the sky, a convenient way to proceed is to introduce the window function $W(\Omega)$, with $W = 1$ in the region covered by the catalog, $W = 0$ elsewhere. The function $W(\Omega)\,Y_l^m(\Omega)$ can be expanded as a sum over the spherical harmonics,

$$W(\Omega)\,Y_l^m(\Omega) = \sum_{l',m'} W_{ll'}^{mm'}\,Y_{l'}^{m'}(\Omega), \tag{46.28}$$

where

$$W_{ll'}^{mm'} = \int d\Omega W(\Omega)\,Y_l^m(\Omega)\,Y_{l'}^{-m'}(\Omega), \tag{46.29}$$

an integral over the survey region. Thus if equation (46.28) is summed over the angular positions of all objects in the whole sky, the sum on the right hand side is the transform a_l^m for the whole sky, the left hand side is the sum over the objects in the survey region Ω,

$$b_l^m = \sum_{j \in \Omega} Y_l^m(j), \tag{46.30}$$

and the result is the convolution,

$$b_l^m = \sum_{l'm'} W_{ll'}^{mm'} a_{l'}^{m'}. \tag{46.31}$$

This means the observed transform b_l^m is a running average of the whole sky transform a_l^m.

The term $l' = 0$ in equation (46.31) should be treated separately because a_0^0 typically is much larger than any other component. Thus the ensemble average of equation (46.31) is, by equation (46.13),

$$\begin{aligned} \langle b_l^m \rangle &= \langle a_0^0 \rangle W_{l0}^{m0} = \mathcal{N}\Omega I_l^m, \\ I_l^m &= \int d\Omega\, W(\Omega) Y_l^m(\Omega)/\Omega. \end{aligned} \tag{46.32}$$

The mean is eliminated by writing the transform as

$$\begin{aligned} c_l^m &= \sum_{\Omega} (Y_l^m(j) - I_l^m) = b_l^m - N I_l^m \\ &= \sum_{l' \neq 0} W_{ll'}^{mm'} a_{l'}^{m'} - I_l^m (N - a_0^0 \Omega/(4\pi)^{1/2}), \end{aligned} \tag{46.33}$$

where N is the number of objects found in Ω. The second term generally is negligible. Dropping it and using equation (46.17), one finds

$$\langle |c_l^m|^2 \rangle = \sum_{l'm'} |W_{ll'}^{mm'}|^2 \langle |a_{l'}^{m'}|^2 \rangle. \tag{46.34}$$

The $W_{ll'}^{mm'}$ peak up at $l = l'$, $m = m'$, and, according to equation (46.11), they satisfy

$$\sum_{l'm'} |W_{ll'}^{mm'}|^2 = \int d\Omega\, W(\Omega) |Y_l^m(\Omega)|^2 \equiv J_l^m. \tag{46.35}$$

If the true power spectrum does not vary too rapidly with l', it can be taken out of the sum in equation (46.34) to get

$$\langle |c_l^m|^2\rangle = J_l^m \langle |a_l^m|^2\rangle. \tag{46.36}$$

That is,

$$|\sum_{\Omega} (Y_l^m(i) - I_l^m)|^2/J_l^m = (u_l(e) + 1)N/\Omega \tag{46.37}$$

is an estimate of the true spectrum.

Finally, of particular interest is the small-angle limit $\theta \ll 1$ radian, where the clustering length covers a small part of the sky. An integral expression for the Legendre polynomial is (Jahnke and Emde 1945)

$$P_l(\cos\theta) = \frac{2}{\pi}\int_0^\theta \frac{dx \cos(l + \tfrac{1}{2})x}{(2(\cos x - \cos\theta))^{1/2}}. \tag{46.38}$$

When $\theta \ll 1$ radian, the denominator is $(\theta^2 - x^2)^{1/2}$, so the integral can be rewritten as

$$\begin{aligned} P_l &\simeq \frac{2}{\pi}\int_0^{\pi/2} d\psi \cos[(l + \tfrac{1}{2})\,\theta\cos\psi] \\ &= J_0((l + \tfrac{1}{2})\theta), \end{aligned} \tag{46.39}$$

where J_0 is the Bessel function of order 0. Thus if most of the contribution to the integral over $w(\theta)$ in equation (46.16) is at $\theta \ll 1$, the relation to the power spectrum becomes

$$\begin{aligned} u_l &= \langle |a_l^m|^2\rangle/\mathcal{N} - 1 \\ &= \mathcal{N}\int 2\pi\theta\, d\theta\, w(\theta) J_0((l + \tfrac{1}{2})\theta). \end{aligned} \tag{46.40}$$

This same result is obtained by approximating a small section of the sky as a flat two-dimensional square area Ω and representing the distribution of objects in Ω by the two-dimensional Fourier transform,

$$\delta_{\mathbf{k}} = \sum_j e^{i\mathbf{k}\cdot\boldsymbol{\theta}_j}/\Omega^{1/2}, \tag{46.41}$$

where $\boldsymbol{\theta}_j$ is the position of the jth object. Following the standard calculation, one sees that if the clustering length is much less than the size of Ω the power spectrum is

$$\langle|\delta_{\mathbf{k}}|^2\rangle/\mathcal{N} - 1 = \mathcal{N}\int d^2\theta w(\theta)e^{i\mathbf{k}\cdot\boldsymbol{\theta}}, \tag{46.42}$$

for $\mathbf{k} \neq 0$. The angle integral brings in the Bessel function (eq. 46.39) leaving

$$\langle|\delta_{\mathbf{k}}|^2\rangle/\mathcal{N} - 1 = \mathcal{N}\int 2\pi\theta\, d\theta w(\theta)J_0(k\theta), \tag{46.43}$$

which can be compared to equation (46.40).

The major virtue of the power spectrum over its transform $w(\theta)$ is that the spectrum separates density fluctuations on different angular scales. This is illustrated in Figure 46.1, which shows estimates of

$$u_l = \sum_{m=-l}^{+l} |a_l^m|^2/(2l + 1) \tag{46.44}$$

for three latitude zones in the Lick catalog,

$$\begin{aligned} &\text{A: } b > 55°, \\ &\text{B: } 40 < b < 55°, \\ &\text{C: } b < -40°, \qquad \delta > -23°. \end{aligned} \tag{46.45}$$

Because of the window function effect, the statistical fluctuations of the u_l at neighboring l are correlated. Taking this into account, one would conclude that there is good reproducibility of the clustering seen in different parts of the sky at $l \gtrsim 40$ ($\theta \lesssim 4°$), moderate scatter at $10 \lesssim l \lesssim 40$ ($15° \gtrsim \theta \gtrsim 4°$), and considerable difference among zones at $l \lesssim 10$. The standard presumption is that this scatter at large angular scale is due to variable absorption in the galaxy, though an unknown part could be true large-scale clustering. The scatter in $w(\theta)$ at small θ is appreciably more than the scatter in u_l at the same angular resolution because w at small θ is affected by density fluctuations on small and large scales.

For further discussions of the power spectrum see Yu and Peebles (1969), Peebles (1973b), and Webster (1976a). Examples of the window function $W_{ll'}^{mm'}$ are given by Hauser and Peebles (1973). Examples of the exponential distribution of the u_l are given there and in Peebles and Hauser (1974). This latter paper also gives further details on the small l part of u_l for the Lick sample.

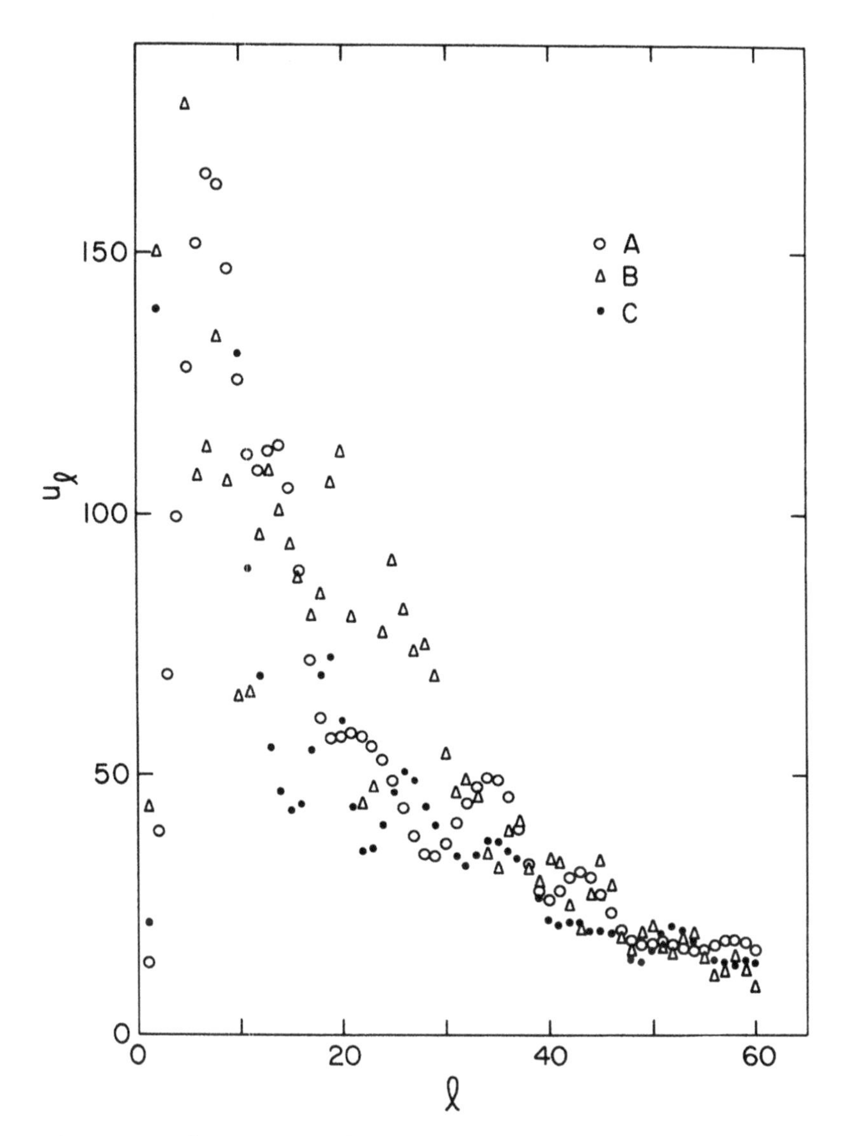

FIG. 46.1. Angular power spectra for three zones in the Lick catalog (from Peebles and Hauser 1974; published by The University of Chicago Press; copyright 1974 by the American Astronomical Society).

47. ESTIMATING $w(\theta)$

The angular correlation function $w(\theta)$ is estimated from the list of angular positions of the objects found in some section of the sky. If N objects are found in a survey region that subtends solid angle Ω, an estimate of the density is

$$\mathcal{N} = N/\Omega. \tag{47.1}$$

One sees from equation (45.4) that the expected count of pairs at separation θ to $\theta + \delta\theta$ is

$$n_p = \tfrac{1}{2}\mathcal{N}^2\Omega\langle\delta\Omega\rangle[1 + w(\theta_1)], \tag{47.2}$$

where θ_1 is between θ and $\theta + \delta\theta$, depending on the shape of $w(\theta)$, and $\langle\delta\Omega\rangle$ is the mean value of the solid angle subtended by the ring θ to $\theta + \delta\theta$ and within Ω for ring centers randomly placed in Ω. The factor of two takes account of the fact that each pair represents two neighbors. An estimate of the correlation function is then

$$1 + w(\theta_1) = 2n_p\Omega/(N^2\langle\delta\Omega\rangle), \tag{47.3}$$

where N and n_p are the counts of objects and pairs at separation θ in the sample. (One might prefer $N(N - 1)$ to N^2 here and in the following equations; however, N usually is large so the difference is negligible and can be ignored.)

Another way to arrive at equation (47.3) is to start from the number n of neighbors at distance θ to $\theta + \delta\theta$ from a particular one of the objects in the sample. Since the expected value is

$$\langle n\rangle = \mathcal{N}[1 + w(\theta_1)]\delta\Omega, \tag{47.4}$$

where $\delta\Omega$ is the solid angle of the ring within Ω, an estimate of the correlation function is

$$1 + w(\theta_1) = n\Omega/(N\delta\Omega). \tag{47.5}$$

If $\delta\Omega$ is small, so typically $n = 0$ and rarely is n greater than one, the variance of n is proportional to $\delta\Omega$ and the variance of $n/\delta\Omega$ is proportional to $(\delta\Omega)^{-1}$. The average of equation (47.5), weighted by $\delta\Omega$, for all the objects in the sample is

$$1 + w(\theta_1) = 2n_p\Omega/(N^2\langle\delta\Omega\rangle), \tag{47.6}$$

where now $\langle\delta\Omega\rangle$ is the solid angle of the ring within Ω, averaged over the positions of all the objects in the sample.

If the objects are assigned labels such as morphological types (Davis and

Geller 1976), or apparent magnitudes (Soneira 1978b), or if the objects come from different catalogs (Seldner and Peebles 1977a), then an estimate of the two-point cross correlation function for objects with different labels (§ 44) is

$$1 + w_{ab}(\theta_1) = n_p\Omega/(N_a N_b \langle\delta\Omega\rangle). \tag{47.7}$$

Here n_p is the count of pairs ab, and N_a, N_b are the counts of objects. There is no factor of 2 here because an a in $\delta\Omega_1$ and a b in $\delta\Omega_2$ is different from a b in $\delta\Omega_1$ and an a in $\delta\Omega_2$.

For like objects the total number of pairs in a sample of N is $N(N-1)/2$, so one sees from equation (47.6) that

$$\begin{aligned}\int_\Omega d\Omega_1 d\Omega_2[1 + w(\theta_{12})] &\simeq \Omega \sum (1 + w)\langle\delta\Omega\rangle \\ &= \Omega^2(1 - 1/N).\end{aligned} \tag{47.8}$$

The situation is like that discussed in Section 32: since the estimate of $\mathcal{N}$ is based on the number of objects in the sample, there is an integral constraint on the estimate of w. Of course, the constraint does not apply to the ensemble w discussed in Section 45. If the latter is positive at small θ and very close to zero at large θ, equation (47.6) may yield accurate estimates of w at small θ, but then the estimates at large θ will be biased low. This only reflects the point that one cannot hope to estimate true w at θ from a survey area with size comparable to θ.

Several methods of computing $\langle\delta\Omega\rangle$ have been used. Hauser and Peebles (1973) selected the N_1 interior objects at distances $>\theta$ from the boundary and for each of them counted the number of neighbors from the full sample at distance θ to $\theta + \delta\theta$. The count n_p' is twice the number of distinct pairs among interior objects plus the number of pairs one of which is an interior object, the other near the boundary. Equation (47.6) is modified to

$$1 + w(\theta_1) = \frac{n_p'\Omega}{NN_1\delta\Omega}, \qquad \delta\Omega = 2\pi\delta(\cos\theta). \tag{47.9}$$

This simplifies the computation, but it does eliminate some information. A second method is to compute directly the solid angle subtended by $\theta, \theta + \delta\theta$, around each object and within the survey region. If $\theta \ll 1$, $\delta\theta \ll \theta$, and the boundary close to straight on the scale of θ, the solid angle subtended by an object at distance ψ from the boundary is (Seldner 1977)

$$\delta\Omega = 2(\pi - \cos^{-1}\psi/\theta)\theta\delta\theta, \qquad \psi < \theta. \tag{47.10}$$

In the approximation of this equation, the expected fraction of objects at distances ψ to $\psi + d\psi$ from the boundary is

$$df = \Theta d\psi/\Omega, \tag{47.11}$$

where Θ is the length of the boundary, and so the mean of equation (47.10) is

$$\langle \delta\Omega \rangle \simeq [1 - \Theta\theta/(\pi\Omega)] 2\pi\theta\delta\theta. \tag{47.12}$$

If the boundaries of the catalog are two parallel small circles of fixed declination δ_1 and δ_2, as in the 4C catalog, then (Seldner 1977)

$$\begin{gathered} \Theta = 2\pi(\cos\delta_1 + \cos\delta_2), \\ \langle \delta\Omega \rangle = (1 - C\theta) 2\pi\theta\delta\theta, \qquad C = \frac{1}{\pi}\cot\left(\frac{\delta_1 - \delta_2}{2}\right). \end{gathered} \tag{47.13}$$

Finally, it is easy to derive $\langle \delta\Omega \rangle$ by Monte Carlo integration: place N_t points at random in the survey area; let $n_p(t)$ be the number of pairs among these trial points at separation θ to $\theta + \delta\theta$, and let n_p be the corresponding number of pairs in the real catalog of N objects. Since $w = 0$ for the trial points, the estimate of w for the data is (eq. 47.3)

$$1 + w(\theta_1) = \frac{n_p}{n_p(t)} \frac{N_t^2}{N^2}. \tag{47.14}$$

This was used in the analysis of the Zwicky catalog (Peebles and Hauser 1974).

In some cases such as the Lick sample, one has counts of objects for an array of cells rather than individual object positions. The variance of the cell counts is an integral over w (eq. 45.6); this was used by Rubin (1954). If the variance is known for counts in cells of various sizes, one can deconvolve the integral to get $w(\theta)$. Another approach is to use the correlation of counts in disjoint cells. If all cells have the same size Ω_c, then

$$\begin{aligned} \langle n \rangle &= \mathcal{N}\Omega_c, \\ \langle (n_1 - \langle n \rangle)(n_2 - \langle n \rangle) \rangle &= \langle n \rangle^2 \int \frac{d\Omega_1 d\Omega_2}{\Omega_c^2} w(\theta_{12}) \\ &= \langle n \rangle^2 w(\theta'_{12}). \end{aligned} \tag{47.15}$$

The mean count per cell is $\langle n \rangle$. In the second equation one wants the average over all pairs of cells with chosen and fixed relative position; n_1 and n_2 are the counts in the two (disjoint) cells, and the integral is over $d\Omega_1$ in cell 1, $d\Omega_2$ in cell 2. As indicated the integral can be written as $w(\theta'_{12})$, where the argument is very nearly equal to the distance between cell centers. Equation (47.15) was first applied by Limber (1954) and Neyman, Scott, and Shane (1956) to the analysis of the Lick sample. The effects of counting errors in the statistics $\langle n_1 \rangle$ and $\langle n_1 n_2 \rangle$ and the correction for variable depth of the telescope field are discussed by Groth and Peebles (1977).

48. Statistical uncertainty in the estimate of $w(\theta)$

The statistical uncertainty in the estimate of w generally proves to be unimportant compared to the systematic errors; the main problem in galaxy catalogs is that the effective depth of the survey may vary across the sky because of variable obscuration in the galaxy. Such systematic errors might be negligible in a few cases, as the 4C radio source catalog. If, as in the 4C catalog, the distribution of objects also is close to random, the standard deviation of w may be computed as follows.

The estimate of w from the count n_p of pairs at separation θ to $\theta + \delta\theta$ satisfies the equation (eq. 47.3)

$$(\mathcal{N}^2 \Omega \delta\Omega) w = 2n_p - N^2 \delta\Omega / \Omega. \tag{48.1}$$

There are N objects in the survey region of solid angle Ω. The solid angle of the ring θ to $\theta + \delta\theta$ is $\delta\Omega$, and the boundary correction will be ignored. The dominant statistical uncertainty in w arises from the difference of the two large and nearly equal terms on the right hand side of this equation. The density $\mathcal{N}$ on the left side can be replaced with the ensemble average value.

As in Section 36 a convenient way to proceed is to divide Ω into infinitesimal cells with $n_1 = 0$ or 1 objects in the element $\delta\Omega_1$, so

$$N = \Sigma\, n_i, \qquad 2n_p = \Sigma\, n_i n_j \Theta(\theta_{ij}), \tag{48.2}$$

where the window function is

$$\Theta(\theta') = 1, \qquad \theta < \theta' < \theta + \delta\theta, \tag{48.3}$$

$\Theta = 0$ otherwise. Equation (48.1) is then

$$(\mathcal{N}^2 \Omega \delta\Omega) w = \Sigma\, n_i n_j [\Theta(\theta_{ij}) - \delta\Omega/\Omega]. \tag{48.4}$$

Since the correlations among object positions are assumed to be negligible, we have

$$\langle n_1 \rangle = \mathcal{N}\delta\Omega_1, \qquad \langle n_1 n_2 \rangle = \mathcal{N}^2 \delta\Omega_1 \delta\Omega_2, \tag{48.5}$$

and so on. Thus the expectation value of equation (48.4) is

$$\begin{aligned}(\mathcal{N}^2\Omega\delta\Omega)\langle w \rangle &= \Sigma \langle n_1 n_2 \rangle(\Theta_{12} - \delta\Omega/\Omega) - \Sigma \langle n_1^2 \rangle \delta\Omega/\Omega \\ &= \mathcal{N}^2 \int d\Omega_1 d\Omega_2 (\Theta_{12} - \delta\Omega/\Omega) - \mathcal{N} \int d\Omega\delta\Omega/\Omega \\ &= -\mathcal{N}\delta\Omega.\end{aligned} \tag{48.6}$$

In the first equation the terms $i \neq j$ and $i = j$ are written separately: the window function vanishes at $i = j$ to exclude pairs at zero separation. The sum $i \neq j$ vanishes leaving the bias in w that agrees with equations (47.8) and (32.7). The variance of w is given by the mean of the square of equation (48.4),

$$(\mathcal{N}^2\Omega\delta\Omega)^2 \langle w^2 \rangle = \Sigma \langle n_i n_j n_k n_l \rangle (\Theta_{ij} - \delta\Omega/\Omega)(\Theta_{kl} - \delta\Omega/\Omega). \tag{48.7}$$

As in equation (48.6) one must treat separately all the various cases where two or more of the indices are equal. Most such sums vanish. For example, the terms where three indices are different with $i = j$ give

$$-(\langle\delta\Omega\rangle/\Omega)\mathcal{N}^3 \int d\Omega_1 d\Omega_2 d\Omega_3 (\Theta_{23} - \langle\delta\Omega\rangle/\Omega) = 0. \tag{48.8}$$

The largest contribution is from the terms with $i = k, j = l$ and $i = l, j = k$:

$$2\Sigma \langle n_1^2 n_2^2 \rangle (\Theta_{12} - \delta\Omega/\Omega)^2 \simeq 2\mathcal{N}^2 \int d\Omega_1 d\Omega_2 \Theta_{12}^2 = 2\mathcal{N}^2\Omega\delta\Omega, \tag{48.9}$$

since $\delta\Omega/\Omega \ll 1$. Thus equation (48.7) becomes

$$\langle w^2 \rangle = 2/(\mathcal{N}^2\Omega\delta\Omega). \tag{48.10}$$

Since $\langle w \rangle^2$ (eq. 48.6) is smaller than this by the factor $\delta\Omega/\Omega$, it can be ignored, so equation (48.10) is the expected mean square uncertainty in w. Since the mean number of pairs is (eq. 48.1)

$$n_p = \mathcal{N}^2\Omega\delta\Omega/2, \tag{48.11}$$

the standard deviation of the w estimate is just (Peebles 1973b)

$$\delta w = n_p^{-1/2}, \tag{48.12}$$

where n_p is the count of (distinct) pairs used in the estimate of w.

This result assumes correlations among the objects can be ignored. One sees from equation (48.7) that in general δw depends on the four-point correlation function. Some estimates and models for this are available, but the effect on δw will not be written down because it seems doubtful it would be of any use. Of greater interest is the problem of testing for the possible reality of clustering on small angular scales when the density varies on large angular scales due to variable obscuration in the galaxy. Here one would have more confidence in estimates of the expected variance of the power spectrum because the spectrum decouples density variations on large and small angular scales.

49. Relation between angular and spatial two-point correlation functions

A general model for the relation between the angular and spatial correlation functions describing the galaxy distribution is set up as follows. The probability that a galaxy with absolute magnitude M in the range δM is found in the randomly placed volume element δV is (compare eq. 31.1)

$$\delta P = \Phi(M)\delta M\delta V. \tag{49.1}$$

This defines the luminosity function Φ. The probability that a galaxy with magnitude M_1 in the range δM_1 is found in the element δV_1 and a second galaxy with magnitude M_2 in the range δM_2 is found in δV_2 at distance r_{12} from the first is (eq. 31.4)

$$\delta P = [\Phi(M_1)\Phi(M_2) + \Gamma(M_1, M_2, r_{12})]\delta V_1\delta V_2\delta M_1\delta M_2. \tag{49.2}$$

If the galaxies were uncorrelated in position and magnitude, Γ would vanish. The functions ϕ and Γ integrated over magnitude are the number density and two-point spatial correlation function defined in equations (31.1) and (31.4),

$$n = \int \Phi\, dM, \qquad n^2\xi(r) = \int dM_1 dM_2 \Gamma(M_1, M_2, r). \tag{49.3}$$

The functions corresponding to Φ and Γ for correlations among angular position and apparent magnitude are found by integrating equations (49.1) and (49.2) along lines of sight. It will be assumed here that the effects of redshift, space curvature, and absorption may be neglected (the former two are discussed in § 56), so the apparent magnitude of a galaxy at distance r (Mpc) with absolute magnitude M is

$$m = M + 5 \log r + 25. \tag{49.4}$$

Then the probability that a galaxy appears in the sky in the element of solid angle $\delta\Omega$ at apparent magnitude m to $m + \delta m$ and at distance r to $r + \delta r$ from the observer is

$$\delta P = r^2 \delta r \delta\Omega \delta m \, \Phi(m - 5 \log r - 25). \tag{49.5}$$

The result of integrating this over distance is the probability of finding a galaxy of apparent magnitude m in $\delta\Omega$,

$$\delta P = \delta\Omega \delta m \int_0^\infty r^2 dr \, \Phi(m - 5 \log r - 25). \tag{49.6}$$

With the change of variables

$$r = s 10^{0.2m}, \tag{49.7}$$

this becomes

$$\begin{gathered} \delta P = (d\mathcal{N}/dm) \delta m \delta\Omega, \\ d\mathcal{N}/dm = 10^{0.6m} \int_0^\infty s^2 \, ds \Phi(-5 \log s - 25), \end{gathered} \tag{49.8}$$

which is the usual number-magnitude relation.

The probability of finding a galaxy with apparent magnitude m_1 to $m_1 + \delta m_1$ in $\delta\Omega_1$ and a second galaxy with apparent magnitude m_2 to $m_2 + \delta m_2$ in $\delta\Omega_2$ at distance θ_{12} from the first is found by integrating the probability in equation (49.2) along the two lines of sight:

$$\begin{gathered} \delta P = \delta\Omega_1 \delta\Omega_2 \delta m_1 \delta m_2 \left[\frac{d\mathcal{N}}{dm_1} \frac{d\mathcal{N}}{dm_2} + g(m_1, m_2, \theta_{12}) \right], \\ g(m_1, m_2, \theta) = \int_0^\infty r_1^2 \, dr_1 r_2^2 dr_2 \Gamma(M_1, M_2, r_{12}), \\ r_{12}^2 = r_1^2 + r_2^2 - 2 r_1 r_2 \cos\theta, \\ M_1 = m_1 - 5 \log r_1 - 25, \qquad M_2 = m_2 - 5 \log r_2 - 25. \end{gathered} \tag{49.9}$$

The first equation defines the two-point correlation function in apparent magnitude and angular separation. One sees in the second equation that g is a linear integral over the corresponding spatial function Γ. It is the simplicity of this relation between the angular and spatial statistics that makes the correlation functions particularly convenient for practical application.

50. Small separation approximation and the scaling relation

In the case of greatest interest one assumes the maximum scale of appreciable galaxy clustering is small compared to the typical distances of the galaxies in the sample, so it is indeed a fair sample of the universe. Here the contribution to the integral g (eq. 49.9) is appreciable only when the objects are nearly at the same distance, $|r_1 - r_2| \ll r_1$, and the angular separation is much less than one radian.

When $\theta_{12} \ll 1$, the spatial separation of the two galaxies is

$$\begin{aligned} r_{12}^2 &= r_1^2 + r_2^2 - 2r_1r_2(1 - \theta^2/2) \\ &= (r_1 - r_2)^2 + r_1r_2\theta^2, \end{aligned} \tag{50.1}$$

and with

$$u = r_2 - r_1, \qquad r = (r_1 + r_2)/2, \tag{50.2}$$

the separation is

$$r_{12}^2 = u^2 + (r\theta)^2. \tag{50.3}$$

In this approximation the angular two-point function is

$$\begin{aligned} g(m_1, m_2, \theta) &= \int_0^\infty r^4 dr \int_{-\infty}^{\infty} du \Gamma(M_1, M_2, r_{12}), \\ M_\alpha &= m_\alpha - 5 \log r - 25. \end{aligned} \tag{50.4}$$

This expression is simplified by introducing some more notation. The mean magnitude and magnitude difference for the pair are

$$m = (m_1 + m_2)/2, \qquad \Delta m = m_2 - m_1. \tag{50.5}$$

Galaxies selected by apparent magnitude are found to have fairly definite absolute magnitudes, a typical value being M^*, say, with a standard deviation about this of perhaps one magnitude. M^* and m define a characteristic distance (eq. 49.4)

$$D = 10^{0.2(m-M^*)-5}\ \text{Mpc}. \tag{50.6}$$

With

$$r = Dy, \tag{50.7}$$

equation (50.4) becomes the scaling relation

$$g = D^5 \hat{g}(\Delta m, \theta D), \tag{50.8}$$

where

$$\begin{gathered} \hat{g}(\Delta m, x) = \int_0^\infty y^4 dy \int_{-\infty}^\infty du \Gamma(M_1, M_2, r_{12}), \\ M_1 = M^* - \Delta m/2 - 5 \log y, \qquad M_2 = M^* + \Delta m/2 - 5 \log y, \\ r_{12} = (u^2 + (xy)^2)^{1/2}. \end{gathered} \tag{50.9}$$

The single galaxy distribution of equation (49.6) with r replaced with yD is

$$\delta P = \delta\Omega \delta m D^3 \int_0^\infty y^2 dy \Phi(M^* - 5 \log y). \tag{50.10}$$

Equations (50.8) and (50.10) show how the one- and two-point functions scale with the effective depth D (or m). These scaling relations reflect the assumed geometry and the assumed statistical homogeneity of the distribution. The variation of $\hat{g}$ with Δm depends on the shape of the galaxy luminosity function through equation (50.9).

Some catalogs list angular positions of the galaxies brighter than a limiting magnitude m_0 rather than individual apparent magnitudes. Following Neyman, Scott, and Shane, one can model the random errors in the selection of galaxies by assuming the catalog includes a galaxy of apparent magnitude m with probability $f(m - m_0)$. Then the two-point correlation function for the catalog is the result of multiplying g (eq. 50.4) by f for each galaxy and integrating over the magnitudes,

$$\begin{gathered} \mathcal{N}^2 w(\theta) = D^5 \int y^4 dy du\, dM_1 dM_2 \Gamma(M_1, M_2, r_{12}) f_1 f_2, \\ f_\alpha = f(M_\alpha - M^* + 5 \log y), \qquad r_{12}{}^2 = u^2 + (yD\theta)^2. \end{gathered} \tag{50.11}$$

Here the characteristic survey depth is

$$D = 10^{0.2(m_0 - M^*) - 5}\ \text{Mpc}. \tag{50.12}$$

Since $\mathcal{N} \propto D^3$ (eq. 49.8), w varies with the depth of the survey according to the relation

$$w(\theta) = D^{-1} W(\theta D). \tag{50.13}$$

One can understand the scaling relation (50.13) as follows. The correlation function w measures the ratio of the number of neighbors in excess of random to the number expected for a uniform distribution. If one counts neighbors in solid angle $\delta\Omega \propto D^{-2}$ at angular distance θD from a galaxy at distance D from the observer, one is looking at a fixed projected area at the galaxy and a fixed projected distance from the galaxy, so the number of correlated neighbors seen is independent of D while the number of accidental neighbors from the foreground and background is proportional to D. Thus $w \propto D^{-1}$ at fixed θD.

The function W does depend on the shape of the observer function f, which will be different for different catalogs. However, it is found that if f has a reasonably small spread, W is not sensitive to its shape because f effectively broadens the luminosity function, which is broad already; for examples see Peebles and Hauser (1974). Thus it appears to be reasonable to ignore the difference in f in different catalogs.

The scaling relation for the angular power spectrum follows from equation (46.40), which with equation (50.13) is

$$u_l = (\mathcal{N}/D^3) \int 2\pi x\, dx W(x) J_0(xl/D) \tag{50.14}$$

for $\theta \ll 1$ radian, $l \gg 1$. Since $\mathcal{N} \propto D^3$, the relation is

$$u_l = U(l/D). \tag{50.15}$$

This is as expected because u_l measures the number of neighbors in excess of random within distance $\theta \propto l^{-1}$ (§ 46), and if this angle scales as $\theta \propto D^{-1}$, the number in excess of random should be independent of D.

The scaling relations (50.13) and (50.15) have played an important role in testing that the angular correlations of galaxies in the catalogs do reflect the clustering in space rather than systematic errors like the effect of patchy obscuration in the galaxy. The scaling relation was first applied to the study of clustering among positions of Abell clusters (Hauser and Peebles 1973). Recent results are summarized in Section 57.

51. Decoupling of Magnitude and Position

Equation (50.11) for w still is quite complicated for practical application since it involves a function of three variables. Fortunately it can be considerably simplified in an approximation that seems to be reasonably good. The assumption is that the absolute magnitude of a galaxy is statistically independent of its position relative to other galaxies, so the joint distribution Γ in M_1, M_2 and r_{12} (eq. 49.2) can be written as a product of distributions in magnitude and separation,

$$\Gamma(M_1, M_2, r_{12}) = \Phi(M_1)\Phi(M_2)\xi(r_{12}), \tag{51.1}$$

where Φ is the galaxy luminosity function of equation (49.1) and ξ is the two-point correlation function (eq. 31.4). In this approximation equation (50.11) is

$$\begin{gathered} \mathcal{N}^2 w(\theta) = D^5 n^2 \int_0^\infty y^4 \phi(y)^2\, dy \int_{-\infty}^\infty du\,\xi(r), \\ r^2 = u^2 + (yD\theta)^2, \end{gathered} \tag{51.2}$$

where

$$n = \int dM\Phi(M) \tag{51.3}$$

is the mean space number density of galaxies and

$$\phi(y) = \int_{-\infty}^\infty dM(\Phi(M)/n)\, f(M - M^* + 5\log y) \tag{51.4}$$

is the probability that a galaxy at distance yD will be in the catalog. The selection function ϕ thus measures the combined effects of the luminosity function and the observer function $f(m - m_0)$. In terms of the selection function the mean number density in the catalog is (eq. 49.6)

$$\mathcal{N} = nED^3, \qquad E = \int_0^\infty y^2 dy \phi(y). \tag{51.5}$$

By equations (51.2) and (51.5),

$$w(\theta) = D^{-1} W(\theta D), \tag{51.6}$$

where the characteristic depth is given by equation (50.12) and

$$W(x) = \frac{\int_0^\infty y^4 \phi(y)^2 dy \int_{-\infty}^\infty du \xi((u^2 + (xy)^2)^{1/2})}{\left[\int_0^\infty y^2 dy \phi(y)\right]^2}. \tag{51.7}$$

This is the integral equation that has been used in most estimates of the spatial galaxy correlation function starting with Rubin (1954) and Limber (1954).

Equation (51.1) for Γ in equation (50.9) gives the two-point distribution in apparent magnitudes and angular separation,

$$\begin{aligned} g(\theta, m, \Delta m) &= D^5 \hat{g}(\Delta m, \theta D), \\ \hat{g}(\Delta m, x) &= \int_0^\infty y^4 dy \Phi_1 \Phi_2 \int_{-\infty}^\infty du \xi((u^2 + (xy)^2)^{1/2}), \\ \Phi_{1,2} &= \Phi(M^* \pm \Delta m/2 - 5 \log y). \end{aligned} \tag{51.8}$$

We know that equation (51.1) is only a convenient but crude approximation because galaxy morphological type is correlated with position (early type galaxies tend to appear in dense concentrations) but it is consistent with the available statistical tests. In an unpublished study Soneira (1978b) found that equation (51.8) with conventional models for $\Phi(M)$ and $\xi(r)$ gives a reasonable fit to the joint distribution in θ and in Δm in the Zwicky catalog. Lake and Tremaine (1980) found that equation (51.1) agrees with Holmberg's (1969) counts of faint companions of spiral galaxies.

52. Relation between ξ and w: some examples

The relation between spatial and angular correlation functions ξ and w is illustrated by two convenient analytic models: ξ equal to a Gaussian or a power law. In the first case,

$$\xi = Ae^{-(r/r_0)^2}, \tag{52.1}$$

the integral over u in equation (51.7) can be evaluated leaving

$$w(\theta) = \frac{\pi^{1/2} A r_0}{DE^2} \int_0^\infty y^4 \phi(y)^2 \, dy \, e^{-(\theta D y/r_0)^2}. \tag{52.2}$$

The part $y^4\phi^2$ is fairly sharply peaked at $y \sim 1$, small at small y because of the geometrical factor y^4, and small at large y because the integral galaxy luminosity function sharply cuts off at the bright end. This allows us to separate the behavior of $w(\theta)$ into two limiting cases. If $\theta \lesssim r_0/D$, the exponential varies with y less rapidly than does $y^4\phi^2$, so equation (52.2) is

$$w(\theta) \simeq \frac{\pi^{1/2} A r_0}{DE^2} \left[\int_0^\infty y^4 \phi^2 dy \right] e^{-(\theta D/r_0)^2}. \tag{52.3}$$

The angular function looks like a Gaussian near the peak: a flat top of width r_0/D. If $\theta \gg r_0/D$, the exponential is the more rapidly varying term, and it cuts off the integral at $y \sim r_0/(\theta D)$ leaving

$$w(\theta) \propto y^5 \phi(y)^2, \qquad y = r_0/(\theta D). \tag{52.4}$$

The correlation function has a tail at $\theta \gg r_0/D$ that varies about as θ^{-5} (because the integral luminosity function ϕ varies only slowly with y at the faint end). The θ^{-5} tail comes from clusters with sizes $\sim r_0$ but much closer to the observer than typical galaxies in the sample. The observed $w(\theta)$ for galaxies does approximate a power law, but the slope is a good deal shallower, $w \propto \theta^{-0.8}$, so there is no danger that the present observations have been confused by this effect.

The second convenient model is a power law,

$$\xi(r) = Br^{-\gamma}, \qquad \gamma > 1. \tag{52.5}$$

This form in equation (51.7) gives

$$w(\theta) = \frac{B}{DE^2} \int_0^\infty y^4 \phi^2 dy \int_{-\infty}^{\infty} du \, [u^2 + (\theta D y)^2]^{-\gamma/2}. \tag{52.6}$$

With

$$u = \theta D y x, \tag{52.7}$$

one finds

$$w = A\theta^{1-\gamma}, \qquad A = \frac{B}{D^\gamma} \frac{H_\gamma}{E^2} \int_0^\infty y^{5-\gamma} \phi(y)^2 dy, \tag{52.8}$$

where the number H_γ from the integral over u can be reduced to a product of Gamma functions,

$$H_\gamma = \int_{-\infty}^{\infty} dx (1 + x^2)^{-\gamma/2} = \Gamma(\tfrac{1}{2})\Gamma((\gamma - 1)/2)/\Gamma(\gamma/2). \tag{52.9}$$

Equation (52.8) says that the angular function is a power law with index reduced by one unit from the power law index γ of ξ. This was derived by Totsuji and Kihara (1969). If $\gamma \leq 1$ the integral H diverges. Here the small angle assumption of Section 50 fails because most of the correlated pairs of galaxies seen at small angular separation are at very different distances from the observer. If $\gamma \geq 6$, the integral in equation (52.8) diverges at $y \to 0$. Here the correlation function ξ is decreasing so rapidly with increasing separation that the dominant contribution to $w(\theta)$ is from the cluster nearest the observer. The observed galaxy correlation function approximates a power law with $\gamma \approx 1.8$, well removed from either of these cases.

The angular power spectrum for either of these models may be computed using equation (46.40). In the power law model

$$w = A\theta^{1-\gamma} \tag{52.10}$$

the result is (Peebles and Hauser 1974)

$$u_l = 2\pi \mathcal{N} A \int_0^\infty x^{2-\gamma} dx J_0(x)/l^{3-\gamma}. \tag{52.11}$$

If $\gamma > 3$, the integral diverges at small separation. This is because the spectrum measures the number of neighbors within separation $\sim l^{-1}$, and if $\gamma > 3$, almost all the neighbors are closer than some small-scale cut-off of the $r^{-\gamma}$ power law. Thus u_l remains almost constant if $\gamma > 3$ and l^{-1} exceeds the cut-off of the power law model. If $\gamma < 3$, u_l varies as a power of l. The dimensionless integral in equation (52.11) is evaluated by Fall (1979).

In the Gaussian model the angular correlation function can be approximated as

$$w(\theta) = Ce^{-(\theta D/r_0)^2}. \tag{52.12}$$

The $\gamma = 5$ tail of w can be ignored because, as has just been remarked, it has little effect on the shape of the spectrum. The simplest way to evaluate u_l here is to use equation (46.39) to write equation (46.40) as

$$\begin{aligned} u_l &= \mathcal{N} C \int d\theta_1 d\theta_2 e^{-(\theta_1^2+\theta_2^2)D^2/r_0^2 + il\theta_1} \\ &= \pi \mathcal{N} C (r_0/D)^2 e^{-(lr_0/2D)^2}. \end{aligned} \tag{52.13}$$

Since the observed ξ for galaxies is close to a power law, it is interesting to see how accurate the small angle approximation is for this case at reasonable values of the angle θ. In Figure 52.1 the upper dashed straight line is given by equations (52.8) and (52.9) with

$$\xi(r) = (D/r)^2, \tag{52.14}$$

close to the observed shape for galaxies and with the selection function (Abell 1962)

$$\begin{aligned} \phi(y) &= y^{-5\beta}, \qquad \beta = 0.25, \qquad y < 1, \\ &= y^{-5\alpha}, \qquad \alpha = 0.75, \qquad 1 < y < y_0, \\ &= 0, \qquad y > y_0, \\ &\qquad y_0^{-5\alpha} = 0.01, \end{aligned} \tag{52.15}$$

which is thought to be fairly realistic. The upper solid curve is the result of numerically integrating the equation

$$\begin{aligned} w(\theta) &= E^{-2}\int_0^\infty (y_1 y_2)^2 dy_1 dy_2 \phi(y_1)\phi(y_2)\xi(r), \\ r &= D(y_1^2 + y_2^2 - 2y_1 y_2 \cos\theta)^{1/2}, \end{aligned} \tag{52.16}$$

with ξ given by equation (52.14) and ϕ by equation (52.15). This is the exact equation (49.9) in the case that magnitudes and positions decouple (eq. 51.1). One sees that the power law approximation to w is quite good even at $\theta = 30°$.

The bottom two curves in the figure show how $w(\theta)$ responds to a feature in $\xi(r)$. The spatial correlation function is a broken power law,

$$\begin{aligned} \xi &= (D/r)^2, \qquad r < r_b = 0.0052D; \\ \xi &= r_b(D/r)^3, \qquad r > r_b. \end{aligned} \tag{52.17}$$

The lower solid curve in the figure is the result of integrating equation (52.16) with the selection function of equation (52.15) and ξ given by equation (52.17). The lower dashed line is computed from equations (52.8) and (52.9), that is, it assumes $\xi = r_b(D/r)^3$ for all r and it uses the small angle approximation. The sharp break in the slope of ξ has been placed at the effective angle

$$r_b/D = 0.30° \tag{52.18}$$

This is indicated by the arrow in the figure.

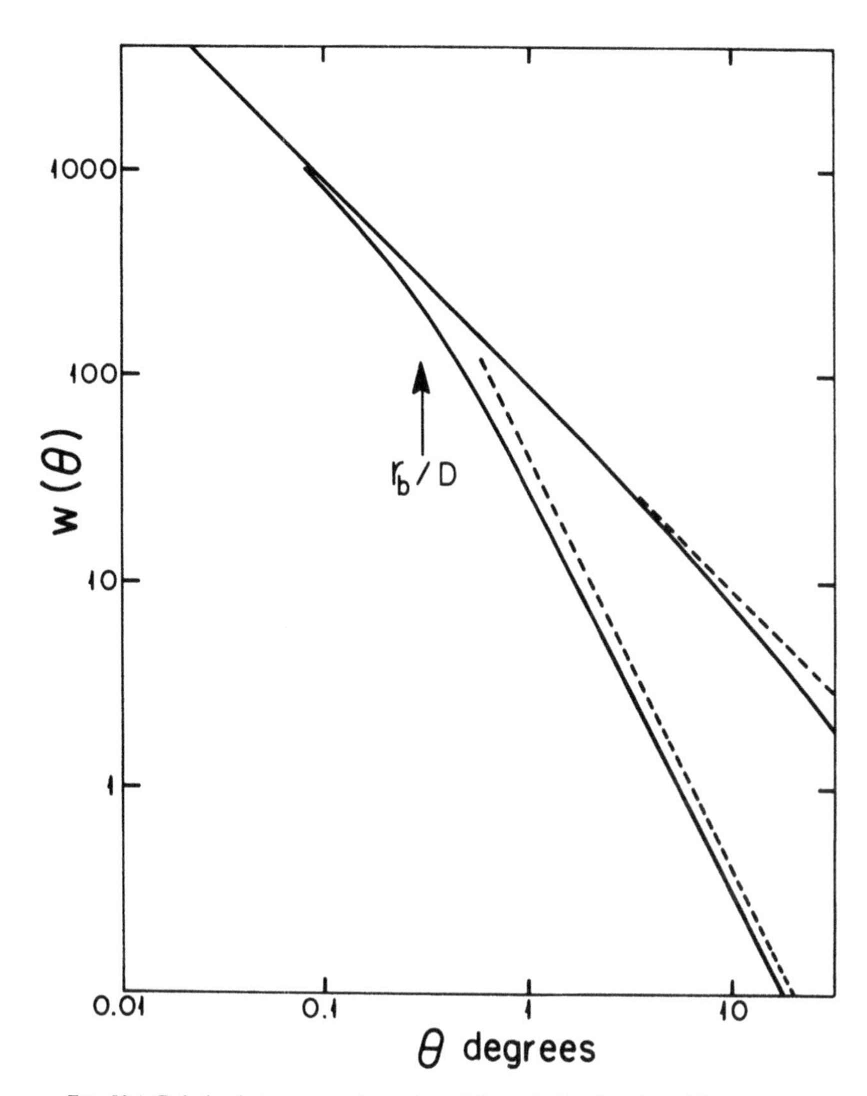

FIG. 52.1. Relation between angular and spatial correlation functions. The upper two curves both correspond to a power law ξ (eq. 52.14). The straight dashed line is the small separation approximation, the solid line the result of numerical integration of Limber's equation. The bottom two curves correspond to a two-power-law ξ (eq. 52.17). The solid curve is the result of numerical integration, and the dashed curve is computed from the small separation approximation with ξ taken to be a pure power law.

The feature certainly is less pronounced in $w(\theta)$ than in $\xi(r)$. It is smoothed mainly by the projection effect: galaxies at projected separation r are at true separations $>r$ with the distribution in true separation determined by ξ (the integral over u in eq. 51.7). It is smoothed also because the distances from the observer for the galaxies in the sample have an appreciable dispersion around D. Nevertheless, $w(\theta)$ does show a fairly prominent change of slope at about the angle expected from the naive application of equation (52.8). This is so because with the assumed selection function ϕ the distribution of distances from us of galaxies brighter than m_0, $\propto y^2\phi(y)$, is fairly tight, and the distribution of distances of correlated pairs, $\propto y^4\phi(y)^2$, is tighter still. Of course, if ξ had a narrow feature, perhaps an abrupt decrease in value going from r to $r + \Delta r$ with $\Delta r < r$, the projection effect (integral over u) would smooth this step to the width $\Delta r'$ comparable to r. Because of this projection effect, one should not find that $w(\theta)$ varies appreciably on angular scales smaller than $\Delta\theta \sim \theta$ (Fall and Tremaine 1977).

Finally it might be noted that if the selection function were such that $y^4\phi(y)^2$ is appreciable over a range of y much bigger than a factor of two or so, the major smoothing of the correlation function would be caused by the spread of distances of the galaxies, and this would substantially limit the amount of information one could hope to derive from $w(\theta)$.

53. Inversion of the equation

The result of changing the variable of integration in equation (51.7) from u to r and then exchanging the order of integration is

$$W(x) = \frac{2}{E^2}\int_0^\infty dr\xi(r)\int_0^{r/x}\frac{y^4dy\phi(y)^2}{(1-(xy/r)^2)^{1/2}}. \tag{53.1}$$

The relation thus is of the form

$$W(x) = E^{-2}\int_0^\infty dr\xi(r)F(x/r), \tag{53.2}$$

where the window function is

$$F(z) = 2\int_0^{z^{-1}}\frac{y^4dy\phi(y)^2}{(1-(yz)^2)^{1/2}}. \tag{53.3}$$

Equation (53.2) shows that W simply is a convolution (in logarithmic variables) of ξ. Fall and Tremaine (1977) and Parry (1977) have shown that this form can be inverted in a remarkably simple and elegant way.

Since the window function depends on the ratio rather than difference of arguments, one uses the Mellin rather than Fourier transform. The following properties of the transform are used. The definition is

$$\tilde{f}(s) = \int_0^\infty f(x) x^{s-1}\, dx. \tag{53.4}$$

Second is the inversion formula: if the integral $\tilde{f}$ is absolutely convergent for $\alpha < s < \beta$, then

$$f(x) = \frac{1}{2\pi i} \int_{\sigma - \infty i}^{\sigma + \infty i} x^{-s} \tilde{f}(s)\, ds, \tag{53.5}$$

where $\alpha < \sigma < \beta$ (Courant and Hilbert 1953, p. 103). Third is the analog of the convolution theorem: for the function

$$G(x) = \int_0^\infty dy x^p y^q f(xy) g(y), \tag{53.6}$$

the Mellin transform is

$$\begin{aligned} \tilde{G}(s) &= \int_0^\infty dy dx x^{p+s-1} y^q f(xy) g(y) \\ &= \tilde{f}(p+s)\tilde{g}(q-p-s+1), \end{aligned} \tag{53.7}$$

where the second line follows from the change of variables $xy = z$. In a similar way one finds that for the function

$$K(x) = \int_0^\infty dy x^p y^q f(x/y) g(y), \tag{53.8}$$

the Mellin transform is

$$\tilde{K}(s) = \tilde{f}(p+s)\tilde{g}(q+p+s+1). \tag{53.9}$$

Finally, appearing in the window function of equation (53.3) is the function

$$\begin{aligned} b(x) &= 2/(1-x^2)^{1/2}, \qquad 0 \le x < 1, \\ b(x) &= 0, \qquad x \ge 1. \end{aligned} \tag{53.10}$$

The Mellin transform of this is

$$\tilde{b}(s) = 2\int_0^1 x^{s-1}dx/(1 - x^2)^{1/2}$$
$$= 2\int_0^\infty dy(1 + y^2)^{(s+1)/2}, \tag{53.11}$$

where the second line follows from $x = (1 + y^2)^{-1/2}$. This function appeared in equation (52.9):

$$\tilde{b}(s) = \Gamma(1/2)\Gamma(s/2)/\Gamma((s + 1)/2). \tag{53.12}$$

Now the solution proceeds as follows. One notes that equation (53.2) is the same form as equation (53.8), so the convolution is

$$E^2\tilde{W}(s) = \tilde{F}(s)\tilde{\xi}(s + 1). \tag{53.13}$$

Second, equations (53.3) and (53.10) are

$$F(z) = \int_0^\infty dyy^4\phi^2(y)b(yz), \tag{53.14}$$

which is the same form as equation (53.6), so

$$\tilde{F}(s) = \tilde{b}(s)\tilde{\phi}^2(5 - s). \tag{53.15}$$

The transform of ξ is then

$$\tilde{\xi}(s) = E^2\tilde{W}(s - 1)/(\tilde{\phi}^2(6 - s)\tilde{b}(s - 1)). \tag{53.16}$$

Using

$$\Gamma(1/2) = \pi^{1/2}, \qquad \Gamma(n + 1) = n\Gamma(n), \tag{53.17}$$

one can write $\tilde{b}$ as

$$\frac{1}{\tilde{b}(s - 1)} = \frac{s - 1}{2\pi}\frac{\pi^{1/2}\Gamma(s/2)}{\Gamma((s - 1)/2)(s - 1)/2} = \frac{s - 1}{2\pi}\tilde{b}(s), \tag{53.18}$$

and on setting

$$E^2/(2\pi\tilde{\phi}^2(6 - s)) = \tilde{P}(s), \tag{53.19}$$

we have

$$\tilde{\xi}(s) = (s - 1)\tilde{b}(s)\tilde{W}(s - 1)\tilde{P}(s). \tag{53.20}$$

On setting

$$\tilde{W}(s-1)\tilde{b}(s) \equiv \tilde{H}(s), \tag{53.21}$$

we get from equations (53.8) and (53.9)

$$H(x) = \frac{2}{x}\int_0^1 dy W(x/y)(1-y^2)^{-1/2}, \tag{53.22}$$

and with

$$\tilde{J}(s-1) \equiv \tilde{H}(s)\tilde{P}(s), \tag{53.23}$$

equations (53.8) and (53.9) give

$$J(r) = 2\int_0^\infty dx P(x)\int_0^1 dy \frac{W(r/xy)}{(1-y^2)^{1/2}}, \tag{53.24}$$

where equation (53.20) is now

$$\tilde{\xi}(s) = (s-1)\tilde{J}(s-1). \tag{53.25}$$

From the definition of the transform one finds finally that this states

$$\xi(r) = -dJ/dr. \tag{53.26}$$

This with equation (53.24) is the inversion: the spatial correlation function ξ is given as a double integral of a derivative of the observed angular correlation function. Fall and Tremaine have used this solution to compute ξ from the w estimate for the Lick sample: the results agree well with the trial and error method of Section 52.

54. Angular three-point correlation function

The treatment of the three-point correlation function parallels that of the two-point function in Sections 49 to 51.

A. General relation of the reduced spatial and angular functions

The probability of finding three galaxies with magnitudes M_1, M_2 and M_3 in the volume elements δV_1, δV_2 and δV_3 respectively is written as

$$\begin{aligned}\delta P = [&\Phi(M_1)\Phi(M_2)\Phi(M_3) + \Phi(M_1)\Gamma(M_2, M_3, r_{23}) \\ &+ \Phi(M_2)\Gamma(M_1, M_3, r_{13}) + \Phi(M_3)\Gamma(M_1, M_2, r_{12}) \\ &+ \Gamma_3(M_1, M_2, M_3, r_{12}, r_{23}, r_{31})]\delta V_1\delta V_2\delta V_3\delta M_1\delta M_2\delta M_3,\end{aligned} \tag{54.1}$$

where Γ and Φ are defined in equation (49.2). The form of this equation follows equation (34.1), and as in equation (49.3) one sees that the integral of Γ_3 over magnitudes is

$$n^3\zeta(r_{12}, r_{23}, r_{31}) = \int dM_1 dM_2 dM_3 \Gamma_3. \tag{54.2}$$

The probability of finding galaxies with apparent magnitudes m_1, m_2, m_3 in the elements of solid angle $\delta\Omega_1$, $\delta\Omega_2$, $\delta\Omega_3$ respectively is found by integrating (54.1) along the three lines of sight. This gives

$$\begin{aligned}\delta P = \delta\Omega_1 \delta\Omega_2 \delta\Omega_3 \delta m_1 \delta m_2 \delta m_3 \Bigg[&\frac{d\mathcal{N}}{dm_1}\frac{d\mathcal{N}}{dm_2}\frac{d\mathcal{N}}{dm_3} + \frac{d\mathcal{N}}{dm_1} g(m_2, m_3, \theta_{23}) \\ &+ \frac{d\mathcal{N}}{dm_2} g(m_1, m_3, \theta_{13}) + \frac{d\mathcal{N}}{dm_3} g(m_1, m_2, \theta_{12}) \\ &+ g_3(m_1, m_2, m_3, \theta_{12}, \theta_{23}, \theta_{31})\Bigg],\end{aligned} \tag{54.3}$$

where $d\mathcal{N}/dm$ is the result of integrating the single point distribution Φ along the line of sight (eq. 49.8), and g is the result of integrating the reduced two-point function Γ (eq. 49.9). If the maximum scale of appreciable clustering is small compared to the depth of the catalog, the terms involving $d\mathcal{N}/dm$ and g describe triplets seen close together in the sky but generally at quite different distances from us, and in the integral over Γ_3 all three galaxies are at very nearly the same distance. As in Section 50 we can use this to simplify the integral g_3.

B. Small separation approximation

The effects of space curvature and expansion will be neglected here. With

$$M_\alpha = m_\alpha - 5 \log r_\alpha - 25, \tag{54.4}$$

the reduced three-point function is

$$\begin{gathered} g_3(m_1 \ldots \theta_{31}) = \int_0^\infty (r_1 r_2 r_3)^2\, dr_1 dr_2 dr_3 \Gamma_3(M_1 \ldots r_{31}), \\ r_{\alpha\beta}^2 = r_\alpha^2 + r_\beta^2 - 2 r_\alpha r_\beta \cos\theta_{\alpha\beta}. \end{gathered} \tag{54.5}$$

In the small separation approximation discussed in Section 50 with

$$\begin{gathered} r = r_1, \qquad u = r_2 - r_1, \qquad v = r_3 - r_1, \\ m = m_1, \qquad \Delta m_a = m_2 - m_1, \qquad \Delta m_b = m_3 - m_1, \end{gathered} \tag{54.6}$$

and the dimensionless distance variable (eqs. 50.6, 50.7)

$$y = r/D, \qquad D = 10^{0.2(m - M^*)-5}\ \text{Mpc}, \tag{54.7}$$

equation (54.5) becomes

$$g_3(m_1 \ldots \theta_{31}) = D^7 \hat{g}_3(\Delta m_a, \Delta m_b, D\theta_{12}, D\theta_{23}, D\theta_{31}), \tag{54.8}$$

where

$$\begin{gathered}
\hat{g}_3(\Delta m_a, \Delta m_b, x_{12}, x_{23}, x_{31}) = \int_0^\infty y^6 dy \int_{-\infty}^\infty du dv \Gamma_3(M_1 \ldots r_{31}), \\
M_1 = M^* - 5 \log y, \qquad M_2 = M_1 + \Delta m_a, \qquad M_3 = M_1 + \Delta m_b, \\
r_{12}{}^2 = u^2 + (yx_{12})^2, \qquad r_{13}{}^2 = v^2 + (yx_{13})^2, \\
r_{23}{}^2 = (u - v)^2 + (yx_{23})^2.
\end{gathered} \tag{54.9}$$

Equation (54.8) shows how g_3 scales with m or the effective distance D. The variation with magnitude difference Δm_a, Δm_b, of course, is more complicated, depending on the luminosity function.

In a catalog that lists angular positions of galaxies brighter than some fixed nominal limiting magnitude m_0, the three-point angular correlation function z is defined by the expression

$$\delta P = \mathcal{N}^3(1 + w_{12} + w_{23} + w_{31} + z)\delta\Omega_1 \delta\Omega_2 \delta\Omega_3, \tag{54.10}$$

which can be compared to equation (34.1). One sees from equation (54.1) that the reduced part is

$$\mathcal{N}^3 z = \int dM_1 dM_2 dM_3 (r_1 r_2 r_3)^2 dr_1 dr_2 dr_3 f_1 f_2 f_3 \Gamma_3, \tag{54.11}$$

where

$$f_\alpha = f(M_\alpha + 5 \log r + 25 - m_0) \tag{54.12}$$

is the observer function, as in equation (50.11). In the small separation approximation this becomes

$$\begin{gathered}
D^2(nE)^3 z = \int_0^\infty y^6 dy \int_{-\infty}^\infty du dv \int_{-\infty}^\infty dM_1 dM_2 dM_3 f_1 f_2 f_3 \Gamma_3, \\
f_\alpha = f(M_\alpha - M^* + 5 \log y), \qquad D = 10^{0.2(m_0 - M^*)-5}\ \text{Mpc}, \\
r_{12}{}^2 = u^2 + (yD\theta_{12})^2, \text{ etc.,}
\end{gathered} \tag{54.13}$$

where the constant E is defined in equation (51.5).

Equation (54.13) yields the scaling relation

$$z(\theta_{12}, \theta_{23}, \theta_{31}) = D^{-2}\hat{z}(D\theta_{12}, D\theta_{23}, D\theta_{31}). \tag{54.14}$$

One can interpret this in the same way as for equation (50.13). The function $\mathcal{N}^2 z \propto D^6 z$ measures the probability in excess of random of finding neighbors in $\delta\Omega_2$ and $\delta\Omega_3$ at distances θ_{12} and θ_{13} from a randomly chosen object in the catalog. When the angles scale with the depth of the sample as $\theta_{\alpha\beta} \propto D^{-1}$ and the solid angles $\delta\Omega_2$ and $\delta\Omega_3$ vary as D^{-2}, this probability is constant because the projected area and separation at the object are constant. Therefore $z \propto D^{-2}$ at fixed $D\theta_{\alpha\beta}$.

Under the assumption that galaxy positions and magnitudes are uncorrelated (§ 51), Γ_3 can be written as the product

$$\Gamma_3 = \Phi(M_1)\Phi(M_2)\Phi(M_3)\zeta(r_{12}, r_{23}, r_{31}), \tag{54.15}$$

where ζ is defined in equation (54.2). Then the integral over each M in equation (54.13) yields $n\phi(y)$, where ϕ is the selection function (eq. 51.4), and equation (54.13) becomes (Peebles and Groth 1975)

$$\begin{gathered} z(\theta_{12}, \theta_{23}, \theta_{31}) = \frac{1}{D^2}\frac{\int_0^\infty y^6 dy \phi(y)^3 \int_{-\infty}^{\infty} du dv \zeta}{\left(\int_0^\infty y^2 dy \phi(y)\right)^3}, \\ r_{12}^2 = u^2 + (yD\theta_{12})^2, \qquad r_{13}^2 = v^2 + (yD\theta_{13})^2, \\ r_{23}^2 = (u - v)^2 + (yD\theta_{23})^2. \end{gathered} \tag{54.16}$$

This equation has been used in all estimates of the spatial three-point correlation function.

C. Model for ζ

As it happens a model for ζ that considerably simplifies equation (54.16) matches the data remarkably well. The assumption is that ζ, which is a function of three variables, can be written in terms of a function of one variable,

$$\zeta(r_a, r_b, r_c) = Q[\xi(r_a)\xi(r_b) + \xi(r_b)\xi(r_c) + \xi(r_c)\xi(r_a)], \tag{54.17}$$

where Q is a constant. The new function has been written as $\xi(r)$, the two-point correlation function, because that agrees with what is observed, but that is not required. It will be noted that equation (54.17) satisfies two necessary conditions: that ζ is symmetric in its three arguments and that $\zeta \to 0$ when one point is well away from the other two. The form of equation (54.17) appears in theories of liquid physics and turbulence where the full three-point correlation function (34.1) occasionally is written as the Kirkwood superposition approximation (for example, Ichimaru 1973)

$$\delta P = n^3(1 + \xi_a)(1 + \xi_b)(1 + \xi_c)\delta V_1 \delta V_2 \delta V_3, \tag{54.18}$$

with ξ the reduced two-point function. This generates equation (54.17) with $Q = 1$, quite close to the observed value, but it adds to ζ the term

$$\xi_a \xi_b \xi_c. \tag{54.19}$$

In the conventional applications one often assumes $\xi \ll 1$, so this term can be neglected. However, the opposite limit applies here: in the range of separations in most of the empirical studies of the galaxy distribution $\xi \gg 1$, so if the term (54.19) were present, it would have dominated (54.17) and would have made the variation of the angular function z with θ distinctly different from what is observed. Thus the Kirkwood superposition approximation is not relevant here.

The model for ζ (eq. 54.17) in equation (54.16) yields

$$z = \frac{Q}{E^3}\left[\int_0^\infty y^6 dy \phi(y)^3 \Upsilon(yD\theta_a)\Upsilon(yD\theta_b) + \text{cycl.}\right],$$
$$\Upsilon(yD\theta) = \int_{-\infty}^{\infty} du\, \xi((u^2 + (yD\theta)^2)^{1/2})/D. \tag{54.20}$$

The sum is over the three terms with arguments $\theta_a\theta_b$, $\theta_b\theta_c$, and $\theta_c\theta_a$. As in Section 52 there are two convenient models for ξ: a power law, which agrees with the observations, and a Gaussian, which gives at least some impression of how sensitive z might be to the shape of ζ.

In the Gaussian model,

$$\xi = Ae^{-(r/r_0)^2}, \qquad \Upsilon = \pi^{1/2} A r_0 D^{-1} e^{-(y\theta D/r_0)^2}. \tag{54.21}$$

As discussed in Section 52 the function $y^6\phi(y)^3$ is sharply peaked near $y = 1$, so if $\theta D/r_0$ is not much greater than unity $y^6\phi^3$ acts like a delta function, reducing equation (54.20) to

$$z \propto e^{-(\theta_a^2 + \theta_b^2)(D/r_0)^2} + \text{cycl.} \tag{54.22}$$

When $\theta D/r_0 \gg 1$, the integral over y is cut off at $y \ll 1$ by the Υ factors, and equation (54.20) says z varies with the size θ of the triangle as

$$z \propto (\theta)^{-7}. \tag{54.23}$$

Thus as in Section 52 the angular function reflects the Gaussian shape of ξ near the peak but has a power law tail that comes from the clusters of size $\sim r_0$ at distances from the observer much less than the typical distance D. The observed correlation function varies as a power of the triangle size, $z \propto \theta^{-1.6}$; since this is considerably shallower than the θ^{-7} tail, there is no danger the observations have been confused by this near cluster effect.

In the power law model,

$$\xi = Br^{-\gamma}, \qquad \gamma > 1, \qquad \Upsilon = (y\theta)^{1-\gamma}BD^{-\gamma}H_\gamma. \tag{54.24}$$

The dimensionless integral H_γ is given in equation (52.9). Thus equation (54.20) becomes

$$z = P[w(\theta_a)w(\theta_b) + w(\theta_b)w(\theta_c) + w(\theta_c)w(\theta_a)], \tag{54.25}$$

where (Peebles and Groth 1975)

$$\frac{P}{Q} = \frac{\left(\int_0^\infty y^{8-2\gamma}dy\phi(y)^3\right)\left(\int_0^\infty y^2 dy\phi(y)\right)}{\left(\int_0^\infty y^{5-\gamma}dy\phi(y)^2\right)^2}, \tag{54.26}$$

and $w(\theta)$ is the angular correlation function belonging to $\xi(r)$ (eq. 52.8). That is, we have the convenient result in the model that if ξ varies as a power of r, the same model applies to z in terms of $w(\theta)$. The value of P/Q is determined by integrals over ϕ: it is independent of D in the Newtonian approximation and not very sensitive to the shape of ϕ or the details of the relativistic corrections (fig. 56.1 below). What is more, this model for z agrees with the observations in considerable detail.

D. *Methods of estimating z*

Two approaches have been used. In the Monte Carlo method one observes that the expected count of triplets defining triangles with some chosen range of sizes and shapes is

$$\langle DDD \rangle = N^3(\Omega_2/\Omega^2)[1 + w_a + w_b + w_c + z]. \tag{54.27}$$

There are $N \gg 1$ objects in the survey region of size Ω, and Ω_2 is the product of two solid angles fixed by the range of sizes and shapes of the triangles and by the shape of the boundary of Ω. If N_t points are placed at random in the survey region, the expected count of triplets among them is

$$\langle RRR \rangle = N_t^3 \Omega_2 / \Omega^2, \tag{54.28}$$

and the expected count of triplets in which real objects define two sides of the triangle and the triangle is completed by one of the random objects is

$$\langle DDR \rangle = N_t N^2 (\Omega_2/\Omega^2)(3 + w_a + w_b + w_c). \tag{54.29}$$

Thus an estimate of z is

$$z = \frac{DDD - DDR\, N/N_t}{RRR(N/N_t)^3} + 2. \tag{54.30}$$

Examples of the numerical values of the quantities in this expression are given by Peebles and Groth (1975).

The second method uses the counts N_i of objects in cells: one sees from equations (54.10) and (47.15) that estimates of z are (Peebles 1975)

$$\begin{aligned} z &= \frac{\langle N_i N_j N_k \rangle}{\langle N_i \rangle \langle N_j \rangle \langle N_k \rangle} - \frac{\langle N_i N_j \rangle}{\langle N_i \rangle \langle N_j \rangle} - \frac{\langle N_j N_k \rangle}{\langle N_j \rangle \langle N_k \rangle} - \frac{\langle N_k N_i \rangle}{\langle N_k \rangle \langle N_i \rangle} + 2 \\ &= \frac{\langle (N_i - \langle N_i \rangle)(N_j - \langle N_j \rangle)(N_k - \langle N_k \rangle) \rangle}{\langle N_i \rangle \langle N_j \rangle \langle N_k \rangle}. \end{aligned} \tag{54.31}$$

The averages are over the set of disjoint cells ijk whose centers define triangles in some chosen range of size and shape. As discussed in Section 47 the arguments of z usually can be taken to be the distances between the cell centers.

Estimation of the joint three-point distribution in angular separation and apparent magnitude in principle is a straightforward extension of these methods and would be interesting as a test of the assumed decoupling of magnitude and position.

55. Angular four-point correlation function

This closely parallels the discussion in the last section. Following equations (35.1) and (54.1), we write the probability of finding galaxies of magnitudes M_1, M_2, M_3, and M_4 in the volume elements δV_1, δV_2, δV_3, and δV_4 as

$$\begin{aligned}\delta P = [&\Phi(M_1)\Phi(M_2)\Phi(M_3)\Phi(M_4)\\ &+ \Phi(M_1)\Phi(M_2)\Gamma(M_3, M_4, r_{34}) + \cdots \text{ (6 terms)}\\ &+ \Gamma(M_1, M_2, r_{12})\Gamma(M_3, M_4, r_{34}) + \cdots \text{ (3 terms)}\\ &+ \Phi(M_1)\Gamma_3(M_2, M_3, M_4, r_{23}, r_{34}, r_{42}) + \cdots \text{ (4 terms)}\\ &+ \Gamma_4(M_1 \ldots r_{41})]\delta M_1 \ldots \delta V_4.\end{aligned} \tag{55.1}$$

The probability of finding four galaxies of apparent magnitudes m_1 to m_4 in the elements of solid angle $\delta\Omega_1$ to $\delta\Omega_4$ is found by integrating this along the four lines of sight. As discussed in Section 54A the result is the same form with Φ replaced by $d\mathcal{N}/dm$ and Γ_α replaced by g_α, and g_4 is an integral over Γ_4. Therefore, if the maximum scale of appreciable clustering is small compared to the depth of the survey, we can assume that the four distances from the observer in the integral over Γ_4 all are nearly equal. In this approximation the reduced four-point correlation function in apparent magnitude and angular separation is

$$g_4(m_1 \ldots \theta_{41}) = D^9 \hat{g}_4(\Delta m_a, \Delta m_b, \Delta m_c, D\theta_{12}, \ldots), \tag{55.2}$$

where

$$\begin{aligned}\hat{g}_4(\Delta m_a, \Delta m_b, \Delta m_c, x_{12}, \ldots) &= \int_0^\infty y^8 dy \int_{-\infty}^\infty dt du dv \Gamma_4,\\ M_1 = M^* - 5 \log y, &\qquad M_2 = M_1 + \Delta m_a, \text{ etc.};\\ r_{12}^2 = t^2 + (yx_{12})^2, &\qquad r_{23}^2 = (u - t)^2 + (yx_{23})^2, \text{ etc.}\end{aligned} \tag{55.3}$$

Equation (55.2) shows how g_4 scales with effective depth (eq. 54.7).

If the catalog lists angular positions of all objects brighter than apparent magnitude m_0, the angular correlation function u is defined in analogy with equation (35.1) by the equation

$$\begin{aligned}\delta P = \mathcal{N}^4(1 &+ w_{12} + w_{13} + w_{14} + w_{23} + w_{24} + w_{34}\\ &+ w_{12}w_{34} + w_{13}w_{24} + w_{14}w_{23}\\ &+ z_{123} + z_{124} + z_{134} + z_{234} + u)\delta\Omega_1\delta\Omega_2\delta\Omega_3\delta\Omega_4.\end{aligned} \tag{55.4}$$

The reduced part u is found by multiplying equation (55.2) by the observer function for each galaxy, as in equation (54.13), and integrating over magnitudes:

$$D^3(nE)^4u = \int_0^\infty y^8 dy \int_{-\infty}^\infty dtdudv \int_{-\infty}^\infty dM_1 dM_2 dM_3 dM_4 f_1 f_2 f_3 f_4 \Gamma_4,$$
$$f_\alpha = f(M_\alpha - M^* + 5\log y), \qquad D = 10^{0.2(m_0 - M^*)-5},$$
$$r_{12}^2 = t^2 + (yD\theta_{12})^2, \text{etc.} \tag{55.5}$$

This yields the scaling relation

$$u(\theta_{12}, \ldots) = D^{-3}U(D\theta_{12}, D\theta_{23}, \ldots). \tag{55.6}$$

Under the assumption that magnitudes and positions decouple, Γ_4 is a product of $\Phi(M_\alpha)$'s and the four-point function η defined in equation (35.1). As in equation (54.16) equation (55.5) can be reduced to

$$u = \frac{1}{D^3}\frac{\int_0^\infty y^8\, dy\phi(y)^4 \int_{-\infty}^\infty dtdudv\,\eta}{\left(\int_0^\infty y^2\, dy\phi(y)\right)^4}, \tag{55.7}$$

$$r_{12}^2 = t^2 + (yD\theta_{12})^2, \text{etc.}$$

A convenient model for η is (Fry and Peebles 1978)

$$\begin{aligned}\eta_{1234} = {} & R_a[\xi_{12}\xi_{23}\xi_{34} + \text{sym. (12 terms)}] \\ & + R_b[\xi_{12}\xi_{13}\xi_{14} + \text{sym. (4 terms)}],\end{aligned} \tag{55.8}$$

where R_a and R_b are constants. In the first line there are 12 terms corresponding to all different ways of joining the four points by an unbroken line, and in the second there are 4 terms for the four ways of joining three points to a common fourth. This expression is symmetric in all its arguments, and it approaches zero if one or two points are well removed from the others. This model is a natural guess given the observational success of the analogous model for ζ, and a further theoretical justification of sorts is given in Sections 61 and 62; however, its main importance is that the model greatly simplifies discussion of the relation between angular and spatial functions and does not disagree with the crude data available on the four-point function.

As in equation (54.20) one sees that with equation (55.8) the integrals over t, u, and v in equation (55.7) all are of the form of $\Upsilon(yD\theta_{\alpha\beta})$ so (55.7) becomes a sum of terms like

$$\frac{\int_0^\infty y^8 dy \phi(y)^4 \Upsilon(yD\theta_{12})\Upsilon(yD\theta_{23})\Upsilon(yD\theta_{34})}{\left(\int_0^\infty y^2 dy \phi(y)\right)^4}. \tag{55.9}$$

If $\xi \propto r^{-\gamma}$, then $\Upsilon \propto \theta^{1-\gamma}$ (eq. 54.24), and u has the same form as η,

$$u = a[w_{12}w_{23}w_{34} + \text{sym.}] + b[w_{12}w_{13}w_{14} + \text{sym.}], \tag{55.10}$$

where w and ξ are related by equations (52.5, 52.8) and

$$\frac{a}{R_a} = \frac{b}{R_b} = \frac{\left[\int_0^\infty y^{11-3\gamma}\phi(y)^4 dy\right]\left[\int_0^\infty y^2\phi(y)dy\right]^2}{\left[\int_0^\infty y^{5-\gamma}dy\phi(y)^2\right]^3}. \tag{55.11}$$

All estimates of u so far have used the products of counts in cells: if N_1, N_2, N_3, and N_4 are the counts of galaxies in four disjoint cells, then the expected value of the product of the four counts is

$$\langle N_1 N_2 N_3 N_4 \rangle = \int dP, \tag{55.12}$$

where dP is given by equation (55.4) and the integral is over $d\Omega_1$ in cell 1, $d\Omega_2$ in cell 2, and so on. On using (55.4) to evaluate this integral, one finds

$$\begin{aligned}
&\langle (N_1 - \langle N_1\rangle)(N_2 - \langle N_2\rangle)(N_3 - \langle N_3\rangle)(N_4 - \langle N_4\rangle)\rangle \\
&\quad - \langle (N_1 - \langle N_1\rangle)(N_2 - \langle N_2\rangle)\rangle\langle (N_3 - \langle N_3\rangle)(N_4 - \langle N_4\rangle)\rangle \\
&\quad - \langle (N_1 - \langle N_1\rangle)(N_3 - \langle N_3\rangle)\rangle\langle (N_2 - \langle N_2\rangle)(N_4 - \langle N_4\rangle)\rangle \\
&\quad - \langle (N_1 - \langle N_1\rangle)(N_4 - \langle N_4\rangle)\rangle\langle (N_2 - \langle N_2\rangle)(N_3 - \langle N_3\rangle)\rangle \\
&= \mathcal{N}^4 \int d\Omega_1 d\Omega_2 d\Omega_3 d\Omega_4 u \\
&\simeq \langle N_1\rangle\langle N_2\rangle\langle N_3\rangle\langle N_4\rangle u_{1234}.
\end{aligned} \tag{55.13}$$

As indicated the integral can be approximated as the product of the mean counts in each cell with u evaluated at the separations of the cell centers. The left-hand side of equation (55.13) can be estimated by averaging the products of counts over the configurations of four cells identical up to a translation, rotation, or inversion (Fry and Peebles 1978).

56. Correction for curvature and expansion

The galaxy surveys reach fairly high redshifts—a typical value for galaxies in the Lick catalog is $Z \sim 0.08$ and the deep surveys go well beyond that—so it is useful to see how the relations between angular and spatial correlation functions are affected by the cosmological model (Fall 1976a, Groth and Peebles 1977, Dautcourt 1977, Phillipps et al. 1978).

The standard Friedman-Lemaître model will be used and it will be supposed that $\Lambda = p = 0$; a generalization to nonzero pressure or cosmological constant is straightforward but uninteresting because at practical survey depths the results are much more sensitive to the uncertainties in luminosity function and K-correction than to the acceleration parameter. The line element will be taken to be the Robertson-Walker form

$$ds^2 = c^2dt^2 - \frac{a^2dx^2}{1 - (x/Rc)^2} - a^2x^2(d\theta^2 + \sin^2\theta d\phi^2). \qquad (56.1)$$

This neglects local fluctuations in space curvature, that is, deflection of the line of sight by individual mass concentrations. The expansion rate is

$$H^2 = (\dot{a}/a)^2 = {}^8\!/\!_3\, \pi G\rho - 1/(aR)^2, \qquad (56.2)$$

and the density parameter is

$$\Omega_0 = 8\pi G\rho_0/3H_0^2, \qquad (56.3)$$

so we can write

$$R^{-2} = a_0^2 H_0^2(\Omega_0 - 1). \qquad (56.4)$$

The cosmological redshift of an object whose light is received now, at epoch t_0, and was emitted from the object at epoch t is

$$1 + Z = a_0/a(t). \qquad (56.5)$$

In the coordinates of equation (56.1) the coordinate distance of an object at redshift $Z(t)$ is fixed by the equation

$$c\int_t^{t_0} dt/a(t) = \int_0^x dx/(1 - x^2/R^2c^2)^{1/2}, \qquad (56.6)$$

which with equation (56.2) can be reduced to (Mattig 1958)

$$H_0a_0x/c = 2[(\Omega_0 - 2)(1 + \Omega_0 Z)^{1/2} + 2 - \Omega_0 + \Omega_0 Z]/[\Omega_0^2(1 + Z)]. \quad (56.7)$$

Another convenient form is the reciprocal of this relation, the redshift as a function of coordinate distance:

$$1 + Z = 2[2y(1 - \Omega_0) + \Omega_0 + (2 - \Omega_0)(1 + y^2(1 - \Omega_0))^{1/2}]/(2 - y\Omega_0)^2, \quad (56.8)$$
$$y = H_0a_0x/c.$$

The flux of energy received from an object with luminosity $\mathcal{L}$ at redshift Z is

$$f = \frac{\mathcal{L}}{4\pi a_0^2 x^2}(1 + Z)^{-2}. \quad (56.9)$$

The denominator of the first factor is the area over which the radiation now has spread, and the redshift factor takes account of the loss of energy of each photon and the diminuition of the rate of reception of photons. This equation translates to the distance modulus

$$m - M = 5 \log [a_0x(1 + Z)] + 25 + \kappa Z, \quad (56.10)$$

where a_0x is measured in megaparsecs. The K-correction is (Oke and Sandage 1968, Pence 1976)

$$\Delta m \cong \kappa Z. \quad (56.11)$$

This takes account of the fact that the magnitude m measured in a fixed wavelength band measures a wavelength and bandwidth at the object that are functions of redshift. As indicated the K-correction will be approximated as κZ; for $Z \lesssim 0.5$ the best value of the constant of proportionality is thought to be between 3 and 5.

Now we can write down expressions for the correlation functions. The element of solid angle $\delta\Omega$ subtends proper area

$$\delta A = a(t)^2x^2\delta\Omega \quad (56.12)$$

at coordinate distance x, and the increment δx in distance corresponds to proper radial interval (eqs. 56.1 and 56.4)

$$\delta r = a(t)\delta x/F(x),$$
$$F = [1 - (H_0 a_0 x/c)^2(\Omega_0 - 1)]^{1/2}. \tag{56.13}$$

The probability of finding a galaxy of apparent magnitude m in $\delta\Omega$ is then

$$\delta P = \delta\Omega\delta m d\mathcal{N}/dm,$$
$$d\mathcal{N}/dm = \int_0^\infty x^2\, dx F(x)^{-1} a^3 \Phi(M, t). \tag{56.14}$$

Here Φ is the luminosity function, the mean number of galaxies per unit proper volume and magnitude interval at epoch t. The absolute magnitude M is given in terms of m and x by equations (56.8) and (56.10). The joint probability of finding galaxies with apparent magnitudes m_1 and m_2 in $\delta\Omega_1$ and $\delta\Omega_2$ is (eq. 49.9)

$$\delta P = \delta m_1 \delta m_2 \delta\Omega_1 \delta\Omega_2 \left[\frac{d\mathcal{N}}{dm_1}\frac{d\mathcal{N}}{dm_2} + g\right],$$
$$g = \int_0^\infty x_1{}^2 x_2{}^2\, dx_1 dx_2 F_1{}^{-1} F_2{}^{-1} a_1{}^3 a_2{}^3 \Phi_1 \Phi_2 \xi. \tag{56.15}$$

Following Section 51 it has been assumed that the two-point correlation function can be written as a product of the distributions in separation and magnitude (eq. 51.1). The luminosity functions Φ_1 and Φ_2 are functions of x_1, m_1 and x_2, m_2 as in equation (56.14). In general ξ is a function of the spatial coordinate separation and the cosmic times at the two points; however, it appears that in any realistic situation the integral is dominated by points whose cosmic times are very nearly the same so that ξ can be taken to be a function of one epoch and of the proper separation of the points at that epoch. In this small separation approximation we can rewrite equation (56.15) as

$$g(m_1, m_2, \theta) = \int_0^\infty x^4 dx F(x)^{-2} a^6 \Phi_1 \Phi_2 \int_{-\infty}^\infty du\, \xi(r, t),$$
$$r = a(u^2/F^2 + x^2\theta^2)^{1/2}, \qquad u = x_2 - x_1, \tag{56.16}$$
$$\Phi_\alpha = \Phi(t, M = m_\alpha - 5 \log [a_0 x(1 + Z)] - 25 - \kappa Z).$$

The proper separation r has been computed as in equations (56.12) and (56.13).

In a catalog of angular positions of galaxies brighter than limiting magnitude m_0, the probability that a galaxy at distance x is brighter than the cutoff is the selection function

$$\phi(x) = n^{-1}\int_{-\infty}^{M} \Phi dM = \Psi(M(x), t)/n,$$
$$M(x) = m_0 - 5\log[a_0 x(1+Z)] - 25 - \kappa Z. \tag{56.17}$$

The observer function has been ignored (eq. 51.4). The function Ψ is the mean number of galaxies per unit proper volume more luminous than M at epoch t. The proper number density of galaxies will be taken to vary as

$$n(t) = n_0(a_0/a(t))^3, \tag{56.18}$$

which means galaxies are not being created or destroyed. The density and two-point angular correlation function are then

$$\mathcal{N} = n_0 a_0^3 \int_0^\infty x^2\, dx\phi(x)/F(x),$$
$$\mathcal{N}^2 w(\theta) = n_0^2 a_0^6 \int_0^\infty x^4\, dx\phi^2 F^{-2}\int_{-\infty}^{\infty} du\xi(r,t). \tag{56.19}$$

A convenient model for ξ that fits the observations quite well is the power law (eq. 52.5)

$$\xi(r,t) = Br^{-\gamma}(1+Z)^{-(3+\epsilon)}, \tag{56.20}$$

where B is constant. The last factor models possible evolution of the clustering: if $\epsilon = 0$, $n(t)\,\xi(r,t)$ at fixed proper separation r is constant, so the clustering measured in proper coordinates is not changing (eq. 31.7 with $\xi \gg 1$). The u integral can be evaluated for this form for $\xi(r)$, as in equations (52.6) to (52.9), giving

$$w = A\theta^{1-\gamma},$$

$$A = BH_\gamma a_0^{-\gamma}\frac{\int_0^\infty x^{5-\gamma}\, dx\phi(x)^2(a/a_0)^{3+\epsilon-\gamma}/F(x)}{\left(\int_0^\infty x^2 dx\phi(x)/F(x)\right)^2}. \tag{56.21}$$

When the redshift is small this reduces to equation (52.8). The cosmological model enters through the volume factor F (eq. 56.13), the redshift factor a_0/a, and the variation of the selection function ϕ with distance x (eq. 56.17).

The relation between angular and spatial three-point functions is written down in the same way. In the small separation approximation the reduced three-point function (eq. 54.3) is

$$g_3(m_1, m_2, m_3, \theta_{12}, \theta_{23}, \theta_{31})$$
$$= \int_0^\infty x^6 dx F(x)^{-3} a^9 \Phi_1 \Phi_2 \Phi_3 \int_{-\infty}^{\infty} du dv \zeta(r_{12}, r_{23}, r_{31}, t). \quad (56.22)$$

The Φ_α and $r_{\alpha\beta}$ are defined as in equation (56.16). For a catalog of angular positions of objects brighter than limiting magnitude m_0 this becomes

$$\mathcal{N}^3 z(\theta_{12}, \theta_{23}, \theta_{31}) = n_0^3 a_0^9 \int_0^\infty x^6 dx \phi(x)^3 F(x)^{-3} \int_{-\infty}^{\infty} du dv \zeta. \quad (56.23)$$

As in Section 54 this is considerably simplified if

$$\zeta = Q[\xi_a \xi_b + \xi_b \xi_c + \xi_c \xi_a]. \quad (56.24)$$

When the ξ are given by equation (56.20) and Q is a constant, the result is (Groth and Peebles 1977)

$$z = P[w_a w_b + w_b w_c + w_c w_a], \quad (56.25)$$

where $w(\theta)$ is given by equation (56.21) and

$$\frac{P}{Q} = \frac{\left(\int_0^\infty x^{8-2\gamma} dx \phi(x)^3 a^{6+2\epsilon-2\gamma}/F(x)\right)\left(\int_0^\infty x^2 dx \phi/F\right)}{\left(\int_0^\infty x^{5-\gamma} dx \phi(x)^2 a^{3+\epsilon-\gamma}/F\right)^2}. \quad (56.26)$$

The treatment of the four-point function is just the same although the equations are still longer. For a catalog of objects brighter than m_0 the reduced four-point angular function in the small separation approximation is

$$\mathcal{N}^4 u = n_0^4 a_0^{12} \int_0^\infty x^8 dx (\phi/F)^4 \int_{-\infty}^{\infty} dt du dv \eta, \quad (56.27)$$

where the arguments of η are computed as in equation (56.16). Of particular interest because it is simple is the model discussed in Section 55:

$$\eta = R_a[\xi_{12}\xi_{23}\xi_{34} + \text{sym.}] + R_b[\xi_{12}\xi_{13}\xi_{14} + \text{sym.}], \quad (56.28)$$

with ξ given by equation (56.20) and R_a and R_b constants. This form in equation (56.27) yields

$$u = a[w_{12}w_{23}w_{34} + \text{sym.}] + b[w_{12}w_{13}w_{14} + \text{sym.}], \qquad (56.29)$$

where w is given by equation (56.21) and (Fry and Peebles 1978)

$$\frac{a}{R_a} = \frac{b}{R_b} = \frac{\left(\int_0^\infty x^{11-3\gamma}\, dx\phi^4 a^{9+3\epsilon-3\gamma}/F\right)\left(\int_0^\infty x^2\, dx\phi/F\right)^2}{\left(\int_0^\infty x^{5-\gamma}\, dx\phi^2 a^{3+\epsilon-\gamma}/F\right)^3}. \qquad (56.30)$$

Some examples of the results of numerical integration of these equations are shown in Figure 56.1. The model for the luminosity function agrees with equation (52.15):

$$\begin{aligned} \Psi(M) &= C\,\text{dex}\,[\beta(M - M^*)], \qquad M > M^*; \\ &= C\,\text{dex}\,[\alpha(M - M^*)], \qquad M_0 < M < M^*; \\ &= 0, \qquad M < M_0; \\ \phi_0 &= \text{dex}\,[\alpha(M_0 - M^*)]. \end{aligned} \qquad (56.31)$$

It has been assumed that $\Psi \propto a^{-3}$ at fixed M, that is, galaxy evolution has been ignored. Conventional values for the parameters are

$$\begin{gathered} \alpha = 0.75, \qquad \beta = 0.25, \qquad \phi_0 = 0.01, \\ M^*_{pg} = -18.6 + 5 \log h. \end{gathered} \qquad (56.32)$$

In the curves labeled 1 in the figure these parameters are used together with density parameter $\Omega_0 = 1$, $\kappa = 3$ in the model for the K-correction (eq. 56.11), and $\epsilon = 0$ in the model ξ (eq. 56.20). In each of the other sets of curves just one parameter has been changed from these standard values. In model 2 $\beta = 0$, so there are no galaxies fainter than M^*. In model 3 $\phi_0 =$ 0.0001, so the luminosity function extends to brighter (and very rare) galaxies. In model 4, the K-correction has been changed to $5Z$. Only the two-point function for this model is shown; the other quantities all are very close to the results for model 1. The results change by $\lesssim 5$ percent if the density parameter is reduced to $\Omega_0 = 0.1$. Changing ϵ to -1.25, which makes ξ constant at fixed comoving separation x, makes the two-point function at the right side of the graph larger by 20 percent, with like changes in P/Q and a/R_a. In all the models $\gamma = 1.75$, close to the observed value.

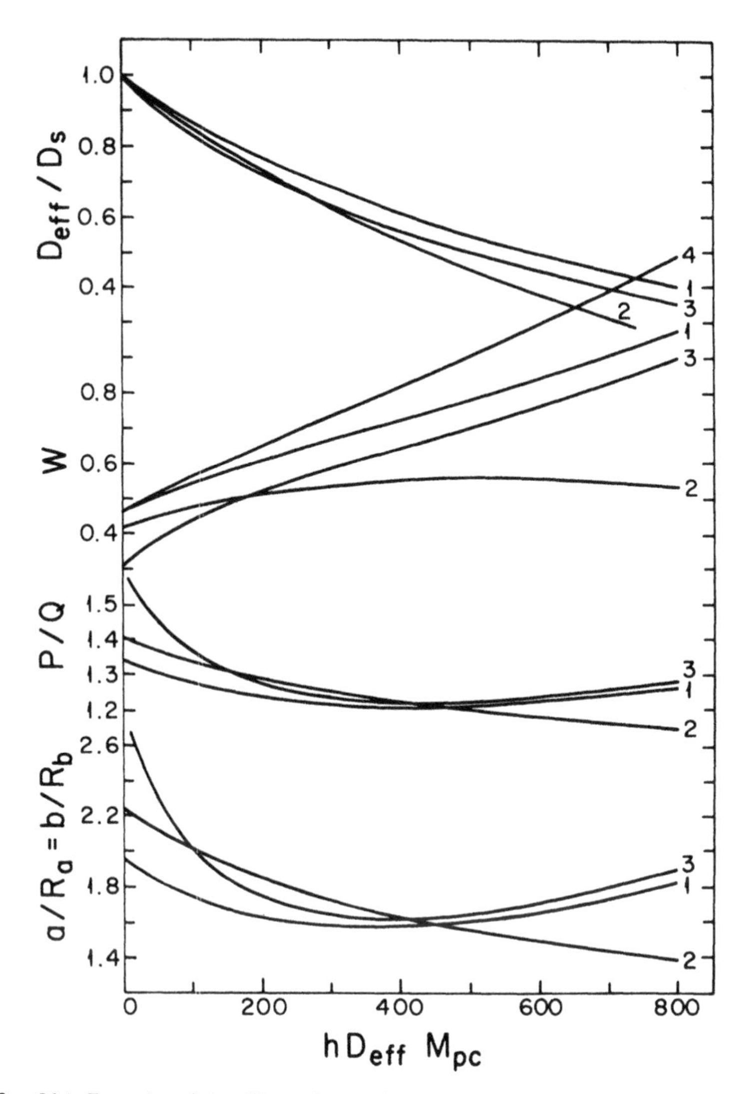

FIG. 56.1. Examples of the effects of expansion and curvature. The top graph is the ratio of two effective depths, D_{eff} and D_s (eqs. 56.33). W measures the deviation from the Euclidean scaling law for $w(\theta)$ (eq. 56.38). P/Q and a/R_a are the conversion factors for the models for the three- and four-point correlation functions. The independent variable is the effective catalog depth (eqs. 56.33 and 56.34). The labels on the curves correspond to different parameter choices: model 1 uses the parameters in equation (56.32) with $\Omega_o = 1$, $\kappa = 3$, $\epsilon = 0$, and $\gamma = 1.75$. One parameter is changed from this standard set in each of the other curves: in (2), $\beta = 0$, in (3), $\phi_o = 0.0001$, in (4), $\kappa = 5$.

The independent variable in the graph is the effective depth D_{eff} defined by the equations

$$\left(\frac{D_{eff}}{D_s}\right)^3 = \frac{a_0^3 \int_0^\infty x^2 dx F^{-1} \phi(M = m_0 - 5 \log [a_0 x(1+Z)] - 25 - \kappa Z)}{\int_0^\infty r^2 dr \phi(M = m_0 - 5 \log r)},$$

$$D_s = \text{dex}\,[0.2(m_0 - M^*) - 5]\ \text{Mpc}, \qquad (56.33)$$

where m_0 is the catalog limiting magnitude. Thus

$$D_{eff} = (\mathcal{N}(m_0)/n_0 E)^{1/3}, \qquad (56.34)$$

where $\mathcal{N}(m_0)$ is the expected density in the catalog (eq. 56.19) and $n_0 D_s^3 E$ (eqs. 50.12 and 51.5) is the density that would have been expected at limiting magnitude m_0 in the absence of expansion and curvature corrections. The parameter D_{eff} has proved to be a useful measure of survey depth. The limiting magnitude is the more fundamental quantity but in some cases it is not well known so it is more convenient to normalize to the density. Thus with the standard parameters listed above an estimate of the effective depth for the Zwicky catalog (Zwicky et al. 1961–1968) is (Groth and Peebles 1977)

$$m_0 = 14.9, \qquad \mathcal{N}_Z = 2050\ \text{sr}^{-1}, \qquad D_{eff} = 47.2h^{-1}\ \text{Mpc}, \quad (56.35)$$

so for the sample in the Jagellonian field (Rudnicki et al. 1973), where the density is

$$\mathcal{N}_J = 1.10 \times 10^6\ \text{sr}^{-1}, \qquad (56.36)$$

the effective depth would be

$$D_{eff}(J) = D_{eff}(Z)(\mathcal{N}_J/\mathcal{N}_Z)^{1/3} = 380\ h^{-1}\ \text{Mpc}. \qquad (56.37)$$

One reads from the top graph $D_{eff}/D_s = 0.62$, which with equation (56.33) would say the limiting magnitude is $m_0(pg) \sim 20.3$.

The second graph in the figure shows how the two-point correlation function varies with depth in the power law model for ξ. Equation (56.21) for $w(\theta)$ has been written as

$$w(\theta) = H_\gamma (B/D_{eff}^{\gamma}) W \theta^{1-\gamma}, \qquad (56.38)$$

so W measures the deviation from the Euclidean scaling law $w \propto D^{-\gamma}$ (eq. 52.8). The third and fourth graphs show the factors P/Q (eq. 56.26) and a/R_a (eq. 56.30) for the conversion from spatial to angular models for the three- and four-point functions.

One sees from the figure that the main problems at the moment are the luminosity function and K-correction. For a deep survey, $D_{eff} \gtrsim 500\ h^{-1}$ Mpc, the expected value of $w(\theta)$ scaled from shallower catalogs can change a factor ~2 under moderate variations of the luminosity function and K-correction. Until these are better understood the parameters in the cosmology, as Ω_0 and ϵ, probably will not much matter. Evolution of galaxy luminosities has not been discussed separately because that just changes the best effective K-correction, which is so uncertain anyway.

As has been discussed by Phillipps et al. (1978) a full treatment of the effect of redshift must take account of the fact that different morphological types of galaxies have different K-corrections. Also, one must expect that the detection efficiency (f in eq. 50.11) is a function of angular size as well as apparent magnitude, and that this function may be substantially different for different ways of detecting the galaxies. Of course, these points raise the further problem that morphology is known to be systematically different in compact clusters and the field. This effect is described in terms of correlation functions by Davis and Geller (1976). It means that the appearance of clustering at high redshift may be affected by the fact that galaxies in compact clusters tend to have larger K-corrections than do average galaxies (which would tend to make the apparent distribution too smooth by suppressing the numbers of galaxies visible in the tighter concentrations). It will likely be some time before we have a reliable treatment of all these effects. Meanwhile, in a very deep survey, $D_{eff} \sim 500 - 1000\ h^{-1}$ Mpc, the expected value of w scaled from shallower surveys is uncertain by a factor perhaps as large as 2. The numbers P/Q and a/R_a seem to be much less affected by these uncertainties.

57. Summary of numerical results

The galaxy two-point angular correlation function at small separations is well approximated by the power law model

$$w(\theta) = A\theta^{-\delta}. \tag{57.1}$$

The best estimates of the index δ come from the Zwicky and Lick galaxy catalogs. The Zwicky catalog (Zwicky et al. 1961) lists angular positions and magnitudes of galaxies brighter than $m \sim 15.5$ at declination $\delta > 0$. For this sample at $m = 15$ the effective depth defined in equation (56.34) is

$\sim 50\ h^{-1}$ Mpc. Peebles and Hauser (1974) found $w(\theta)$ in this catalog, and Soneira (1978b) examined the joint distribution in angular position and apparent magnitude. The Lick catalog (Shane and Wirtanen 1967) lists counts of galaxies brighter than $m \sim 18.9$ in $10'$ by $10'$ cells at $\delta > -23°$. The effective depth is $\sim 220\ h^{-1}$ Mpc. Limber (1954) and Neyman, Scott, and Shane (1956) first estimated the galaxy autocorrelation function here. Groth and Peebles (1977) found $w(\theta)$ for the new reduction of the data by Seldner et al. (1977). From the two samples Groth and Peebles found the power law index to be

$$\delta = 0.77 \pm 0.04. \tag{57.2}$$

The scaling of $w(\theta)$ with the depth of the sample is important as a test for possible systematic errors in the data due to variable local obscuration and the limited sample sizes. Groth and Peebles (1977) discussed the scaling relation for the Zwicky, Lick, and Jagellonian catalogs. The catalog of galaxies in the Jagellonian field (Rudnicki et al. 1973) goes to an effective depth $\sim 400\ h^{-1}$ Mpc in a 6° by 6° field. The sample subtends 40 by 40 h^{-1} Mpc, comparable to the Zwicky catalog but at a depth 8 times greater. The Durham group (Ellis 1980, Shanks et al. 1980) have extended the scaling test to $D_{eff} \sim 600\ h^{-1}$ Mpc. In the Zwicky and Lick samples relativistic effects are small, and so the expected variation of $w(\theta)$ with D_{eff} is insensitive to the luminosity function and K-correction. At fixed θ the values of w in the two samples differ by a factor of about 10 and this agrees with the computed ratio based on the ratio of number densities to 10 percent accuracy. The amplitude of w in the Jagellonian sample is 20 percent below the number computed from the shallower catalogs, and the amplitude of w in the deepest Durham samples is ~ 50 percent high. Both are within the uncertainties due to the luminosity function and sampling fluctuations. This success of the scaling relation shows that we have a reliable measure of the clustering of galaxies not seriously affected by variable obscuration in our galaxy or by the limited sizes of the samples.

The power law shape of $w(\theta)$ is reproduced by a power law model for the spatial two-point correlation function,

$$\xi = (r_0/r)^\gamma, \qquad \gamma = \delta + 1 = 1.77 \pm 0.04. \tag{57.3}$$

One can estimate the characteristic length r_0 from the amplitude of $w(\theta)$ and a model for the selection function $\phi(r)$. Where ξ is a power law, the relation is given by equations (52.8) and (52.9). The method was first used by Rubin (1954) and Limber (1954) and was applied in the power law model by Totsuji and Kihara (1969). Most recently Groth and Peebles (1977) found $r_0 = 4.7\ h^{-1}$ Mpc. The main uncertainty in this number comes from the selection function $\phi(r)$. That problem is avoided if redshifts are known for a fair sample of galaxies so the actual distribution in distances is known. Davis, Geller, and Huchra (1978) found the selection function for the bright galaxies at $m \lesssim 13$ where an almost complete set of redshifts is available; they obtained $r_0 = 2.4\ h^{-1}$ Mpc for the bright galaxies in the northern galactic hemisphere and $r_0 = 3.6$ for the bright galaxies in the southern hemisphere. However, the sample is shallow and it is doubtful that it is representative. The best present estimate of r_0 comes from the sample of Kirshner, Oemler, and Schechter (1979), who measured redshifts of the 166 galaxies brighter than apparent magnitude $J = 15$ in eight well-separated fields each about 4° by 4° square. The number of galaxies is small but fortunately the eight fields are well spread across the sky and so the clustering is fairly well sampled. The methods described in section 76 below yield (Peebles 1979b)

$$r_0 = 4.23 \pm 0.26\ h^{-1}\ \text{Mpc}. \tag{57.4}$$

The standard deviation is based on the scatter of results among the eight fields. The sample is biased against rich clusters so we might suspect that r_0 is low. There is in addition a highly uncertain sampling error from the limited depth of the survey. Since we do yet have a firmly established value for r_0, a round number will be adopted in the following calculations,

$$r_0 = 4h^{-1}\ \text{Mpc}. \tag{57.5}$$

The power law form $r^{-\gamma}$ for $\xi(r)$ might be expected to fail at small r because dynamic drag causes tight pairs of galaxies to move together. Ostriker and Turner (1979) found some evidence that the number of galaxy pairs at separations $r \lesssim 40\ h^{-1}$ kpc is less than would be expected from the power law model for $\xi(r)$. However, Gott and Turner (1979) showed that in the Zwicky catalog $w(\theta)$ is a good approximation to a power law at $10'' \lesssim \theta \lesssim 3°$ with $\delta = 0.79$ (eq. 57.2). The lower limit corresponds to projected separation

$$r_{mn} = \theta D_{eff} \sim 3\ h^{-1}\ \text{kpc}. \tag{57.6}$$

Lake and Tremaine (1980) used the counts of dwarf companions of spiral galaxies to test the size of $\xi(r)$ at small r and again concluded that the power law model is a good approximation to $r \sim r_{mn}$.

In the Lick sample there is a feature in $w(\theta)$ at projected separation

$$r_{mx} \sim 10\, h^{-1}\ \mathrm{Mpc}. \tag{57.7}$$

It appears that $\xi(r)$ is slightly larger than the power law at $r \sim r_{mx}$ and that the logarithmic derivative of ξ at $r > r_{mx}$ is greater than $\gamma \sim 1.8$ (Davis, Groth, and Peebles 1977, Fall and Tremaine 1978). A similar effect is found in the deep Durham samples (Ellis 1980, Shanks et al. 1980) but at the smaller projected separation $\sim 3\, h^{-1}$ Mpc. Further evidence of a break in ξ at about r_{mx} comes from attempts to make computer model distributions of galaxies that reproduce the visual texture of the galaxy distribution in the Lick sample (Soneira and Peebles 1978). If the power law variation of ξ extends much beyond r_{mx}, it puts too many galaxies in each clump, and so makes the model galaxy map look too patchy. In all three cases the effect is close to the noise and should be taken with caution. It may be an interesting coincidence that r_{mx} is comparable to r_0, which means the change of slope of ξ is at $\xi \sim 1$. The possible theoretical significance of the feature is discussed in Chapter IV.

The two-point correlation function is fairly reliably known to be greater than zero at $r \lesssim 15\, h^{-1}$ Mpc in the Lick sample and is lost in the noise at larger separations. Though structures do extend to larger scales, as in the enhanced density of galaxies around rich clusters that is detected to $r \sim 40$ h^{-1} Mpc, we have no believable evidence on whether the galaxy two-point correlation function $\xi(r)$ is positive or negative at $r \gtrsim 15\, h^{-1}$ Mpc.

The three-point correlation functions in the Zwicky, Lick, and Jagellonian samples have been discussed by Peebles and Groth (1975), Groth and Peebles (1977), and Peebles (1975). In each sample the power law model in equation (56.25) gives a good fit to the variation of z with the triangle size and shape. The model is tested from $r_1 \sim 50\, h^{-1}$ kpc on the short side to $r_3 \sim 5\, h^{-1}$ Mpc on the long side. The parameter P in the model for z should be very nearly independent of catalog depth (fig. 56.1). The values of P derived from different parts of the Lick catalog are consistent to $\sim$25 percent, and the values from the Zwicky, Lick, and Jagellonian samples agree to $\sim$50 percent. Groth and Peebles (1977) concluded that the parameter Q in the model for the spatial function ζ (eq. 54.17) is

$$Q = 1.29 \pm 0.21. \tag{57.8}$$

In the complete redshift sample of bright galaxies at galactic latitude $b > +30°$, apparent magnitude $B < 13.2$ and redshift $>$1500 km s^{-1} (Huchra, Davis, and Geller 1979) the parameter is (Peebles 1980a)

$$Q = 0.80 \pm 0.07. \tag{57.9}$$

The error here is internal and Q probably is low because the sample is chosen to contain no prominent clusters.

The four-point function u has been estimated for the Lick and Zwicky samples (Fry and Peebles 1978). The estimates for the Lick sample are not very accurate because u is small and quite close to the systematic errors; the estimates for the Zwicky sample are even cruder because although u is much larger, the noise is larger still. The power law model in equation (56.29) gives an adequate description of the estimates for both catalogs, but that is not a very serious test of the model. The best fit parameters in the model are

$$\begin{aligned} a + b/3 &= 6.4 \pm 0.6, \qquad & a - b/3 &= 1.8 \pm 1.1, \\ a &= 4.1 \pm 0.7, \qquad & b &= 6.8 \pm 1.7. \end{aligned} \tag{57.10}$$

The combination $a + b/3$ measures the overall size of u when the separations of the four points all are about comparable (because there are 12 a-terms, 4 b-terms) and so is better determined than the combination $a - b/3$. It appears that the assumptions $a = 0$ or $b = 0$ are ruled out. Two other interesting cases $a = b$ and $a = b/3$ both seem to be allowed. The conversion to the model for the spatial function η gives

$$\begin{aligned} R_a + R_b/3 &= 4.0 \pm 0.6, \qquad & R_a - R_b/3 &= 1.1 \pm 0.7, \\ R_a &= 2.5 \pm 0.6, \qquad & R_b &= 4.3 \pm 1.2 \end{aligned} \tag{57.11}$$

58. Power spectrum of the extragalactic light

If the galaxy correlation functions measure the distribution of all luminous matter they can be used to predict the fluctuations in the extragalactic contribution to the brightness of the sky, in much the same way Ambartsumian (1944) and Chandrasekhar and Münch (1952) discussed the fluctuations in the brightness of the Milky Way. Of considerable interest is the point made by Gunn (1965) and Shectman (1973): if the dimensionless spatial autocorrelation function ξ of the luminosity density is known from galaxy clustering, then from the autocorrelation function of the extragalactic light one can derive the mean space luminosity density. This number is important in discussions of the mean mass density, and it is not very well known (*PC*, Chapter IVb; Dube, Wickes, and Wilkinson 1977). Shectman (1974) has measured the power spectrum of the extragalactic light. The relation to the spatial correlation function was discussed by Geller (1975).

As in Section 56 the line element will be taken to be the Robertson-Walker form (eq. 56.1). The flux of energy from a source at redshift Z is given by equation (56.9). The flux per unit frequency interval is

$$f(\nu_0) = \frac{\mathcal{L}(\nu)}{4\pi x^2}\frac{a}{a_0^{\,3}}, \qquad \nu = \nu_0\frac{a_0}{a} = \nu_0(1 + Z). \tag{58.1}$$

The observed frequency is ν_0, corresponding to frequency ν at the source; the factor a_0/a in the conversion from equation (56.9) to equation (58.1) comes from the redshift of the bandwidth.

A convenient way to proceed is to write the flux received from the direction Ω in the element of solid angle $\delta\Omega$ as

$$i(\Omega, \nu_0)\delta\Omega = \Sigma n_l f_l(\nu_0), \tag{58.2}$$

where $n_l = 1$ if there is a galaxy in $\delta\Omega$ at coordinate distance x_l in the range of proper distance $c\delta t$, the galaxy luminosity being $\mathcal{L}(\nu)_l$ in the range $\delta\mathcal{L}$; and $n_l = 0$ otherwise (§ 36). The factor f_l is the flux received if there is a galaxy. The ensemble average value of n_l is

$$\langle n_l\rangle = \delta V\delta\mathcal{L}\Phi(t, \nu, \mathcal{L}), \qquad \delta V = a^2x^2c\delta t\delta\Omega, \tag{58.3}$$

where Φ is the luminosity function at epoch t and frequency ν, as in equations (49.1) and (56.14), with the variable changed from magnitude M to $\mathcal{L}$. The mean value of equation (58.2) is then

$$\begin{gathered}\langle i(\nu_0)\rangle = \frac{c}{4\pi}\int_0^{t_0} dt\, n\langle\mathcal{L}(\nu, t)\rangle(a/a_0)^3, \\ n\langle\mathcal{L}\rangle = \int \mathcal{L}\,d\mathcal{L}\,\Phi.\end{gathered} \tag{58.4}$$

The second equation defines the mean luminosity per object where n is the mean proper number density of objects at epoch t. The first equation is the usual expression for the brightness of the sky as an integral over the luminosity density $j_\nu = n\langle\mathcal{L}\rangle$ (for example, *PC*, chapter IVb).

The mean of the product of two different n's in equation (58.2) is

$$\langle n_1 n_2\rangle = \delta V_1\delta V_2\delta\mathcal{L}_1\delta\mathcal{L}_2[1 + \xi(r_{12}, t)]\Phi(\mathcal{L}_1)\Phi(\mathcal{L}_2). \tag{58.5}$$

It is assumed that the distribution in $\mathcal{L}$ is statistically independent of position (§ 51) and that the difference of cosmic times at the two points in

ξ can be ignored (§ 56). Then in the small separation approximation (§ 50), the autocorrelation function of the brightness of the extragalactic light is, for $\theta \neq 0$,

$$C(\theta) = \langle \delta i(\Omega_a, \nu_0) \delta i(\Omega_b, \nu_0) \rangle$$

$$= \frac{c}{16\pi^2} \int_0^{t_0} dt \int_{-\infty}^{\infty} du\, \xi(r, t) \langle \mathcal{L}(\nu, t) \rangle^2 n^2 (a/a_0)^6, \quad (58.6)$$

$$r^2 = u^2 + (ax\theta)^2.$$

Here $\delta i(\Omega_a, \nu_0)$ is the difference of the brightness in the direction Ω_a from the mean (eq. 58.4). In the power law model,

$$\xi = B(t)[u^2 + (ax\theta)^2]^{-\gamma/2}, \quad (58.7)$$

the u integral can be evaluated, giving

$$C(\theta) = K\theta^{1-\gamma},$$

$$K = \frac{cH_\gamma}{16\pi^2} \int_0^{\infty} dt\, (ax)^{1-\gamma} [j(t, \nu(t))]^2 (a/a_0)^6 B(t). \quad (58.8)$$

The dimensionless integral H_γ is evaluated in equation (52.9) and the mean luminosity density at epoch t is

$$j(t, \nu) = n \langle \mathcal{L} \rangle. \quad (58.9)$$

A useful approximation to equation (58.8) is obtained if the observed frequency ν_0 is in the visible part of the spectrum. Then the integrand sharply decreases with increasing $t_0 - t$, which is increasing distance from us because the spectrum j_ν for a galaxy rapidly decreases toward the blue. This means that, to a good approximation, we can ignore the time variation of $a(t)$ and B and write the distance from us in terms of the redshift,

$$c(t_0 - t) \simeq a_0 x \simeq cZ/H_0, \quad (58.10)$$

where H_0 is the present value of Hubble's constant. In this approximation equation (58.8) becomes

$$C(\theta) = K\theta^{1-\gamma},$$

$$K = \frac{BH_\gamma}{16\pi^2} \left(\frac{c}{H_0}\right)^{2-\gamma} \int_0^{\infty} dZ\, Z^{1-\gamma} [j(t_0, \nu_0(1 + Z))]^2. \quad (58.11)$$

The power spectrum, in the normalization used by Shectman (1974), is

$$\begin{aligned} s(\eta) &= (2\pi)^{-2} \int C(\theta) e^{i\eta\cdot\theta}\, d^2\theta \\ &= (2\pi)^{-1} \int \theta\, d\theta\, C(\theta) J_0(\eta\theta), \end{aligned} \tag{58.12}$$

where J_0 is the Bessel function of zero order (eq. 46.39). For the power law of equation (58.11) the spectrum is

$$s(\eta) = \frac{K}{2\pi} \eta^{\gamma-3} \int_0^\infty x^{2-\gamma}\, dx\, J_0(x) = 0.12\, K\eta^{\gamma-3}. \tag{58.13}$$

This is the wanted expression for the power spectrum of the extragalactic light in the power law model for ξ.

It might be noted that equation (58.6) for $C(\theta)$ does not apply at $\theta = 0$ because as in Section 36 there is an extra contribution from the $n_1{}^2$ terms. This adds a delta function to $C(\theta)$ (or more accurately a spike with angular size typical of the galaxies contributing to the light) and so adds a constant term to $s(\eta)$. Following the above method and using $\langle n_1{}^2 \rangle = \langle n_1 \rangle$, one finds that this constant is

$$s_0 = \frac{c}{2^6 \pi^4} \int_0^\infty dt \frac{a^4}{a_0{}^6 x^2} n \langle \mathcal{L}^2 \rangle. \tag{58.14}$$

The last factor is the mean square luminosity of a galaxy. The integral diverges as x^{-1}: the main effect is from the nearest galaxies in the sample. This contribution to the power spectrum would seem to be uninteresting because of the problem of separating it from the white noise from stars.

To evaluate K in equation (58.11), we need the dimensionless integral

$$I = \int_{Z_c}^\infty dZ\, Z^{1-\gamma} [\, j(\nu_0(1 + Z))/j(\nu_0)]^2. \tag{58.15}$$

The lower limit is greater than zero if the nearby bright galaxies are removed from the sample: in Shectman's measurements galaxies brighter than $m_R \sim 18$ were removed, corresponding roughly to a cutoff at $Z_c \sim 0.15$. The spectrum $j(\nu)/j(\nu_0)$ might be expected to have the same shape as for a giant galaxy. The measurements of Oke and Sandage (1968) with $\lambda_0 = 6500$ A give

$$I = 0.73. \tag{58.16}$$

Half of this integral comes from $0.15 \leq Z \lesssim 0.25$, and there is negligible contribution beyond $Z \sim 0.6$. The approximation $Z \ll 1$ leading from equation (58.8) to equation (58.11) thus is somewhat crude but the results should be good to better than a factor of 2.

Equations (58.11), (58.13), and (58.16) with equation (57.5) for $B(t_0)$ give

$$s(\eta) = 1.5 \times 10^{48} h^{-2} \eta^{-1.23} [j(\nu_0, t_0)]^2 \qquad \mathrm{erg}^2 \mathrm{cm}^{-4} \mathrm{s}^{-2} \mathrm{Hz}^{-2} \mathrm{ster}^{-2} (\mathrm{rad})^2, \tag{58.17}$$

where the units of the wave number η are radians of phase per radian of angle. Shectman found

$$s \sim 4.7 \times 10^{-49}, \qquad \eta = 1000, \tag{58.18}$$

in the units of equation (58.17). This yields

$$\begin{aligned} j(\nu_0) &= 4 \times 10^{-47} h \,\mathrm{ergs\,cm}^{-3} s^{-1} \mathrm{Hz}^{-1} \\ &= 2 \times 10^8 h \mathcal{L}_\odot \mathrm{Mpc}^{-3} \end{aligned} \tag{58.19}$$

at 6500 A. Standard estimates based on galaxy counts are in the range (*PC*. chapter IV)

$$j \sim 1 - 3 \times 10^8 h \mathcal{L}_\odot \mathrm{Mpc}^{-3}. \tag{58.20}$$

The agreement of equations (58.19) and (58.20) is remarkable, certainly as good as one could have hoped for.

At the characteristic depth $Z \sim 0.25$ the angular wavelength $2\pi/\eta$ for $\eta = 1000$ (eq. 58.18) subtends wavelength

$$\lambda = (2\pi/\eta) D = 4.7 h^{-1} \mathrm{Mpc}. \tag{58.21}$$

If ξ falls below the power law $r^{-\gamma}$ at $r \sim 10 h^{-1}$ Mpc as it appears to do in the Lick sample, one would look for a break in the spectrum at $\eta \sim 500$. No such effect is seen in Shectman's results, but it could be lost in the large noise at small η. At $\eta \gtrsim 1000$ the spectrum should vary as $\eta^{-1.23}$. Shectman's estimate decreases with increasing η appreciably more rapidly than that. It is not clear whether this is a problem: at $\eta \gtrsim 1000$ the correction for the white noise background from stars is large and could have been overestimated.

The agreement of j from Shectman's measurement and from the conventional galaxy counts is an important check of consistency of three points: the assumption that the covariance function for the luminosity density j is the same as the galaxy correlation function ξ; the normalization of ξ at $r \sim 4h^{-1}$Mpc; and the conventional estimate of the mean space luminosity density. If appreciable amounts of luminous matter have been missed, this missed part has to be distributed a good deal more uniformly than are the galaxies. An improved measurement capable of testing the variation of s with η would be of considerable interest and would appear to be difficult.

59. Moments of the Number of Neighbors

An immediate and interesting application of the numerical results in Section 57 is to the calculation of the number of neighbors of a galaxy. The mean number of galaxies within distance r of a randomly chosen one is given by equation (36.23),

$$\langle N \rangle_p = nV + n\int_0^r \xi \, dV, \qquad V = \frac{4}{3}\pi r^3, \tag{59.1}$$

where n is the mean number density. In the power law model the integral is

$$\int_0^r \xi \, dV = V(r_0/r)^\gamma K_1, \qquad K_1 = 3/(3-\gamma), \qquad \xi(r) = (r_0/r)^\gamma. \tag{59.2}$$

The mean square number of neighbors within distance r of a galaxy is given by equation (36.25). Two more integrals are needed. The first is

$$\begin{aligned} \int_0^r dV_1 dV_2 \, \xi(|\mathbf{r}_1 - \mathbf{r}_2|) &= V^2 (r_0/r)^\gamma J_2, \\ J_2 &= \left(\frac{3}{4\pi}\right)^2 \int_0^1 d^3x_1 d^3x_2 / x_{12}^\gamma \\ &= 72/[(3-\gamma)(4-\gamma)(6-\gamma)2^\gamma]. \end{aligned} \tag{59.3}$$

The dimensionless integral is given by Peebles and Groth (1976). The integral over the model (54.17) for the three-point function is

$$\begin{aligned} \int_0^r dV_1 dV_2 \, \zeta(x_1, x_2, |\mathbf{x}_1 - \mathbf{x}_2|) &= Q(K_1^2 + 2K_2)(r_0/r)^{2\gamma} V^2, \\ K_2 &= \left(\frac{3}{4\pi}\right)^2 \int_0^1 d^3x_1 d^3x_2 (x_1 x_{12})^{-\gamma}. \end{aligned} \tag{59.4}$$

Groth (1977) has reduced K_2 to the form ($\gamma < 2$)

$$[2(2-\gamma)^2(3-\gamma)/9]K_2 =$$
$$F(\gamma-3, 2-\gamma; 3-\gamma; -1) - F(\gamma-3, 2-\gamma; 3-\gamma; 1)$$
$$-\frac{1}{(4-\gamma)}F(\gamma-4, 2-\gamma; 3-\gamma; -1) \tag{59.5}$$
$$+\frac{1}{(4-\gamma)}F(\gamma-4, 2-\gamma; 3-\gamma; 1),$$

where F is the hypergeometric function. Some of his numerical results are listed in Table 59.1. For the observed power law index,

$$K_2 = 5.02, \qquad \gamma = 1.77. \tag{59.6}$$

The second moment is then

$$\begin{aligned}\langle N^2\rangle_p &= \langle N\rangle_p + (nV)^2[1 + (J_2 + 2K_1)(r_0/r)^\gamma \\ &\quad + Q(2K_2 + K_1{}^2)(r_0/r)^{2\gamma}] \\ &= \langle N\rangle_p + (nV)^2[1 + 6.70(r_0/r)^{1.77} + 20.6(r_0/r)^{3.54}],\end{aligned} \tag{59.7}$$

for $\gamma = 1.77$.

The mean of the cube of the number of neighbors is given by equation (36.28). In the power law models for ξ, ζ, and η the result is

$$\begin{aligned}\langle N^3\rangle_p = {}&(nV)^3(r_0/r)^{3\gamma}[6R_a(K_{3a} + K_1K_2) + R_b(K_1{}^3 + 3K_{3b})] \\ &+ (nV)^3(r_0/r)^{2\gamma}[3QJ_3 + 3K_1J_2] + C,\end{aligned} \tag{59.8}$$

TABLE 59.1
EVALUATION OF K_2*

γ	K_2
0.0	1.0
0.5	1.28
1.0	1.88
1.5	3.31
2.0	7.82
2.5	34.0

*Eq. 59.4.

where C is a sum over the lower moments,

$$C = 3\langle N^2\rangle_p(1 + nV) - \langle N\rangle_p[2 + 3nV + 3(nV)^2] + (nV)^3, \qquad (59.9)$$

and the dimensionless integrals are

$$K_{3a} = \left(\frac{3}{4\pi}\right)^3 \int_0^1 d^3x_1 d^3x_2 d^3x_3 (x_1 x_{12} x_{23})^{-\gamma} = 9.5,$$
$$K_{3b} = \left(\frac{3}{4\pi}\right)^3 \int_0^1 d^3x_1 d^3x_2 d^3x_3 (x_1 x_{12} x_{13})^{-\gamma} = 10.3, \qquad (59.10)$$
$$J_3 = \left(\frac{3}{4\pi}\right)^3 \int_0^1 d^3x_1 d^3x_2 d^3x_3 (x_{12} x_{23})^{-\gamma} = 3.38.$$

The values are the results of numerical integration for $\gamma = 1.77$. With these numbers equation (59.8) is

$$\langle N^3\rangle_p = (nV)^3[520(r_0/r)^{5.31} + 26(r_0/r)^{3.54} + 1] + C. \quad (59.11)$$

Some results are listed in Table 59.2. The calculation assumes magnitude and position are uncorrelated. If this is so, we can choose a magnitude limit M_0, estimate the mean density of galaxies brighter than M_0, and with this n compute from the equations the moments of the numbers of neighbors brighter than M_0 around a galaxy. The numbers are most reliable if we concentrate on galaxies at $M \sim M^*$ because they are most

TABLE 59.2
MOMENTS OF THE NUMBER OF NEIGHBORS

$h^{-3}n$ (Mpc^{-3})	hr (Mpc)	nV	$\langle N\rangle$	$\langle N^2\rangle^{1/2}$	$\langle N^3\rangle^{1/3}$	M_2*	M_3†
0.03	5.0	16	42	61	94	1990	—
0.03	1.0	0.13	3.7	7.0	12	35	1300
0.03	0.1	1.3×10^{-4}	0.21	0.60	1.0	0.32	0.79
0.01	5.0	5.2	14	20	32	230	—
0.01	1.0	0.04	1.2	2.5	4.3	4.8	58
0.01	0.1	4×10^{-5}	0.07	0.30	0.51	0.082	0.12
0.10	0.1	4×10^{-4}	0.70	1.6	2.6	1.91	14

*$M_2 = \langle(N - \langle N\rangle)^2\rangle$.
†$M_3 = \langle(N - \langle N\rangle)^3\rangle$.

common in a catalog selected by apparent magnitude (and they are thought to be the galaxies that contain most of the mass). The first three entries in the table have $n = 0.03\ h^3\ \mathrm{Mpc}^{-3}$, corresponding roughly to $M \lesssim M^*$. Also listed are some numbers for brighter and fainter magnitude limits.

For $n = 0.03\ h^3\ \mathrm{Mpc}^{-3}$, a galaxy generally has more than one bright neighbor at $r \leq 1\ h^{-1}$ Mpc, while a randomly placed sphere of this radius contains such a galaxy with 13 percent probability. The scatter in the number of neighbors is considerably larger than Poisson: given $\langle N \rangle = 3.7$ the expected second and third central moments of N both are 3.7 for a Poisson distribution, and the last two columns of the table indicate the moments are considerably larger than this. If r is reduced to $100\ h^{-1}$ kpc, the mean number of bright neighbors, $M \lesssim M^*$, is about 0.2, compared to 1.3×10^{-4} for a randomly placed sphere. The rms number of neighbors is 0.6 and the mean from the third moment, $\langle N^3 \rangle^{1/3}$, is about unity. A reasonable fit to these numbers is given by the following distribution in the number of neighbors at $r < 100\ h^{-1}$ kpc:

$$
\begin{gathered}
P_0 = 0.86, \qquad P_1 = 0.09, \\
P_2 = 0.03, \qquad P_3 = 0.02, \\
P_N = 0, \qquad N > 3; \qquad r < 100\ h^{-1}\ \mathrm{kpc}, \qquad M \lesssim M^*.
\end{gathered}
\tag{59.12}
$$

60. Model for P_N

Suppose a sphere of radius r is placed at random. What is the probability that it contains some chosen number of galaxies or some chosen amount of mass? One way to compute this was described in Section 39 but it is not useful here because for interesting values of n and r the series strongly oscillates. Another approach is a conjecture based on the moments of counts in a randomly placed sphere.

The sphere will have radius

$$
r < r_0, \tag{60.1}
$$

so $\xi(r) > 1$, and it will be supposed that the correlation functions apply to faint enough galaxies (or even nucleons; §63) so that the mean number density satisfies

$$
nV(r_0/r)^\gamma > 1, \qquad V = \tfrac{4}{3}\pi r^3, \tag{60.2}
$$

for some appreciable range of $r < r_0$. This eliminates the shot noise from the moments.

The second moment of the count of objects within the randomly placed sphere is (eq. 36.6)

$$\begin{aligned}\langle N^2\rangle &= nV + (nV)^2 + n^2\int_0^r dV_1 dV_2 \xi_{12}\\ &= nV + (nV)^2 + (nV)^2 (r_0/r)^\gamma J_2,\end{aligned} \tag{60.3}$$

where the second line follows from the power law model for ξ and the number J_2 is evaluated in equation (59.3). Under the conditions of equations (60.1) and (60.2) the dominant term is the last one:

$$\langle N^2\rangle = J_2 (nV)^2 (r_0/r)^\gamma. \tag{60.4}$$

In the same way one finds from equations (36.13) and (36.20)

$$\begin{aligned}\langle N^3\rangle &= 3J_3 Q(nV)^3 (r_0/r)^{2\gamma},\\ \langle N^4\rangle &= 12(J_{4a}R_a + 1/3\, J_{4b}R_b)(nV)^4 (r_0/r)^{3\gamma},\end{aligned} \tag{60.5}$$

where the dimensionless integrals are

$$\begin{aligned}J_3 &= \left(\frac{3}{4\pi}\right)^3 \int_0^1 dV_1 dV_2 dV_3 (x_{12}x_{23})^{-\gamma} = 3.38,\\ J_{4a} &= \left(\frac{3}{4\pi}\right)^4 \int_0^1 dV_1 dV_2 dV_3 dV_4 (x_{12}x_{23}x_{34})^{-\gamma} = 6.3,\\ J_{4b} &= \left(\frac{3}{4\pi}\right)^4 \int_0^1 dV_1 dV_2 dV_3 dV_4 (x_{12}x_{13}x_{14})^{-\gamma} = 6.3.\end{aligned} \tag{60.6}$$

The integrals are evaluated by numerical integration for $\gamma = 1.77$. With the parameters listed in equations (57.5), (57.8) and (57.11) the moments are

$$\begin{aligned}\langle N\rangle &= nV,\\ \langle N^2\rangle &= 1.82\,(nV)^2 (r_0/r)^\gamma,\\ \langle N^3\rangle &= 13.1\,(nV)^3 (r_0/r)^{2\gamma},\\ \langle N^4\rangle &= (300 \pm 45)\,(nV)^4 (r_0/r)^{3\gamma}.\end{aligned} \tag{60.7}$$

These expressions show an interesting regularity: apart from the numerical factor each moment is a fixed multiple of the next lower one. It is easy to find a model for the distribution of N that reproduces this. Suppose the sphere is empty with probability $1 - \epsilon$ and contains the fixed number N_0 with probability ϵ. Then the nth moment is

$$\langle N^n \rangle = \epsilon N_0^n. \tag{60.8}$$

this approximates equations (60.7) with

$$N_0 \sim nV(r_0/r)^\gamma, \tag{60.9}$$

though it misses the rapid increase of the numerical factors. We can remedy this and make the model more realistic by introducing some dispersion around N_0. Since N_0 is supposed to be large (eq. 60.2), we can take P_N to be a function of the continuous variable N. The model

$$P(N) = (1 - \epsilon)\delta(N) + \frac{\epsilon}{2N_0} e^{-(N/N_0)^{1/2}} \tag{60.10}$$

gives the moments

$$\langle N^n \rangle = (2n + 1)!\, \epsilon N_0^n, \tag{60.11}$$

and if

$$\langle N \rangle = 6\epsilon N_0 = nV, \qquad N_0 = \frac{nV}{10}\left(\frac{r_0}{r}\right)^\gamma, \tag{60.12}$$

we find

$$\begin{aligned} \langle N^2 \rangle &= 2(nV)^2(r_0/r)^\gamma, \\ \langle N^3 \rangle &= 8.4(nV)^3(r_0/r)^{2\gamma}, \\ \langle N^4 \rangle &= 60(nV)^4(r_0/r)^{3\gamma}. \end{aligned} \tag{60.13}$$

The numerical coefficient for the fourth moment is still too small and could be improved by making the dispersion around N_0 broader but the refinement does not seem interesting.

The correlation functions thus suggest that if a sphere of radius $r \lesssim 4$ h^{-1} Mpc is placed at random, the probability that it contains little or no mass is high:

$$P = 1 - \epsilon \cong 1 - \frac{5}{3}\left(\frac{r}{r_0}\right)^{\gamma} \cong 1 - 0.14\,(\mathrm{hr})^{1.77}. \tag{60.14}$$

That is, it usually does not fall on a group or cluster. If it does happen to fall on a lump of matter, it contains mass typically on the order of $N_0(r)$ though with considerable dispersion around this characteristic value. With the mean mass density

$$\rho = 1.9 \times 10^{-29}\Omega h^2 \mathrm{g\ cm}^{-3}, \tag{60.15}$$

where Ω is the density parameter belonging to the galaxies, the characteristic mass (eq. 60.12) is

$$M_0 = \frac{\rho V}{10}\left(\frac{r_0}{r}\right)^{\gamma} \sim 1.4 \times 10^{12}\Omega h^{-1}(\mathrm{hr})^{1.23}\mathrm{M}_{\odot}, \tag{60.16}$$

with r in megaparsecs. For $r = 100\ h^{-1}$ kpc the mass is

$$M_0 \sim 8 \times 10^{10}\Omega h^{-1}\mathrm{M}_{\odot}, \qquad r = 100h^{-1}\mathrm{kpc}. \tag{60.17}$$

At $r = 10h^{-1}$ kpc, the conventional size of a galaxy, we find

$$M_0 \sim 5 \times 10^{9}\Omega h^{-1}\mathrm{M}_{\odot}, \qquad r = 10h^{-1}\mathrm{kpc}. \tag{60.18}$$

The correlation functions thus suggest the following picture (Peebles 1973d). The matter distribution in effect is viewed with resolution r fixed by the size of the randomly placed sphere. The matter is found in lumps of typical size and mass r and $M_0(r)$. When the resolution is changed to $r' < r$, these lumps must tend to break up into sublumps, for the distribution again appears in lumps of typical size r' and mass $M_0(r') < M_0(r)$. An interesting coincidence is that if the resolution is adjusted all the way to the nominal size of a galaxy, $r \sim 10h^{-1}$ kpc, the lump mass comes to roughly that of a galaxy. The tempting speculation is that this is telling us galaxies are only part of a fairly continuous clustering phenomenon (§ 4B). Some other aspects of this speculation are mentioned in the following sections.

61. Clustering models

Following the program of Neyman and Scott (1952) described in Section 40, we can seek a prescription for placing galaxies in clumps that reproduces the observed statistics of the galaxy distribution. Two patterns are discussed here. One, which serves mainly as a foil, can reproduce ξ but then predicts three- and four-point functions ζ and η that conflict with the observations. The second, a clustering hierarchy, reproduces ξ, ζ, and η and so must be closer to reality. This is particularly interesting because, as discussed in Section 26, under fairly natural initial conditions in an expanding world model the mass distribution would be laid down in a clustering hierarchy.

A. Power law cluster model

The clumps are assumed to be spherically symmetric, the galaxy number density varying as a power of distance from the clump center,

$$\begin{aligned} n(r) &= nAr^{-\epsilon}, \qquad r \leq R, \\ &= 0, \qquad r > R, \qquad 1.5 < \epsilon < 3, \end{aligned} \tag{61.1}$$

where n is the large-scale mean density and the constants ϵ, A, and R are the same for all clumps. The limits on the power law index ϵ assure convergence of the integrals at large and small r. The clumps are placed uniformly at random, so $\xi(r)$ is determined by the mean number of pairs of galaxies at separations r to $r + \delta r$ in a clump:

$$\begin{aligned} 2\delta N_p &= 8\pi^2 r^2 \delta r \int r_1^2 dr_1 d\cos\theta\, n(r_1) n(r_2), \\ r_2^2 &= r_1^2 + r^2 - 2rr_1 \cos\theta. \end{aligned} \tag{61.2}$$

If $3/2 < \epsilon < 3$ the main contribution to the integral is at $r_1 \sim r_2 \sim r$. When $r \ll R$, the boundary can be ignored and the integral becomes

$$\begin{aligned} 2\delta N_p &= 8\pi^2 r^{5-2\epsilon} \delta r n^2 A^2 I, \\ I &= \int_0^\infty x^{2-\epsilon}\, dx \int_{-1}^{+1} d\mu (x^2 + 1 - 2x\mu)^{-\epsilon/2}, \end{aligned} \tag{61.3}$$

where $x = r_1/r$. The correlation function is given by equation (40.5) (Peebles 1974b, McClelland and Silk 1977):

$$\xi(r) = \tfrac{1}{2}(3 - \epsilon) I A r^{3-2\epsilon} R^{\epsilon-3}, \qquad r \ll R. \tag{61.4}$$

This matches the observed power law (eq. 57.3) if

$$2\epsilon - 3 = \gamma = 1.77, \qquad \epsilon = 2.38. \tag{61.5}$$

If the clump mass $\propto A$ is different in each clump, then equation (40.5) says A in equation (61.4) should be replaced with $\langle A^2 \rangle / \langle A \rangle$.

The three-point correlation function is determined by the mean count of triplets in a clump. We can see how this scales with the size of the triangle without bothering to write down the integral by noting that if the integral converges (here we need $1 < \epsilon < 3$), the main contribution is at distances from the clump center on the order of the triangle size r. The count of triplets whose range of sizes and shapes define the six-dimensional element $(\delta V)^2$ is then

$$\delta N_t \sim (nAr^{3-\epsilon})(n^2A^2r^{-2\epsilon}(\delta V)^2). \tag{61.6}$$

The first factor is the number of galaxies within distance $\sim r$ of the center. The second factor is the probability for each such galaxy that the triangle is completed by two other galaxies each at distance $\sim r$ from the center. With equation (40.14) this gives (Peebles and Groth 1975)

$$\zeta \sim A^2R^{\epsilon-3}r^{3-3\epsilon}, \qquad r \ll R. \tag{61.7}$$

The ratio of this to the square of the two-point function at $r \sim$ triangle size is

$$\zeta/\xi^2 \sim (R/r)^{3-\epsilon} \propto r^{-0.6}. \tag{61.8}$$

In the same way one finds for the four-point function

$$\eta/\xi^3 \sim (R/r)^{6-2\epsilon} \propto r^{-1.2}, \qquad r \ll R. \tag{61.9}$$

Since in the small separation approximation

$$w(\theta) \sim \theta\xi(\theta D), \qquad z \sim \theta^2\zeta(\theta D), \qquad u \sim \theta^3\eta(\theta D), \tag{61.10}$$

The above results say

$$\begin{aligned} z &\propto \theta^{5-3\epsilon}, \qquad & z/w^2 &\propto \theta^{\epsilon-3}, \\ u &\propto \theta^{6-4\epsilon}, \qquad & u/w^3 &\propto \theta^{2\epsilon-6}. \end{aligned} \tag{61.11}$$

These relations apply when the ratios of relative positions in the configuration of three or four points are held fixed.

These relations are not consistent with the observations. In the Lick sample z/w^2 is constant to $\sim$25 percent accuracy when θ is in the range of 0.15° to 2°, while the model predicts that z/w^2 should vary by a factor of 5 (Groth and Peebles 1977). In the Zwicky sample the variation of z with triangle size is found to be well approximated by the power law model (Peebles 1979a)

$$z \propto \theta^{-\nu}, \qquad \nu - 2\delta = -0.028 \pm 0.079, \tag{61.12}$$

where $\delta = \gamma - 1$. The model predicts

$$\nu - 2\delta = 3 - \epsilon = 0.6, \tag{61.13}$$

some eight standard deviations high. For the four-point function for the Lick sample the fit to a power law model yields

$$u \propto \theta^{-\mu}, \qquad \mu = 2.40 \pm 0.18, \qquad \delta = 0.741 \pm 0.035, \tag{61.14}$$

so

$$\mu - 3\delta = 0.18 \pm 0.21. \tag{61.15}$$

The model predicts

$$\mu - 3\delta = 6 - 2\epsilon = 1.2, \tag{61.16}$$

well above what is observed.

B. Continuous clustering hierarchy

In a clustering hierarchy matter is concentrated in lumps; the lumps are collected into groups; the groups are in supergroups, and so on through some range of cluster masses. The clusters on each level of the hierarchy could have very different sizes so a group is clearly distinguished from a supergroup and so on, or it could be arranged as in the following examples that the spectrum of clustering is continuous. The galaxy distribution shares features with both cases: Abell clusters are rare extreme concentrations that are clearly distinct objects, but the galaxy correlation functions that measure the general clustering pattern are well approximated by a continuous scale-invariant clustering hierarchy.

One way to construct the hierarchy goes as follows (Soneira and Peebles 1977). The centers of η spheres, each of radius R/λ, are placed at random within a sphere of radius R. Within each of these spheres there are placed at random the centers of η spheres each of radius R/λ^2. This repeats through L levels to η^{L-1} spheres, each of radius R/λ^{L-1}, that might be called galaxies. If

$$\lambda = \eta^{1/(3-\gamma)}, \tag{61.17}$$

the mean density within a sphere in the sequence varies with the sphere radius r as

$$n(r) \propto r^{-\gamma}, \qquad R \gtrsim r \gtrsim r_{\min} = R/\lambda^{L-1}. \tag{61.18}$$

This model agrees with the behavior of P_N discussed in section 60. When a sphere of radius $r < R$ is placed at random, the chances are it will fall in the interstices of the spheres of radii $\lesssim r$ in the clustering hierarchy (assuming the clumps do not strongly overlap), so it contains little matter. If it does happen to overlap a sphere of radius $\sim r$, the mass contained is $\sim n(r)r^3 \propto r^{3-\gamma}$, which agrees with equation (60.12).

The correlation functions in this construction can be understood as follows. Suppose an object is chosen at random. The object is in one of the nested sequences of spheres, so if δV is placed at distance r away, it is outside all the small spheres in the sequence but still inside one of radius $\sim r$ with density $\sim n(r)$. The chance of finding a neighbor in δV is then

$$\delta P \sim n(r)\,\delta V \propto r^{-\gamma}\delta V, \qquad r_{\min} \lesssim r \lesssim R, \tag{61.19}$$

so (eq. 40.5)

$$n\xi(r) \sim n(r) \propto r^{-\gamma}, \tag{61.20}$$

as wanted. In the same way one sees that the chance of finding neighbors in δV_1 and δV_2 at distances r_1 and r_2, with $r_1 \ll r_2$, is

$$\delta P \sim n(r_1)n(r_2)\delta V_1 \delta V_2, \tag{61.21}$$

so the three-point function is

$$n^2\zeta \sim n(r_1)n(r_2), \qquad \zeta \sim \xi(r_1)\xi(r_2). \tag{61.22}$$

This fairly closely approximates the behavior of the power law model for ζ (eqs. 40.12 and 54.17) and the observations. Numerical trials confirm that the three-point function in this construction gives a good aproximation to the observed variation of ζ with triangle shape as well as size. However, if the parameters R, η, and L are the same for each clump, the parameter Q (eq. 54.17) is found to be (compare § 62 below)

$$Q \sim 0.5, \tag{61.23}$$

significantly less than what is observed (eq. 57.8).

One can think of two easy ways to raise Q. The first makes η, the number of subclusters per cluster, a random variable with considerable skewness. This can increase the number of triplets defining roughly equilateral triangles ($\propto \langle \eta^3 \rangle$) compared to the number of pairs ($\propto \langle \eta^2 \rangle$) at that separation so that ζ/ξ^2 increases. (For a detailed treatment see Soneira and Peebles 1978, Appendix A.) However, it does not increase $\zeta/\xi(r_1)\xi(r_2)$ when the triangle is elongated with $r_1 \ll r_2$ because then each of the neighbors comes from a different level of the hierarchy with an independent η. On this plan the function

$$Q = \zeta/(\xi_1\xi_2 + \xi_2\xi_3 + \xi_3\xi_1), \tag{61.24}$$

decreases with increasing triangle elongation, contrary to what is observed. A second approach that avoids this problem is to make the number of levels L different in each clump (with the quantities R and η fixed constants). Then the mean numbers of objects and pairs per clump vary as

$$\langle n_\alpha \rangle \propto \langle \eta^L \rangle, \qquad \langle \delta N_p \rangle \propto \langle n_\alpha{}^2 \rangle, \tag{61.25}$$

so (eq. 40.5)

$$n\xi \propto \langle n_\alpha{}^2 \rangle / \langle n_\alpha \rangle. \tag{61.26}$$

The mean number of triplets varies as $\langle n_\alpha{}^3 \rangle$, so the three-point function ζ is

$$n^2\zeta \propto \langle n_\alpha{}^3 \rangle / \langle n_\alpha \rangle. \tag{61.27}$$

The wanted shape of ζ thus is preserved, and the amplitude relative to ξ is proportional to

$$Q \propto \langle n_\alpha{}^3 \rangle \langle n_\alpha \rangle / \langle n_\alpha{}^2 \rangle^2, \tag{61.28}$$

independent of the mean space density n. In the same way one finds that the amplitude of the four-point function varies as

$$R \propto \langle n_\alpha{}^4 \rangle \langle n_\alpha \rangle^2 / \langle n_\alpha{}^2 \rangle^3. \tag{61.29}$$

For more discussion and examples of distributions in n_α that match the observed Q and R see Soneira and Peebles (1978). This paper also describes the application of this prescription to the construction of computer model galaxy distributions that can be compared to the real world. Such models can reproduce fairly well the texture of the map of the Lick data, though one has the impression that the model is a somewhat fuzzy copy of the real thing.

C. Lack of uniqueness

Even if the correlation functions n, ξ, ζ, and η were known to all accuracy, they would not point to a unique prescription for making clumps. Equations (61.28) and (61.29) illustrate this: within the general prescription the moments of the distribution $F(n_\alpha)$ of the number of galaxies per clump must be adjusted to fit the observed Q and R and, of course, that constrains $F(n_\alpha)$ but does not fix it. For that we would need to know all orders of the correlation functions. Another example is given by McClelland and Silk (1977) who discuss as one possibility the assumption that each clump is a spherically symmetric cluster, as in Part A above, perhaps with a Gaussian shape, with some joint frequency distribution $F(n_\alpha, R_\alpha)$ in clump mass and radius. One can imagine all clumps have similar shapes $n_\alpha/R_\alpha^3 N(r/R_\alpha)$, scaled by n_α and R_α. Then the two- and three-point correlation functions (eq. 40.5 and 40.14) are given by integrals over F of the forms

$$n\xi(r) = \int dn_\alpha dR_\alpha F(n_\alpha, R_\alpha) n_\alpha^2 R_\alpha^{-3} f_2(r/R_\alpha)/\langle n_\alpha \rangle,$$
$$n^2\zeta = \int dn_\alpha dR_\alpha F(n_\alpha, R_\alpha) n_\alpha^3 R_\alpha^{-6} f_3(r_a/R_\alpha, r_b/R_\alpha, r_c/R_\alpha)/\langle n_\alpha \rangle, \tag{61.30}$$

where f_2 and f_3 are integrals over the shape $N(r)$, as in equation (61.2). Given the free function $F(n_\alpha, R_\alpha)$ one has considerable freedom to adjust ξ and ζ more or less separately.

Only two clump prescriptions have been discussed in detail here. The first certainly has some reality because rich clusters of galaxies do resemble such monolithic clumps with power law index ϵ not dissimilar to the value needed to produce the observed index γ of ξ (eq. 61.5). However, it conflicts with the observed ζ and η. The hierarchical clustering model reproduces ξ, ζ, and η, and, what is more, it matches in outline the description astronomers have given of the general galaxy distribution (de Vaucouleurs 1970). One would presume therefore that it is closer to the truth. It certainly is not the whole truth because it misses regular relaxed-looking concentrations like the central parts of the Coma cluster, and it does not quite reproduce the impression of large-scale filaments one sees in the Lick sample. Despite these problems, however, it is remarkable that this simple pattern gives such a good approximation to the galaxy distribution.

62. Continuous Clustering Hierarchy: Mandelbrot's Prescription

Mandelbrot (1975, 1977) has proposed a particularly elegant prescription for a continuous clustering hierarchy. Galaxies are placed at each step of a Rayleigh-Lévy random walk: starting from any one of the galaxies, one places the next in the sequence in a randomly chosen direction at distance l drawn from the distribution

$$\begin{aligned} P(>l) &= (l_0/l)^\alpha, \qquad l \geq l_0, \\ &= 1, \qquad l < l_0, \qquad \alpha > 0. \end{aligned} \tag{62.1}$$

This repeats with l and the direction independently chosen each time through a large (perhaps unlimited) number of steps.

A. Distribution of displacements in the random walk

To compute correlation functions we need the probability that the random walk yields a net displacement r to $r + \delta r$ from a galaxy, after any number of steps, to produce a neighbor at that distance. The result is a power law, as must be expected because there is no characteristic length (at $r \gg l_0$). If $\alpha > 2$, the power law index agrees with the usual result for a random walk with fixed step length; if $\alpha < 2$, the index is larger (eq. 62.15 below) and can be adjusted to the observations.

The probability for net displacement $\mathbf{r}$ in the range d^3r after n steps is

$$dP = f_n(r)d^3r, \tag{62.2}$$

and the generating function belonging to this is

$$\psi(n, k) = \int f_n(r)d^3re^{i\mathbf{k}\cdot\mathbf{r}}. \tag{62.3}$$

For one step, the distribution defined by equation (62.1) is

$$\begin{aligned} f_1(r) &= \frac{\alpha}{4\pi}\frac{l_0^\alpha}{r^{\alpha+3}}, \qquad r \geq l_0 \\ &= 0, \qquad r < l_0, \end{aligned} \tag{62.4}$$

and the generating function for this is

$$\psi(1, k) = \frac{\alpha l_0^\alpha}{k} \int_{l_0}^\infty \frac{\sin kr}{r^{\alpha+2}}\, dr. \tag{62.5}$$

It will be assumed that many steps are needed for interesting net displacements r, so equation (62.5) will be needed for $kl_0 \ll 1$. On integrating twice by parts, we find

$$\psi(1, k) \simeq 1 - \frac{(kl_0)^2}{\alpha + 1}\left(\frac{\alpha}{6} + \frac{1}{2}\right) - \frac{l_0^\alpha k}{\alpha + 1} \int_{l_0}^\infty \frac{\sin kr}{r^\alpha}\, dr, \tag{62.6}$$

for $kl_0 \ll 1$. If $\alpha > 2$, the integral in this last equation is dominated by the part near l_0, so we can set $\sin kr = kr$ and get

$$\psi = 1 - \frac{\alpha}{\alpha - 2} \frac{(kl_0)^2}{6}, \qquad \alpha > 2. \tag{62.7}$$

If $0 < \alpha < 2$, the integral converges at $l_0 \rightarrow 0$ and with $x = kr$ gives

$$\begin{gathered} \psi \simeq 1 - (kl_0)^\alpha I_\alpha, \\ I_\alpha = (\alpha + 1)^{-1} \int_0^\infty x^{-\alpha} \sin x\, dx, \qquad 0 < \alpha < 2, \end{gathered} \tag{62.8}$$

to the lowest nontrivial order.

Since each step is statistically independent, the generating functions multiply (eq. 38.3),

$$\psi(n, k) = [\psi(1, k)]^n. \tag{62.9}$$

For $\alpha > 2$ we have

$$\psi(n, k) \simeq \exp\left[-n \frac{\alpha}{\alpha - 2} \frac{(kl_0)^2}{6}\right]. \tag{62.10}$$

The Fourier transform of this is a Gaussian in r,

$$f_n(r) \propto \exp - \frac{3}{2} \frac{\alpha - 2}{\alpha} \frac{r^2}{nl_0^2}, \qquad \alpha > 2. \tag{62.11}$$

This is the same for an ordinary random walk with fixed step length

$$l = l_0(\alpha/(\alpha - 2))^{1/2}. \tag{62.12}$$

We are interested in the other case $\alpha < 2$. With

$$f_n(r) = (2\pi)^{-3} \int d^3k e^{-i\mathbf{k}\cdot\mathbf{r}}[\psi(1, k)]^n, \tag{62.13}$$

the probability density for displacement $\mathbf{r}$ after any number of steps is

$$\begin{aligned} f(r) &= \sum_n f_n = (2\pi)^{-3} \int d^3k e^{-i\mathbf{k}\cdot\mathbf{r}}[1 - \psi(1, k)]^{-1} \\ &= (2\pi)^{-3} I_\alpha^{-1} l_0^{-\alpha} \int d^3k k^{-\alpha} e^{-i\mathbf{k}\cdot\mathbf{r}}. \end{aligned} \tag{62.14}$$

With $\mathbf{k} = \mathbf{y}/r$ this gives the wanted result (Mandelbrot 1975),

$$\begin{aligned} f(r) &= Cr^{-\gamma}, \qquad \gamma = 3 - \alpha, \qquad \alpha < 2, \\ C &= (2\pi)^{-3} I_\alpha^{-1} l_0^{-\alpha} \int d^3y y^{-\alpha} e^{-iy\cos\theta}. \end{aligned} \tag{62.15}$$

This last integral is evaluated in Section 42.

B. An unbounded continuous clustering hierarchy

Mandelbrot points out that the simplest of all prescriptions, and so the one to try first, would put all galaxies in the universe on a single unlimited Rayleigh-Lévy random walk. The following discussion of this case ignores space curvature and the general expansion of the universe, and it assumes the luminosity of each galaxy is randomly drawn from the distribution Φ.

An observer is in a galaxy. The expected observed number of galaxies per steradian brighter than limiting magnitude m_0 averaged over observers on randomly chosen galaxies is

$$\mathcal{N} = 2 \int_0^\infty r^2\, dr\, f(r)\phi(r/D) = 2CD^{3-\gamma} \int_0^\infty x^{2-\gamma}\, dx \phi(x). \tag{62.16}$$

In the first equation ϕ is the selection function belonging to Φ (eq. 51.4): it is the probability that a galaxy at distance r is bright enough to be in the catalog. The characteristic depth is

$$D = 10^{0.2(m_0 - M^*) - 5}\ \text{Mpc}. \tag{62.17}$$

The factor f is given by equation (62.15) and the factor of 2 takes account of the fact that the observer's galaxy could have been placed before or after one seen at distance r.

Equation (62.16) says the galaxy count increases with apparent magnitude less rapidly than the usual $10^{0.6m}$ relation (eq. 49.8): as the depth of observation D is increased, the observer discovers he (or she) is in clusters of ever higher order of size $\sim D$ with density $n \propto D^{-\gamma}$. Two advantages of this plan were discussed by Fournier d'Albe (1907) and Charlier (1922) (§ 3A above). The mean surface brightness of the sky is

$$i = 2\int r^2 dr f(r)\,\mathcal{L}/4\pi r^2, \tag{62.18}$$

where $\mathcal{L}$ is the mean luminosity of a galaxy. The intergral converges at large r if $\gamma > 1$. This is a resolution of Olbers' paradox that if the space distribution of matter were homogeneous and if stars had always been shining (which is now thought to be a questionable idea since stars presumably have a limited energy supply; Harrison 1974), the integral would not converge. Second, the total Newtonian gravitational acceleration of one galaxy due to all the other matter in the universe seems better defined here than in a homogeneous universe. For example, if the clusters on each level are in a gravitational equilibrium, the typical speed of the subclusters within a cluster of size D is

$$v \sim [GM(D)/D]^{1/2} \propto D^{1-\gamma/2}. \tag{62.19}$$

As noted by Mandelbrot, Fournier d'Albe made the interesting point that if $\gamma = 2$, then v is independent of D, which avoids a divergence of v at large or small D, as one would want for a scale-independent world. This value of γ is remarkably close to the observed index $\gamma \sim 1.8$. More recently de Vaucouleurs (1970) has discussed the evidence that the apparent mean density of the universe may decrease with increasing depth, and he suggested that a reasonable value of the power law index in the clustering hierarchy is $\gamma \sim 1.7$, which is in good agreement with the results from the correlation functions.

There are some observational problems with a simple unlimited scaling plan (for example, Sandage, Tammann, and Hardy 1972); of present interest is the angular correlation function. The probability of finding galaxies in δV_1 and δV_2 at positions $\mathbf{r}_1$ and $\mathbf{r}_2$ relative to a randomly chosen galaxy is, from equation (62.15),

$$\delta P = 2[f(r_1)f(r_{12}) + f(r_2)f(r_{12}) + f(r_1)f(r_2)]\delta V_1 \delta V_2, \quad r_{12} = |\mathbf{r}_1 - \mathbf{r}_2|. \tag{62.20}$$

The first term in the parentheses is the probability that the observer's galaxy came first in the walk; then the galaxy in δV_1 came some steps later; then the galaxy in δV_2 some steps later still. The factor of 2 takes account of this same sequence in the reversed order. The two other cyclic permutations take account of the other possible sequences. The probability that an observer finds galaxies brighter than m_0 in the two elements of solid angle $\delta\Omega_1$ and $\delta\Omega_2$ is found by multiplying equation (62.20) by the selection functions $\phi(r_1/D)\phi(r_2/D)$ and integrating along the lines of sight as in equation (62.16):

$$\delta P = \delta\Omega_1\delta\Omega_2[\mathcal{N}^2/2 + 4C^2 \int r_1^2 r_2^2\, dr_1 dr_2 \phi(r_1/D)\phi(r_2/D)/(r_1 r_{12})^\gamma]. \tag{62.21}$$

If the angle θ_{12} between the two elements is much less than one radian and if $\gamma > 1$, the main contribution is in the integral at $r_1 \sim r_2$. Following equation (52.8), we can write this in the small separation approximation as (Mandelbrot 1975)

$$\delta P = \delta\Omega_1\delta\Omega_2\mathcal{N}^2 A\theta^{1-\gamma}, \qquad A = H_\gamma \frac{\int_0^\infty x^{5-2\gamma}\phi(x)^2\, dx}{\left(\int_0^\infty x^{2-\gamma}\phi(x)\, dx\right)^2}, \tag{62.22}$$

where H_γ is given by equation (52.9).

The amplitude A is independent of the depth of the survey, as expected; this dimensionless number could not very well be a function of D since there is no relevant characteristic length in the model. It means that as the observations go deeper, they always uncover new levels of clustering, so the fractional fluctuations in numbers of galaxies at fixed angular scale do not change.

The two observational problems with equation (62.22) are that A is large and it is independent of D. In the Lick sample the two-point correlation function at small θ gives $A = 0.07$ for θ in degrees, or $A = 3 \times 10^{-3}$ for θ in radians, some two and a half orders of magnitude below the prediction. In testing the variation of A with D, we need some caution, for in a deep survey in a patch of size θ_0 by θ_0 the estimate of A cannot exceed $\sim\theta_0^{\gamma-1}$ if $w(\theta)$ is normalized by the observed count of galaxies in the patch. The model would predict that in well-separated patches the counts of faint galaxies fluctuate as much as the counts of bright galaxies, which seems not to be observed (though the observational limits are not yet nearly as good as they could be). The Zwicky and Lick samples cover about the same amounts of solid angle and so the empirical A for them can be compared. The amplitudes are in the ratio 10:1, in excellent agreement with what is expected under spatial homogeneity and conflicting with equation (62.22) by a factor of 10.

C. Truncated clustering hierarchy

To remedy the problems with equation (62.22), we must lessen the spatial gradients on scales 30 h^{-1} Mpc to at least 300 h^{-1} Mpc. A simple way to do this is to suppose that there are many different Rayleigh-Lévy random walks distributed uniformly at random through space as in the Neyman-Scott program. Individual chains cannot be infinitely long because that would make the mean space density n diverge; let us suppose then that each one has a finite number of steps, n_β, that might be a random variable and that each has the same fixed parameters l_0 and α. Now the distribution is a spatially homogeneous random process, and we can compute the correlation functions by the method of Section 40. The mean number of pairs of galaxies in a clump at separation r is, according to equation (62.15),

$$\delta N_p = 4\pi \langle n_\beta \rangle C r^{2-\gamma} \delta r, \tag{62.23}$$

so the two-point function is (eq. 40.5)

$$n\xi(r) = 2Cr^{-\gamma}. \tag{62.24}$$

This power law applies for $r \gg l_0$ and r much less than the typical clump size. The mean number of triplets per clump that define triangles with sides r_a, r_b, and r_c in the range $\delta^2 V$ is, according to equation (62.20),

$$\delta N_t = 2\langle n_\beta \rangle C^2[(r_a r_b)^{-\gamma} + (r_b r_c)^{-\gamma} + (r_c r_a)^{-\gamma}]\delta^2 V, \tag{62.25}$$

so the three-point function is (eq. 40.14)

$$n^2\zeta = 2C^2[(r_a r_b)^{-\gamma} + (r_b r_c)^{-\gamma} + (r_c r_a)^{-\gamma}]. \tag{62.26}$$

This agrees with the empirical model (54.17) for ζ with

$$Q = \tfrac{1}{2}. \tag{62.27}$$

In the same way one finds that the reduced four-point function is

$$n^3\eta = 2C^3[(r_a r_b r_c)^{-\gamma} + \text{sym. (12 terms)}], \tag{62.28}$$

which agrees with the model (55.8) for η with

$$R_a = 0.25, \qquad R_b = 0. \tag{62.29}$$

One sees that the model requires further adjustment. The parameter Q is too small by the factor ~ 2.6 (eq. 57.8). There is no b-type term in η, though this term does improve the fit to the data; more serious, $R_a + R_b/3$, which is a useful measure of the size of the four-point function, is 4.0 ± 0.6 in the fit to the data (eq. 57.11), some 16 times larger than in the model. Both Q and R_a can be fixed by letting l_0, the step length, be a different constant in each clump. This gives (eqs. 40.5 and 40.14)

$$n\xi = 2\frac{\langle n_\beta C\rangle}{\langle n_\beta\rangle} r^{-\gamma}, \qquad n^2\zeta = 2\frac{\langle n_\beta C^2\rangle}{\langle n_\beta\rangle}[(r_a r_b)^{-\gamma} + \text{cycl.}], \tag{62.30}$$

from which

$$Q = \frac{1}{2}\frac{\langle n_\beta C^2\rangle\langle n_\beta\rangle}{\langle n_\beta C\rangle^2} = \frac{1}{2}\frac{\langle n_\beta l_0{}^{-2\alpha}\rangle\langle n_\beta\rangle}{\langle n_\beta l_0{}^{-\alpha}\rangle^2}. \tag{62.31}$$

The parameter in the four-point function similarly is seen to be

$$R_a = \frac{1}{4}\frac{\langle n_\beta C^3\rangle\langle n_\beta\rangle^2}{\langle n_\beta C\rangle^3}. \tag{62.32}$$

Thus by making the distribution in l_0 broad and skew enough, one can bring Q and R_a up to the wanted values.

It is remarkable that this model, which Mandelbrot proposed independent of the empirical work on the galaxy correlation functions, predicts the form for ζ that fits the data so well and predicts a form for η that at least is a good first approximation. However, some of the simplicity of Mandelbrot's original construction is lost: special adjustment is needed to fix Q and R and to truncate $\xi(r)$. It is possible that another of his fractals (Mandelbrot 1977) would give a more elegant prescription.

63. THE MASS CORRELATION FUNCTIONS

All the above discussion has dealt with the way galaxies are distributed because that is what is observed. To treat dynamics we need the mass correlation functions, ξ_ρ, ζ_ρ, and so on. The assumption in the following equations for these functions is that the mass of a galaxy is statistically independent of its position relative to other galaxies. If so, then at separations large compared to a galaxy, ξ is a fair estimate of ξ_ρ, ζ a fair estimate of ζ_ρ, and so on. The assumption seems reasonable at $r \gtrsim 1$ Mpc but may be questionable on small scales, for at the least morphological type is correlated with the presence of close neighbors (Davis and Geller 1976). There is the added interesting problem and complication that the mass autocorrelation functions at 10 h^{-1} kpc depend on how mass is distributed within galaxies. Finally, one must bear in mind that the universe may well contain appreciable mass in forms like an intergalactic plasma that are not distributed like the galaxies. In that case of course ξ_ρ and ζ_ρ refer only to the part contributed by the galaxies.

The autocorrelation function for the mass density $\rho(\mathbf{r})$ is

$$\xi_\rho(r) = \langle(\rho(\mathbf{x} + \mathbf{r}) - \rho_b)(\rho(\mathbf{x}) - \rho_b)\rangle/\rho_b^2, \qquad \rho_b = \langle\rho(\mathbf{x})\rangle. \quad (63.1)$$

The mass density is

$$\rho(\mathbf{r}) = \Sigma n_1 \rho_\alpha(\mathbf{r} - \mathbf{r}_1), \quad (63.2)$$

where $n_1 = 0$ or 1 is the number of galaxies in the volume element d^3r_1 and ρ_α is the mass distribution within the galaxy. In this notation the mean density is

$$\rho_b = n\langle m_\alpha\rangle, \qquad m_\alpha = \int d^3x \rho_\alpha(\mathbf{x}). \quad (63.3)$$

If the galaxy structure as measured by $\rho_\alpha(\mathbf{r})$ is uncorrelated with its position relative to neighboring galaxies, then

$$\begin{aligned} \langle\rho(\mathbf{x} + \mathbf{r})\rho(\mathbf{x})\rangle &= \Sigma\langle n_1^2\rangle\langle\rho_\alpha(\mathbf{x} + \mathbf{r} - \mathbf{r}_1)\rho_\alpha(\mathbf{x} - \mathbf{r}_1)\rangle \\ &\quad + \Sigma\langle n_1 n_2\rangle\langle\rho_\alpha(\mathbf{x} + \mathbf{r} - \mathbf{r}_1)\rangle\langle\rho_\alpha(\mathbf{x} - \mathbf{r}_2)\rangle, \quad (63.4) \\ \langle n_1^2\rangle = nd^3r_1, &\qquad \langle n_1 n_2\rangle = n^2[1 + \xi(r_{12})]d^3r_1 d^3r_2, \end{aligned}$$

where n is the mean number density of galaxies and ξ is the galaxy two-point correlation function (§ 31). If ξ varies slowly on the scale of a galaxy, $\xi(r_{12})$ can be replaced with $\xi(r)$ and equation (63.4) can be reduced to

$$\langle\rho(\mathbf{x} + \mathbf{r})\rho(\mathbf{x})\rangle \simeq n \int d^3y\langle\rho_\alpha(\mathbf{r} + \mathbf{y})\rho_\alpha(\mathbf{y})\rangle + n^2[1 + \xi(r)]\langle m_\alpha\rangle^2. \quad (63.5)$$

The mass autocorrelation function is then

$$\xi_\rho(r) \simeq \xi(r) + f(r),$$
$$f(r) = \int d^3y \langle \rho_\alpha(\mathbf{r} + \mathbf{y}) \rho_\alpha(\mathbf{y}) \rangle / (\rho_b \langle m_\alpha \rangle), \tag{63.6}$$

where

$$\int d^3r f(r) = \langle m_\alpha^{\ 2} \rangle / (\rho_b \langle m_\alpha \rangle). \tag{63.7}$$

An interesting application is to the second moment of the mass found within a randomly placed cell V. The cell will be supposed to be much larger than a galaxy, so that, following Section 36, we can write the mass within V as

$$M = \Sigma m_1 n_1, \tag{63.8}$$

where $n_1 = 0$ or 1 is the number of galaxies with mass m_1 to $m_1 + \delta m_1$ in the volume element δV_1 and

$$\langle n_1 \rangle = \langle n_1^{\ 2} \rangle = \Phi(m_1) \delta m_1 \delta V_1,$$
$$\langle n_1 n_2 \rangle = \delta V_1 \delta V_2 \delta m_1 \delta m_2 \Phi(m_1) \Phi(m_2) [1 + \xi(r_{12})], \tag{63.9}$$

where Φ is the differential galaxy mass function. The mean mass in V is

$$\langle M \rangle = \int_V dV dm \, m\Phi(m) = \langle m \rangle nV = \rho_b V,$$
$$n \langle m \rangle = \int dm \, m\Phi. \tag{63.10}$$

The second central moment is

$$\langle (M - \langle M \rangle)^2 \rangle = \Sigma \, m_1^{\ 2} \langle (n_1 - \langle n_1 \rangle)^2 \rangle$$
$$+ \Sigma \, m_1 m_2 \, \langle (n_1 - \langle n_1 \rangle)(n_2 - \langle n_2 \rangle) \rangle, \tag{63.11}$$

which with equation (63.9) becomes

$$\langle (M - \langle M \rangle)^2 \rangle = \langle m^2 \rangle nV + n^2 \langle m \rangle^2 \int_V dV_1 dV_2 \xi(r_{12}). \tag{63.12}$$

Of course, this expression makes sense only if V is appreciably larger than the size of a galaxy, so we can take it that each galaxy either is fully inside or fully outside V. Under this assumption equations (63.6) and (63.7) give

$$\int dV_1 dV_2 \xi_\rho(r_{12}) = \int dV_1 dV_2 \xi + V\langle m^2\rangle/(\rho_b\langle m\rangle), \qquad (63.13)$$

so equation (63.12) becomes

$$\langle (M - \langle M\rangle)^2\rangle = \rho_b{}^2 \int_V dV_1 dV_2 \xi_\rho(r_{12}), \qquad (63.14)$$

which is the usual expression for the variance in terms of the mass autocorrelation function.

Equations (63.12) and (63.14) can be compared to the expression for the second moment of the number of galaxies in V (eq. 36.6),

$$\mu_2 = \langle (N - \langle N\rangle)^2\rangle = nV + n^2 \int dV_1 dV_2 \xi(r_{12}). \qquad (63.15)$$

If the galaxy number density n is large, μ_2 is determined by ξ, but for small n, $\mu_2 \sim nV$ independent of ξ: μ_2 is dominated by the shot noise or discreteness noise (Fall 1978). If n were increased by adding the abundant dwarf galaxies that have negligible mass, it could change the character of μ_2, but of course it would not affect the quantities $n\langle m\rangle$ and $n\langle m^2\rangle$ that fix the mass moment. Here the galaxy shot noise is replaced with the structure function f. As discussed in Section 64 it is a matter of conjecture whether this serves to make the mass moment vary as a simple power of the size r of V,

$$\langle (M - \langle M\rangle)^2\rangle \propto r^{6-\gamma}, \qquad (63.16)$$

all the way down to r equal to the nominal size of a galaxy.

The three-point autocorrelation function for the mass density is

$$\zeta_\rho(r, s, |\mathbf{r} - \mathbf{s}|) = \langle (\rho(\mathbf{x} + \mathbf{r}) - \rho_b)(\rho(\mathbf{x} + \mathbf{s}) - \rho_b)(\rho(\mathbf{x}) - \rho_b)\rangle/\rho_b{}^3. \qquad (63.17)$$

The approximate relation to the galaxy three-point function is written as for equations (63.4) and (63.5):

$$\begin{aligned}\zeta_\rho(r_a, r_b, r_c) &\simeq \zeta(r_a, r_b, r_c) + \tfrac{1}{2}\,[\xi(r_a) f(r_b) + \xi(r_b) f(r_a)] + \text{cycl.} \\ &\quad + f_2(r_a, r_b, r_c), \qquad (63.18)\\ f_2(r, s, |\mathbf{r} - \mathbf{s}|) &= \langle \textstyle\int \rho_\alpha(\mathbf{x})\rho_\alpha(\mathbf{x} + \mathbf{r})\rho_\alpha(\mathbf{x} + \mathbf{s}) d^3x\rangle/(\langle m_\alpha\rangle \rho_b{}^2).\end{aligned}$$

The first term is the contribution from the clustering of galaxies. There are three terms of the second sort for the cyclic permutations of the sides. If $r_a \sim$ the size of a galaxy and $r_b \gg r_a$, these three terms reduce to $f(r_a)\xi(r_b)$, which is the contribution from both points on the short side in the same galaxy, the third point in a neighbor. The last term is the contribution from all three points in the same galaxy. One sees from equation (63.3) that

$$\int f_2 d^3r d^3s = \langle m^3 \rangle / (\langle m \rangle \rho_b^2). \tag{63.19}$$

As before an interesting check is the third moment of the mass in a randomly placed cell V. Following the derivation of equation (63.12) one finds

$$\langle (M - \langle M \rangle)^3 \rangle = \langle m^3 \rangle nV + 3\langle m^2 \rangle \langle m \rangle n^2 \int d^2V\xi + \langle m \rangle^3 n^3 \int d^3V\zeta, \tag{63.20}$$

where again V must be large compared to the size of a galaxy. When this is so, equations (63.18) and (63.19) yield

$$\int_V \zeta_\rho d^3V = \int \zeta d^3V + 3\int \xi d^2V \langle m^2 \rangle / (\langle m \rangle \rho_b) + \langle m^3 \rangle V / (\langle m \rangle \rho_b^2), \tag{63.21}$$

which brings equation (63.20) to

$$\langle (M - \langle M \rangle)^3 \rangle = \rho_b^3 \int \zeta_\rho d^3V, \tag{63.22}$$

which is the usual expression from equation (63.17).

Equations (63.6) and (63.18) show how one could in principle compare the mass autocorrelation functions estimated on scales $\gtrsim 10\ h^{-1}$ kpc from the galaxy distribution with the mass autocorrelation functions at $r \lesssim 10\ h^{-1}$ kpc derived from the structures of galaxies. The speculation that ξ_ρ and ζ_ρ might approximate power laws all the way down to the size of a galaxy is discussed next.

64. Clustering hierarchy: continuity speculation

It was shown in Section 61 that the low order galaxy correlation functions agree with a clustering hierarchy of the sort that, it was argued in Section 26, might be laid down in an expanding world model under not unreasonable initial conditions. There does seem to be a break in the scaling at $\sim 10\ h^{-1}$ Mpc, and since ξ is on the order of unity at this break, it could simply mark the transition from linear fluctuations on large scales to nonlinear fluctuations on small. There must also be a feature in the mass correlation functions on the scale of a galaxy because galaxies certainly appear as discrete objects. One line of speculation discussed in Section 4B is that the mass was laid down in a clustering hierarchy extending to scales well below that of a galaxy and that relaxation processes (like dissipation in fluid dynamics) rearranged the pattern at $r < r_g$ to make galaxies. If so, and if there has not been much rearrangement on larger scales, then ξ_ρ, ζ_ρ, η_ρ should vary as the primeval power laws down to $r \sim r_g$ and there join smoothly onto the galaxy structure functions (eqs. 63.6 and 63.18). On the other hand if galaxies were formed by one process and then placed in a clustering pattern by another, one might look for rather abrupt changes in the mass correlation functions at $r \sim r_g$.

An indication of continuity appeared in the discussion of P_N in Section 60. A more direct approach would be to use the method of Section 63 to estimate $\xi_\rho(r_g)$, $\zeta_\rho(r_g)$ and so on. Here again the test is schematic at best because the mass distribution in and around galaxies is not understood, but we can at least write down what the continuity hypothesis would require.

The mass autocorrelation function is given by equation (63.6). At

$$r \lesssim r_g = 10\ h^{-1}\ \text{kpc}, \qquad \xi_\rho(r) \simeq f(r), \tag{64.1}$$

and equation (63.7) says

$$f(r_g) \sim \frac{\langle m^2 \rangle}{\langle m \rangle \rho_b V}, \qquad V = \frac{4}{3}\pi r_g^{\ 3}. \tag{64.2}$$

The continuity hypothesis is that this agrees with the extrapolation of the galaxy two-point correlation function,

$$f(r_g) \sim (r_0/r_g)^\gamma. \tag{64.3}$$

With the parameters for $\xi(r)$ listed in Section 57 and in equation (97.13) for ρ_b, equations (64.2) and (64.3) give

$$\langle m^2 \rangle / \langle m \rangle \sim 5 \times 10^{10}\, \Omega_0 h^{-1}\, M_\odot. \tag{64.4}$$

Most of the light comes from large galaxies. In a typical large spiral the mass within $10\ h^{-1}$ kpc of the center is (Faber and Gallagher 1979, Rubin 1979)

$$M(<10\,h^{-1}\,\text{kpc}) \simeq 1 \times 10^{11}\,h^{-1}\,\text{M}_{\odot}. \qquad (64.5)$$

There is a like mass in the inner 10 h^{-1} kpc of a large elliptical. If the density parameter is $\Omega \sim 1$, then equations (64.4) and (64.5) are remarkably similar, consistent with the continuity hypothesis. There is the problem that if the mass in equation (64.5) is the major contribution to the mass of the universe, as the light within r_g is the major contribution to the total luminosity, then from the abundance of large galaxies, $n \sim 0.02\,h^3$ Mpc^{-3}, we get $\Omega \sim 0.01$, and we find a considerable discrepancy between equations (64.4) and (64.5). However, we know that in some groups and clusters the mass per galaxy is considerably larger than in equation (64.5) (Sections 76 and 77 below), and we know from rotation curves for spiral galaxies that $M(<r)$ varies about as r at $r \sim r_g$, so that there is considerable mass in the dark outer part of some spirals (Rubin 1979). Thus it is possible that $\Omega \sim 1$, the major portion of the mass being outside the bright central parts of galaxies. If that is so, however, it is reasonable to ask whether the galaxy clustering is a good measure of the mass autocorrelation function at $r > r_g$, as has been assumed here. The point can be tested by using the pattern of motions of galaxies as tracers of the gravitational field. That is discussed in Chapter IV below.

In taking account of the mass distribution within galaxies, we are in effect adding a level to the clustering hierarchy. If the mass distribution within each galaxy is taken to be the same, we get $Q \sim 0.5$ at $r \sim r_g$ (eq. 61.23); the dispersion in galaxy masses increases this number. The galaxy three-point correlation function at $r \gtrsim 50\,h^{-1}$ Mpc yields $Q \sim 1.2$ (eq. 57.8). An improved test for continuity of ζ_ρ depends on improved understanding of the mass distribution within galaxies and the correlation among masses of close pairs of galaxies.

65. Remarks on the observations

The characteristic length r_0 (eq. 57.3) marks a division in the nature of the clustering of galaxies. The rms fluctuation in the number density averaged over a sphere of radius r is large compared to the mean if $r \lesssim r_0$ while if $r \gg r_0$ the rms fluctuation represents a small perturbation from the mean. Thus it is natural to look for possible differences in the correlation functions at $r \lesssim r_0$ and $r \gg r_0$. The two-point correlation function is thought to depart from the power law form at $r \sim 10\,h^{-1}$ Mpc, interestingly close to r_0, and ξ is known to be positive to $r \sim 15\,h^{-1}$ Mpc. Structures in the galaxy distribution certainly extend well beyond that limit, as in superclusters (§§ 77 and 78 below), but one could imagine that galaxy positions outside superclusters are mildly anticorrelated on scales of 20 to 40 h^{-1} Mpc making ξ small or negative. Improved measures or bounds on ξ, ζ, and η on scales of 10 to 40 h^{-1} Mpc are within reach and will be of considerable interest.

In the statistical analysis of clustering on scales $\lesssim r_0$ an important uncertainty is the nature of the joint distribution in absolute magnitude and position. The assumption that magnitude and position are independent (eq. 51.1) proves to be a good approximation but it is known that position and morphological type are correlated. That effect with the possible correlation of position and absolute magnitude on small scales is important as a hint to the form of the mass three-point function ζ_ρ on small scales (§ 64) and to the role of relaxation processes in tight groups.

The next big step in the statistical analysis of clustering will be the study of the joint distribution in position and redshift. Even crude redshift measurements provide useful distance measures that reduce the noise in the correlation function estimates caused by the overlap of clusters seen close together in projection. Crude redshift samples to adequate depth will greatly improve the measures of the correlation functions at $r \gtrsim 10\ h^{-1}$ Mpc. For the clustering on scales $\lesssim r_0$ the main point of interest is the measurement of the statistical pattern of peculiar velocities and the interpretation in terms of the mass distribution. The next chapter deals with the analysis of these problems in terms of n-point correlation functions in position and velocity.

IV. DYNAMICS AND STATISTICS

66. GOALS

This chapter deals with attempts to understand how the clustering pattern is behaving now and how it got that way. Just as studies of the nature of the distribution of matter can deal with individual objects or the statistics of a sample, analysis of dynamics can treat objects or the behavior of statistical measures. One certainly can analyze the stability of a galaxy or the central parts of a rich compact cluster, and results such as the mass needed to hold the cluster together generally are accepted with little controversy. The situation is different for groups of galaxies where one must isolate the physical system by removing the background and foreground and the galaxies accidently passing through without removing the high velocity members. One approach that will be discussed here is the resort to statistics.

The second goal is to understand how the present galaxy clustering got that way—to find a scenario that leads from reasonable initial conditions through well-stated and reasonable dynamics to a final state that matches significant features of what is observed. Of course, different people can decide very different features are the significant ones. A considerable stimulus to the thought that the galaxy correlation functions might be important quantities for the theory to reproduce has been the discovery that they can be estimated with good reproducibility and that the clustering pattern as measured by them is remarkably simple; we are well conditioned to believe that where a phenomenon is simple there is a much better chance of accounting for it.

The dynamics used in this section is the BBGKY hierarchy adapted from plasma physics. It deals with the generalization of the n-point correlation functions used in the last chapter to the joint distributions in position and peculiar velocity. The first equation in the hierarchy describes how the probability distribution in the peculiar velocity of a randomly chosen particle is affected by the perturbations of neighbors; this is a relation between the one- and two-point correlation functions. The second equation describes how the abundance and motions of pairs of particles are affected by the perturbations of neighbors; this is a relation between the two- and three-point correlation functions. The sequence continues through an infinite set of equations in all the n-point correlation functions.

As will be described, these equations yield some useful constraints on the behavior of mass clustering in an expanding universe.

The correlation functions used in this section describe the clustering of mass, not galaxies, so here one should think of the particles as mass elements such as nucleons. Some results, the particle conservation equations, apply directly to the galaxy distribution, but the dynamics apply only if the galaxy distribution is a fair measure of the pattern of the mass clustering. As discussed in Sections 63 and 64, it is not clear whether this is so. If it is, it will be an important advantage because it will yield boundary values for the BBGKY theory. Tests of the assumption are discussed in Sections 76 to 78.

67. Definitions of variables and distribution functions

In all this chapter dynamics is treated in the Newtonian limit (§§ 7, 8). Matter is approximated as a collection of identical particles, mass m, that interact only by gravity. Depending on the situation the particles might be atoms, or stars, or galaxies.

As in Section 7 convenient independent variables are the comoving (uniformly expanding) coordinates x^α and the canonical momentum

$$\mathbf{p} = ma(t)\mathbf{v}, \tag{67.1}$$

where $\mathbf{v}$ is the proper peculiar velocity. The correlation functions in $\mathbf{x}$ and $\mathbf{p}$ are defined like the correlation functions in position and magnitude. The probability of finding a particle in the volume element d^3x moving with momentum $\mathbf{p}$ in the range d^3p is

$$dP = b(p, t)d^3xd^3p. \tag{67.2}$$

It is assumed that the galaxy distribution is a spatially homogeneous and isotropic random process so b is independent of $\mathbf{x}$ and of the direction of $\mathbf{p}$. The normalization is

$$\int bd^3p = na^3 = \text{constant}, \tag{67.3}$$

where n is the proper mean number density, na^3 the constant mean density in $\mathbf{x}$ coordinates.

The two-point correlation function is defined by the joint probability of finding a particle in d^3x_1 moving with momentum $\mathbf{p}_1$ in the range d^3p_1 and a second particle in d^3x_2 moving with momentum $\mathbf{p}_2$ in the range d^3p_2:

$$\begin{aligned} dP &= \rho_2(1, 2)d^3x_1d^3x_2d^3p_1d^3p_2 \\ &= [b(1)b(2) + c(1, 2)]d^6xd^6p. \end{aligned} \tag{67.4}$$

The argument 1 means $\mathbf{x}_1, \mathbf{p}_1$. The second equation defines the reduced two-point function c. Its integral over momenta is the spatial correlation function (eq. 31.4),

$$\int c(1, 2)d^6p = n^2a^6\xi(x_{12}, t). \tag{67.5}$$

The three-point function is defined by the joint probability of finding particles in $d^3x_1d^3p_1$, $d^3x_2d^3p_2$, and $d^3x_3d^3p_3$,

$$\begin{aligned} dP &= \rho_3(1, 2, 3)d^9xd^9p \\ &= [b(1)b(2)b(3) + b(1)c(2, 3) + b(2)c(3, 1) \\ &\quad + b(3)c(1, 2) + d(1, 2, 3)]d^9xd^9p. \end{aligned} \tag{67.6}$$

The integral over momenta of the reduced part d is (eq. 34.1)

$$\int d\, d^9p = n^3a^9\zeta(1, 2, 3), \tag{67.7}$$

where ζ is the three-point spatial correlation function for the particles.

The four-point function is defined by the equation

$$dP = \rho_4(1, 2, 3, 4)d^{12}xd^{12}p. \tag{67.8}$$

The reduced part e is written as in equation (35.1) and the integral of e over momenta is

$$\int e\, d^{12}p = n^4a^{12}\eta(1, 2, 3, 4). \tag{67.9}$$

As was noted in Section 56 each point in the correlation function in general has a different cosmic time, but in the Newtonian approximation we only need them for points at common time t. Thus $c(1, 2)$ is a function of t and of the six variables needed to define the separation of the points and the momenta relative to the line joining the points.

68. BBGKY HIERARCHY EQUATIONS

Consider a small patch Γ around $\mathbf{x}$, $\mathbf{p}$ in the six dimensional single particle phase space. If Γ is small enough, there is negligible chance that it contains more than one particle from the same realization in the ensemble, so we can describe the matter in Γ by a single particle distribution function, $f_a(\mathbf{x}, \mathbf{p}, t)$ for the ath realization, with

$$f_a = \delta(\mathbf{x} - \mathbf{x}_a(t))\delta(\mathbf{p} - \mathbf{p}_a(t)) \tag{68.1}$$

if Γ contains a particle, $f_a = 0$ otherwise. Thus Liouville's equation is

$$\left(\frac{\partial}{\partial t} + \frac{dx^\alpha}{dt}\frac{\partial}{\partial x^\alpha} + \frac{dp^\alpha}{dt}\frac{\partial}{\partial p^\alpha}\right) f_a = 0. \tag{68.2}$$

The time derivatives are given by equations (7.10) and (8.5),

$$\frac{dx^\alpha}{dt} = p^\alpha/ma^2, \qquad \frac{dp^\alpha}{dt} = \frac{Gm^2}{a}\sum_i (x_i^\alpha - x^\alpha)/|\mathbf{x}_i - \mathbf{x}|^3. \tag{68.3}$$

The result of averaging equation (68.1) across the ensemble is the one-point function in equation (67.2),

$$\langle f_a(\mathbf{x}, \mathbf{p}, t)\rangle = b(p, t). \tag{68.4}$$

The average of the second term in equation (68.2) is

$$\langle p^\alpha \partial f_a/\partial x^\alpha\rangle = p^\alpha \partial b/\partial x^\alpha = 0. \tag{68.5}$$

For the third term we have

$$\begin{aligned}
&\langle f_a(\mathbf{x}_1, \mathbf{p}_1, t) d\mathbf{p}_1/dt\rangle \\
&\quad = Gm^2/a\langle f_a(\mathbf{x}_1, \mathbf{p}_1, t)\Sigma(\mathbf{x}_i - \mathbf{x}_1)/|\mathbf{x}_i - \mathbf{x}_1|^3\rangle \\
&\quad = Gm^2/a \int d^3x_2 d^3p_2 \rho_2(\mathbf{x}_1, \mathbf{p}_1, \mathbf{x}_2, \mathbf{p}_2)(\mathbf{x}_2 - \mathbf{x}_1)/|\mathbf{x}_2 - \mathbf{x}_1|^3 \\
&\quad = Gm^2/a \int d^3x_2 d^3p_2 c(1, 2)(\mathbf{x}_2 - \mathbf{x}_1)/|\mathbf{x}_2 - \mathbf{x}_1|^3.
\end{aligned} \tag{68.6}$$

In the last equation ρ_2 is replaced with the reduced part (eq. 67.4) and the integral over $b(2)$ is dropped (by the prescription in § 8 that the $\mathbf{x}_2$ integral is first over angles at fixed $|\mathbf{x}_2 - \mathbf{x}_1|$). The ensemble average of equation (68.2) is then

$$\begin{gathered}
\frac{\partial}{\partial t} b(p_1, t) + \frac{Gm^2}{a}\frac{\partial}{\partial p_1{}^\alpha}\int d^3x_2 d^3p_2 c(1, 2) x_{21}{}^\alpha/x_{21}{}^3 = 0, \\
\mathbf{x}_{21} = \mathbf{x}_2 - \mathbf{x}_1.
\end{gathered} \tag{68.7}$$

This is the first BBGKY equation.

For the next equation consider two small separate patches in the six dimensional phase space. If the patches are small enough, we can describe the matter distribution in them by the two particle distribution function, $f_a(1, 2)$ in the ath realization, with $\mathbf{x}_1$, $\mathbf{p}_1$ in one patch, $\mathbf{x}_2$, $\mathbf{p}_2$ in the other. Liouville's equation is extended from equation (68.2) to

$$\frac{\partial f_a}{\partial t} + \frac{dx_1^\alpha}{dt}\frac{\partial f_a}{\partial x_1^\alpha} + \frac{dp_1^\alpha}{dt}\frac{\partial f_a}{\partial p_1^\alpha} + (1 \leftrightarrow 2) = 0. \tag{68.8}$$

The last term indicates that the results of exchanging coordinates 1 and 2 in the second and third terms should be added. The averages across the ensemble for the first two terms are

$$\begin{aligned} \langle \partial f_a/\partial t\rangle &= \partial\rho_2/\partial t, \\ \left\langle \frac{dx_1^\alpha}{dt}\frac{\partial f_a}{\partial x_1^\alpha}\right\rangle &= \frac{p_1^\alpha}{ma^2}\frac{\partial c}{\partial x_1^\alpha}. \end{aligned} \tag{68.9}$$

In the third term, we have

$$\begin{aligned} \langle f_a(1, 2)d\mathbf{p}_1/dt\rangle &= (Gm^2/a)\langle f_a(1, 2)\rangle \mathbf{x}_{21}/x_{21}^3 \\ &\quad + Gm^2/a\langle f_a(1, 2)\sum_i \mathbf{x}_{i1}/x_{i1}^3\rangle \\ &= (Gm^2/a)\rho_2(1, 2)\mathbf{x}_{21}/x_{21}^3 \\ &\quad + Gm^2/a \int d^3x_3 d^3p_3 \rho_3(1, 2, 3)\mathbf{x}_{31}/x_{31}^3, \\ &\qquad \mathbf{x}_{i1} = \mathbf{x}_i - \mathbf{x}_1. \end{aligned} \tag{68.10}$$

The gravitational acceleration caused by the particle at $\mathbf{x}_2$,$\mathbf{p}_2$ has been written separately. The ensemble average of equation (68.8) becomes then

$$\begin{aligned} &\frac{\partial}{\partial t}\rho_2(1, 2) + \frac{p_1^\alpha}{ma^2}\frac{\partial}{\partial x_1^\alpha}\rho_2(1, 2) + \frac{Gm^2}{a}\frac{x_{21}^\alpha}{x_{21}^3}\frac{\partial}{\partial p_1^\alpha}\rho_2(1, 2) \\ &\quad + \frac{Gm^2}{a}\frac{\partial}{\partial p_1^\alpha}\int d^3x_3 d^3p_3\rho_3(1, 2, 3)x_{31}^\alpha/x_{31}^3 \\ &\quad + (1 \leftrightarrow 2) = 0. \end{aligned} \tag{68.11}$$

This is the second BBGKY equation. It can be written in terms of the reduced functions c and d (eqs. 67.4 and 67.6): on using equation (68.7) to eliminate the time derivatives of b and the prescription in Section 8 to eliminate the spatial integrals over b, one finds

$$
\begin{aligned}
\frac{\partial}{\partial t} c(1,2) &+ \frac{p_1{}^\alpha}{ma^2}\frac{\partial}{\partial x_1{}^\alpha} c(1,2) \\
&+ \frac{Gm^2}{a}\frac{x_{21}{}^\alpha}{x_{21}{}^3}\frac{\partial}{\partial p_1{}^\alpha}[b(1)b(2) + c(1,2)] \\
&+ \frac{Gm^2}{a}\frac{\partial b(p_1)}{\partial p_1{}^\alpha}\int d^3x_3 d^3p_3 c(2,3) x_{31}{}^\alpha / x_{31}{}^3 \\
&+ \frac{Gm^2}{a}\frac{\partial}{\partial p_1{}^\alpha}\int d^3x_3 d^3p_3 d(1,2,3) x_{31}{}^\alpha / x_{31}{}^3 \\
&+ (1 \leftrightarrow 2) = 0.
\end{aligned}
\tag{68.12}
$$

Again the equation must be symmetrized by adding the result of interchanging 1 and 2 in all terms save the first.

The third and fourth equations in the hierarchy are obtained in the same way. All that will be needed for the following discussion is to note that the ensemble averages of the Liouville equations are of the forms

$$
\begin{aligned}
&\frac{\partial}{\partial t}\rho_3(1,2,3) + \sum_{i=1,3}\left(\frac{p_i^\alpha}{ma^2}\frac{\partial \rho_3}{\partial x_i^\alpha} + \frac{\partial F^\alpha}{\partial p_i^\alpha}\right) = 0, \\
&\frac{\partial}{\partial t}\rho_4(1,2,3,4) + \sum_{i=1,4}\left(\frac{p_i^\alpha}{ma^2}\frac{\partial \rho_4}{\partial x_i^\alpha} + \frac{\partial G^\alpha}{\partial p_i^\alpha}\right) = 0,
\end{aligned}
\tag{68.13}
$$

where F^α and G^α are the sums of the contributions to the rates of change of the momenta.

The Bogoliubov-Born-Green-Kirkwood-Yvon hierarchy equations have been discussed at length in the theory of plasmas and nonideal gases (for example, Montgomery and Tidman 1964, Ichimaru 1973). The only difference here is that the potential has been adjusted and the coordinates have been changed to the cosmological variables $\mathbf{x}, \mathbf{p}$. Application of the BBGKY method to the behavior of the galaxy distribution was first discussed by van Albada (1960), Gilbert (1965, 1966), Bisnovatyi-Kogan and Zel'dovich (1970) and Saslaw (1972). The BBGKY hierarchy equations in the forms given here were derived by Fall and Severne (1976) and Davis and Peebles (1977).

69. Fluid limit

The fluid limit assumes the gravitational acceleration of a particle typically has substantial contributions from many neighbors. When this is so the first term on the right hand side of equation (68.10) representing the direct gravitational interaction between the particles can be neglected compared to the second, which represents the effect of all the neighbors of the pair. This reduces equation (68.12) to

$$\begin{aligned}
&\frac{\partial}{\partial t} c(1,2) + \frac{p_1^\alpha}{ma^2}\frac{\partial}{\partial x_1^\alpha} c(1,2) \\
&\quad + \frac{Gm^2}{a}\frac{\partial}{\partial p_1^\alpha}\int d^3x_3 d^3p_3 [b(p_1)c(2,3) + d(1,2,3)] x_{31}^\alpha / x_{31}^3 \\
&\quad + (1 \leftrightarrow 2) = 0.
\end{aligned} \tag{69.1}$$

Inagaki (1976) has pointed out that the BBGKY hierarchy in the fluid limit can be derived in a simple way from the Vlasov equation. The matter is described by the continuous single particle distribution function $f(\mathbf{x}, \mathbf{p}, t)$, and the gravitational acceleration is computed from

$$\nabla^2 \phi = 4\pi G \rho a^2, \qquad \rho = ma^{-3}\int f d^3p. \tag{69.2}$$

The particle distribution is random on scales small compared to the coherence length of f: the correlations appear through the fluctuations of f. The correlation functions defined in Section 67 are

$$\begin{aligned}
b(p,t) = \langle f(\mathbf{x}, \mathbf{p}, t)\rangle, \qquad &\rho_2(1,2) = \langle f(1) f(2)\rangle, \\
\rho_3(1,2,3) &= \langle f(1) f(2) f(3)\rangle, \\
\rho_4(1,2,3,4) &= \langle f(1) f(2) f(3) f(4)\rangle,
\end{aligned} \tag{69.3}$$

and so on. The Vlasov equation is

$$\frac{\partial f}{\partial t} + \frac{p^\alpha}{ma^2}\frac{\partial f}{\partial x^\alpha} - m\frac{\partial \phi}{\partial x^\alpha}\frac{\partial f}{\partial p^\alpha} = 0. \tag{69.4}$$

The average of the last term is

$$\begin{aligned}
&\frac{Gm^2}{a}\left\langle \frac{\partial f(\mathbf{x}_1, \mathbf{p}_1, t)}{\partial p_1^\alpha}\int \frac{f(\mathbf{x}_2, \mathbf{p}_2, t) x_{21}^\alpha}{x_{21}^3} d^3x_2 d^3p_2 \right\rangle \\
&\quad = \frac{Gm^2}{a}\frac{\partial}{\partial p_1^\alpha}\int d^3x_2 d^3p_2 \rho_2(1,2) x_{21}^\alpha / x_{21}^3.
\end{aligned} \tag{69.5}$$

Thus the average of equation (69.4) is

$$\frac{\partial b(1)}{\partial t} + \frac{Gm^2}{a} \frac{\partial}{\partial p_1^\alpha} \int d^3x_2 d^3p_2 \rho_2(1, 2) x_{21}^\alpha / x_{21}^3 = 0, \qquad (69.6)$$

which agrees with equation (68.7). The result of differentiating the second of equations (69.3) with respect to time and using equation (69.4) is

$$\begin{aligned} \partial\rho_2/\partial t &= \langle f(2)\partial f(1)/\partial t\rangle + \langle f(1)\partial f(2)/\partial t\rangle \\ &= -\frac{p_1^\alpha}{ma^2}\langle f(2)\partial f(1)/\partial x_1^\alpha\rangle \\ &\quad - \frac{Gm^2}{a}\langle f(2)\partial f(1)/\partial p_1^\alpha \int d^3x_3 d^3p_3 f(3) x_{31}^\alpha/x_{31}^3\rangle \\ &\quad + (1 \leftrightarrow 2) \qquad (69.7) \\ &= -\frac{p_1^\alpha}{ma^2}\frac{\partial\rho_2(1,2)}{\partial x_1^\alpha} \\ &\quad - \frac{Gm^2}{a}\frac{\partial}{\partial p_1^\alpha}\int d^3x_3 d^3p_3 \rho_3(1,2,3) x_{31}^\alpha/x_{31}^3 \\ &\quad + (1 \leftrightarrow 2). \end{aligned}$$

This agrees with equation (68.11) in the fluid limit. In the same way by writing out the time derivatives of the third and fourth of equations (69.3) with the help of equation (69.4), one arrives at equations (68.13) for the third and fourth members of the BBGKY hierarchy.

70. Evolution of the Integral of ξ

When the correlations and the discrete particle interactions both are weak the BBGKY equations can be solved for $\xi(x, t)$ in the approximation of linear perturbation theory. A convenient approach was pointed out by Inagaki (1976): in the fluid limit the problem is to solve the Vlasov equation (69.4) for f, and we know that in the linear perturbation approximation (§ 10) the density varies as

$$\int d^3p\, f(\mathbf{x}, \mathbf{p}, t) = na^3[1 + A(\mathbf{x})D_1(t) + B(\mathbf{x})D_2(t)], \qquad (70.1)$$

where D_1 and D_2 are linearly independent solutions to equation (10.3). Thus we have from equations (67.5) and (69.3) with $\langle A\rangle = \langle B\rangle = 0$

Inagaki's solution

$$\begin{aligned}\xi(r, t) = D_1(t)^2\langle A(\mathbf{x} + \mathbf{r})A(\mathbf{x})\rangle \\ + 2D_1(t)D_2(t)\langle A(\mathbf{x} + \mathbf{r})B(\mathbf{x})\rangle \\ + D_2(t)^2\langle B(\mathbf{x} + \mathbf{r})B(\mathbf{x})\rangle.\end{aligned} \tag{70.2}$$

If the growing mode dominates,

$$\xi(r, t) = D_1(t)^2\xi_0(r). \tag{70.3}$$

Another way to arrive at this solution is to use the relation in equation (41.8) between ξ and the power spectrum (Peebles and Groth 1976). If

$$\langle|\delta_{\mathbf{k}}|^2\rangle \gg 1/nV_u, \tag{70.4}$$

the discrete particle noise can be neglected and

$$\xi(x, t) \propto \int d^3k\langle|\delta_{\mathbf{k}}|^2\rangle(\sin kx)/kx. \tag{70.5}$$

In the linear perturbation approximation $\delta_{\mathbf{k}}(t)$ varies with time as a linear combination of $D_1(t)$ and $D_2(t)$, so equation (70.5) reduces to equation (70.2).

A very useful property of the power spectrum is that the long wavelength part varies according to linear perturbation theory even when the mass distribution on small scales is strongly nonlinear (§ 28). We have from equation (41.8) the relation

$$\int d^3x\xi(x, t)e^{-x^2/2x_0^2} = \frac{V_u x_0^3}{(2\pi)^{3/2}}\int d^3k(\langle|\delta_{\mathbf{k}}|^2\rangle - 1/na^3V_u)e^{-k^2x_0^2/2}. \tag{70.6}$$

If x_0 is larger than the interparticle distance and the clustering length, then the part of the spectrum that appears in this integral, $k \lesssim x_0^{-1}$, is in the linear regime, and we have

$$na^3\int d^3x\xi(x, t)e^{-x^2/2x_0^2} \propto D(t)^2, \tag{70.7}$$

for the mode $D(t)$. This integral constraint is valid even though the clustering may be strongly nonlinear on scales less than x_0.

One interesting special case might be mentioned. Suppose at time t_i the particles are distributed uniformly at random with zero peculiar velocities. Then the long wavelength part of the spectrum is

$$\langle|\delta_{\mathbf{k}}|^2\rangle = D(t)^2/(na^3 V_u), \tag{70.8}$$

where D is the solution to equation (10.3) with initial values $D = 1$, $dD/dt = 0$. This in equation (70.6) gives

$$na^3 \int d^3x\xi(x, t) = D(t)^2 - 1. \tag{70.9}$$

The integral is independent of x_0, when x_0 is large compared to the interparticle distance, consistent with the fact that the power spectrum is flat at long wavelength with ξ negligibly small. This equation shows how the number of neighbors in excess of random grows from the initial value zero in an initially uniform random Poisson process.

Equation (70.7) provides a useful constraint on speculations on the possible rate of growth of clustering. For another derivation of the equation and tests in the spherical model and N-body simulations, see Peebles and Groth (1976).

71. Particle conservation equations

A. Conservation of pairs

When equation (68.12) is integrated over p_1 and p_2, it eliminates all the derivatives with respect to the momenta leaving

$$\begin{aligned} n^2a^6\frac{\partial}{\partial t}\xi(x, t) &+ \frac{\partial}{\partial x_1^\alpha}\int d^6pc(1, 2)p_1^\alpha/ma^2 \\ &+ \frac{\partial}{\partial x_2^\alpha}\int d^6pc(1, 2)p_2^\alpha/ma^2 = 0. \end{aligned} \tag{71.1}$$

The correlation function c has several symmetry properties:

$$\begin{aligned} &\text{homogeneity: } c(1, 2) = c(\mathbf{x}_2 - \mathbf{x}_1, \mathbf{p}_1, \mathbf{p}_2); \\ &\text{exchange: } c(1, 2) = c(2, 1); \\ &\qquad c(\mathbf{x}, \mathbf{p}_1, \mathbf{p}_2) = c(-\mathbf{x}, \mathbf{p}_2, \mathbf{p}_1); \\ &\text{parity: } c(\mathbf{x}, \mathbf{p}_1, \mathbf{p}_2) = c(-\mathbf{x}, -\mathbf{p}_1, -\mathbf{p}_2) = c(\mathbf{x}, -\mathbf{p}_2, -\mathbf{p}_1). \end{aligned} \tag{71.2}$$

With these properties, equation (71.1) can be reduced to

$$n^2a^6\frac{\partial}{\partial t}\xi(x,t) = 2\frac{\partial}{\partial x^\alpha}\int d^6pc(1,2)p_1{}^\alpha/ma^2,$$

$$n^2a^6\frac{\partial}{\partial t}\xi(x,t) + \frac{\partial}{\partial x^\alpha}\int d^6pc(1,2)(p_2{}^\alpha - p_1{}^\alpha)/ma^2 = 0, \tag{71.3}$$

$$\mathbf{x} = \mathbf{x}_2 - \mathbf{x}_1.$$

The mean (ensemble average) value of the peculiar velocity difference of a particle pair at separation $\mathbf{x}$ is (eqs. 67.1 and 67.4)

$$\begin{aligned}\mathbf{v} &= \left(\int d^6p\rho_2(1,2)(\mathbf{p}_2 - \mathbf{p}_1)/ma\right)\Big/\int d^6p\rho_2(1,2) \\ &= \left(\int d^6pc(1,2)(\mathbf{p}_2 - \mathbf{p}_1)/ma\right)/[n^2a^6(1+\xi(x,t))].\end{aligned} \tag{71.4}$$

Since

$$v^\alpha = vx^\alpha/x, \tag{71.5}$$

equations (71.3) and (71.4) yield (Peebles 1976b, Davis and Peebles 1977)

$$\frac{\partial\xi}{\partial t} + \frac{1}{x^2a}\frac{\partial}{\partial x}[x^2(1+\xi)v] = 0. \tag{71.6}$$

This equation just expresses conservation of particle pairs, as can be seen by writing it as

$$\frac{\partial}{\partial t}\left(na^3\int_0^x 4\pi x^2\,dx[1+\xi(x,t)]\right) + 4\pi a^2x^2n(1+\xi)v = 0. \tag{71.7}$$

The first term is the rate of change of the mean number of neighbors within distance x of a particle and the second is the mean flux of neighbors out of the surface x = constant.

At small x where $\xi \gg 1$ the clustering is presumed to be fairly well mixed and so not relaxing very fast. This would say the average relative velocity of particle pairs is close to zero, and since $\mathbf{v} = \mathbf{v}_2 - \mathbf{v}_1$ is computed relative to the uniformly expanding coordinates,

$$v = -\dot{a}x, \tag{71.8}$$

so equation (71.6) in this limit is

$$\frac{\partial}{\partial t}(1+\xi) = \frac{\dot{a}}{x^2 a}\frac{\partial}{\partial x}x^3(1+\xi). \tag{71.9}$$

The solution is

$$1+\xi = a^3 g(a(t)x). \tag{71.10}$$

This means $n\xi$ at fixed proper separation $a(t)x$ is independent of time, as required if the average clustering is not changing.

At large x where $\xi \ll 1$ the weak correlation approximation (70.7) is

$$\int_0^x x^2 dx \xi(x,t) \propto D_1(t)^2, \tag{71.11}$$

if the growing mode $D_1(t)$ dominates. Then equation (71.7) is

$$v = -\frac{2a}{x^2(1+\xi)}\frac{\dot{D}_1}{D_1}\int_0^x x^2 dx \xi(x,t). \tag{71.12}$$

If, for example, $\xi = (x_0/x)^\gamma$, $x < x_{mx}$; $\xi = 0$, $x > x_{mx}$,

$$v = -\frac{2}{3-\gamma}\frac{r_0{}^\gamma r_{mx}{}^{3-\gamma}}{r^2}\frac{\dot{D}_1}{D_1}, \tag{71.13}$$

at $x > x_0$. It is left as an exercise to check that equation (71.12) agrees with equation (14.9). The possibility of observing this mean tendency of well-separated galaxies to approach each other is discussed in Section 76.

The pairs conservation equation is a guide to speculation on the behavior of the correlation function: an assumed $\xi(x,t)$ implies through equation (71.7) a function $v(x,t)$ that should agree with equations (71.8) and (71.12) in the limits of small and large x and interpolate between these limits in a not unreasonable way. A useful approximation to ξ is found by setting $v(x,t)$ equal to the limiting values with a discontinuous jump at some $x_0(t)$. Given v, equation (71.6) can be integrated to get ξ. If the universe is close to cosmologically flat so $a \propto D_1 \propto t^{2/3}$ and if $\xi = 0$ at large separation, this approximation to the relative velocity is

$$\begin{aligned} -v/a &= 2x/3t, \qquad x < x_0 \propto t^{4/9}; \\ &= Ct^{1/3}/x^2, \qquad x > x_0. \end{aligned} \tag{71.14}$$

The solution to equation (71.6) is

$$1 + \xi = (1 + \xi_1)(x_0(t)/x)^{9/5}, \qquad x < x_0(t),$$
$$= 1, \qquad x > x_0(t), \tag{71.15}$$

where the constant ξ_1 is

$$\xi_1 = \frac{9}{10}\frac{Ct^{4/3}}{x_0^{\,3}} - \frac{3}{5}. \tag{71.16}$$

This is the shoulder in $\xi(x)$ at the break in the value of v. A convenient number in the model is the ratio of values of v at either side of x_0,

$$K = v_+/v_- = 3Ct^{4/3}/2x_0^{\,3}, \tag{76.17}$$

in terms of which the shoulder in ξ is

$$\xi_1 = \frac{3}{5}\left(\frac{v_+}{v_-} - 1\right). \tag{71.18}$$

If $K = 1$, v and ξ are continuous. It is sometimes argued that newly formed clusters must collapse to relax to virial equilibrium (Gott and Rees 1975). If this were so, $|v|$ would have to be larger than $\dot{a}x$ to make $v + \dot{a}x$ negative. In the model this means $K > 1$, but that causes a shoulder in ξ at $\xi \sim 1$, contrary to what is observed for the galaxy correlation function. Thus it is difficult to see how there could be a well-defined collapse effect, though, of course, some clusters could be collapsing while others of the same size are expanding so that the mean streaming velocity v yields an acceptable form for $\xi(x)$.

Another application of equation (71.9) is to the growth of ξ in a very low density cosmological model. For a simple model, suppose that at $r < r_s$ the clustering is statistically stable and that at $r > r_s$ there develops a power law tail:

$$\xi = a^3 g(ax), \qquad ax < r_s,$$
$$\xi = a^3 g(r_s)(r_s/ax)^m, \qquad r_s \le ax \le ax_{mx}, \tag{71.19}$$
$$ax_{mx} = r_s(a^3 g(r_s))^{1/m}, \qquad \xi = 0, \qquad x > x_{mx},$$

where g is a function of the one variable r, and r_s is a constant. The last two equations express the assumption that the correlation function is cut off at $\xi = 1$. Now equation (71.11) says that when $\Omega \ll 1$ and $D_1(t)$ stops growing, the integral of ξ stops growing. In the model that means $m > 3$, for otherwise there is a growing contribution to the integral from the tail of ξ at $ax > r_s$. An argument for $m = 3$ was given by Gott and Rees (1975). Figure 52.1 shows that a sharp break, $\xi \propto r^{-2}$ to $\xi \propto r^{-3}$, would be quite prominent in the angular function $w(\theta)$, but none is observed. This is the origin of the problem of accounting for the observed $w(\theta)$ in a low density model (Davis, Groth, and Peebles 1977).

B. Conservation of triplets and quadruplets

The result of integrating the first of equations (68.13) over momenta is the triplets conservation equation,

$$\begin{aligned} &\frac{\partial h_3}{\partial t} + \frac{1}{a}\sum_{i=1,3} \frac{\partial}{\partial x_i^\alpha}(h_3 v_i^\alpha) = 0, \\ h_3(1, 2, 3) &= 1 + \xi(1, 2) + \xi(2, 3) + \xi(3, 1) + \zeta(1, 2, 3) \\ &= \int d^9p\rho_3(1, 2, 3)/n^3a^9, \end{aligned} \tag{71.20}$$

where h_3 is the full spatial correlation function and $\mathbf{v}_i$ is the ensemble average value of the peculiar velocity of the particle at the vertex i of the triplet of particles at separations $\mathbf{x}_2 - \mathbf{x}_1$, $\mathbf{x}_3 - \mathbf{x}_1$:

$$v_i^\alpha = \left(\int d^9p\rho_3 p_i^\alpha/ma\right)\Big/(n^3a^9h_3). \tag{71.21}$$

In terms of the reduced correlation functions of equation (67.6), equations (71.20) and (71.21) are (Davis and Peebles 1977)

$$n^3a^9\frac{\partial\zeta}{\partial t} + \sum \frac{\partial}{\partial x_i^\alpha}\int d^9p\,\frac{p_i^\alpha}{ma^2}d = 0, \tag{71.22}$$

where equation (71.1) has been used to eliminate the time derivatives of ξ.

Equation (71.20) is simplified by writing h_3 and the $\mathbf{v}_i$ as functions of relative position and mean velocity,

$$\begin{aligned} \mathbf{x}_{21} &= \mathbf{x}_2 - \mathbf{x}_1, \qquad \mathbf{x}_{31} = \mathbf{x}_3 - \mathbf{x}_1, \\ \mathbf{v}_{21} &= \mathbf{v}_2 - \mathbf{v}_1, \qquad \mathbf{v}_{31} = \mathbf{v}_3 - \mathbf{v}_1, \end{aligned} \tag{71.23}$$

which gives

$$a\frac{\partial h_3}{\partial t}+\frac{\partial}{\partial x_{21}{}^{\alpha}}v_{21}{}^{\alpha}h_3+\frac{\partial}{\partial x_{31}{}^{\alpha}}v_{31}{}^{\alpha}h_3=0. \qquad (71.24)$$

At small separations the mean relative velocities of the particles are close to zero, so, as in equation (71.8),

$$\mathbf{v}_{21}=-\dot{a}\mathbf{x}_{21}, \qquad \mathbf{v}_{31}=-\dot{a}\mathbf{x}_{31}, \qquad (71.25)$$

which makes equation (71.24)

$$a\frac{\partial h_3}{\partial a}=6h_3+x_{21}{}^{\alpha}\frac{\partial h_3}{\partial x_{21}{}^{\alpha}}+x_{31}{}^{\alpha}\frac{\partial h_3}{\partial x_{31}{}^{\alpha}}. \qquad (71.26)$$

The solution is

$$h_3=a^6g_3(a\mathbf{x}_{21},a\mathbf{x}_{31}). \qquad (71.27)$$

As in equation (71.10) this just means the chance of finding neighbors at fixed proper distances $a\mathbf{x}_{21}$ and $a\mathbf{x}_{31}$ is independent of time.

The equation of conservation of quadruplets is found by integrating the second of equations (68.13) over momenta:

$$\frac{\partial h_4}{\partial t}+\frac{1}{a}\sum_{i=1,4}\frac{\partial}{\partial x_i^{\alpha}}(h_4v_i^{\alpha})=0,$$
$$n^3a^{12}h_4=\int\rho_4d^{12}p. \qquad (71.28)$$

Here v_i^{α} is the ensemble average peculiar velocity of the ith particle in the configuration. With the relative coordinates defined as in equation (71.23), the equation becomes

$$a\frac{\partial h_4}{dt}+\frac{\partial}{\partial x_{21}{}^{\alpha}}h_4v_{21}{}^{\alpha}+\frac{\partial}{\partial x_{31}{}^{\alpha}}h_4v_{31}{}^{\alpha}+\frac{\partial}{\partial x_{41}{}^{\alpha}}h_4v_{41}{}^{\alpha}=0. \qquad (71.29)$$

As before, one sees that at small separations, where $\mathbf{v}_{21}=-\dot{a}\mathbf{x}_{21}$ and so on, the equation is

$$a\frac{\partial h_4}{\partial a}=9h_4+x_{21}{}^{\alpha}\frac{\partial h_4}{\partial x_{21}{}^{\alpha}}+x_{31}{}^{\alpha}\frac{\partial h_4}{\partial x_{31}{}^{\alpha}}+x_{41}{}^{\alpha}\frac{\partial h_4}{\partial x_{41}{}^{\alpha}}, \qquad (71.30)$$

with the solution

$$h_4 = a^9 g_4(a\mathbf{x}_{21}, a\mathbf{x}_{31}, a\mathbf{x}_{41}), \tag{71.31}$$

which can be compared to equations (71.10) and (71.27).

72. Relative peculiar velocity dispersion

Consider a random sample of particle pairs each at separation $\mathbf{x} = \mathbf{x}_2 - \mathbf{x}_1$ with x large enough that the correlation functions are small. The mean peculiar velocity of a particle in this sample was discussed in Section 71. The anisotropy of the relative peculiar velocity dispersion is derived here.

The velocity dispersion appears in the first momentum moment of the second BBGKY equation (68.12). The result of multiplying this equation by $p_2{}^\alpha - p_1{}^\alpha$, integrating over momenta, and simplifying the expression by the methods in section 71 is

$$\begin{aligned}\frac{\partial}{\partial t}(1 + \xi(x))v^\alpha + \frac{\dot{a}}{a}(1 + \xi(x))v^\alpha \\ + \frac{1}{a}\frac{\partial}{\partial x^\beta}(1 + \xi)\langle v_{21}{}^\alpha v_{21}{}^\beta\rangle \\ + \frac{2Gm}{a^2}\frac{x^\alpha}{x^3}(1 + \xi) + 2G\rho_b a\frac{x^\alpha}{x^3}\int_0^x d^3x\xi(x) \\ + 2G\rho_b a\int d^3x_3\zeta(1, 2, 3)x_{31}{}^\alpha/x_{31}{}^3 = 0.\end{aligned} \tag{72.1}$$

Here $v^\alpha = \langle v_{21}{}^\alpha\rangle$ is the mean value of the relative peculiar velocity (eq. 71.4). Since v^α fixes the rate of growth of the two-point function ξ, this equation shows how the evolution of ξ is determined by velocity dispersion and gravity. The first of the gravity terms describes the mutual attraction of the particle pair. For strong clustering this is small compared to the next two that describe the gravitational attraction of all the neighboring particles. When x is large, the dominant gravitational term is the integral over ξ and if $\xi(x)$ is negligibly small, equation (72.1) may be approximated as

$$\frac{\partial v^\alpha}{\partial t} + \frac{\dot{a}}{a}v^\alpha + \frac{1}{a}\frac{\partial}{\partial x^\beta}\langle v_{21}{}^\alpha v_{21}{}^\beta\rangle + 2G\rho_b a\frac{x^\alpha}{x^3}\int_0^x d^3x\xi = 0. \tag{72.2}$$

We can write the relative velocity dispersion as

$$\langle v_{21}{}^{\alpha} v_{21}{}^{\beta} \rangle = (\tfrac{2}{3}\langle v^2 \rangle + \Sigma)\delta_{\alpha\beta} + (\Pi - \Sigma) x^{\alpha} x^{\beta}/x^2, \tag{72.3}$$

where $\langle v^2 \rangle$ is the mean square peculiar velocity of randomly chosen particles and Π and Σ represent the effects of correlated motions on the components of the dispersion parallel and perpendicular to the separation x^{α}. This in equation (72.2) yields

$$\begin{aligned} \frac{\partial v^{\alpha}}{\partial t} + \frac{\dot{a}}{a} v^{\alpha} + \frac{1}{a}\left[\frac{2(\Pi - \Sigma)}{x} + \frac{d\Pi}{dx}\right]\frac{x^{\alpha}}{x} \\ + 2G\rho_b a \frac{x^{\alpha}}{x^3}\int_0^x d^3x \xi = 0. \end{aligned} \tag{72.4}$$

We can estimate Π and Σ by noticing that when x is large the most important contributions to these terms come from the correlation of the peculiar motions of the two particles. For the peculiar accelerations of the particles we have

$$\langle g_1{}^{\alpha} g_2{}^{\beta} \rangle = (Gma)^2 \int d^3x_3 d^3x_4 \frac{x_{31}{}^{\alpha} x_{42}{}^{\beta}}{(x_{31} x_{42})^3} P, \tag{72.5}$$

where $P d^3x_3 d^3x_4$ gives the probability of finding particles at $\mathbf{x}_3$ and $\mathbf{x}_4$ given that there are particles at $\mathbf{x}_1$ and $\mathbf{x}_2$. This is the four-point distribution in equation (35.1). When x is large, the largest part of P is the term $n^2(1 + \xi(x_{34}))$, which represents the clustering among the particles in the neighborhood of particles 1 and 2. To simplify evaluation of the integral in equation (72.5), let us suppose that

$$I = \int_0^x \xi(x)\, d^3x \tag{72.6}$$

converges to a fixed value at $x_0 \ll x$. Then $\mathbf{x}_3 \simeq \mathbf{x}_4$ in the integral and

$$\begin{aligned} \langle g_1{}^{\alpha} g_2{}^{\beta} \rangle &\simeq (G\rho_b a)^2 I J^{\alpha\beta}, \\ J^{\alpha\beta} &= \int d^3x_3 \frac{x_{31}{}^{\alpha} x_{32}{}^{\beta}}{(x_{31} x_{32})^3} = 2\pi\left(\frac{\delta^{\alpha\beta}}{x} - \frac{x^{\alpha} x^{\beta}}{x^3}\right). \end{aligned} \tag{72.7}$$

Now since the correlation between g_1 and g_2 is weak at large x, we can compute the resulting correlation of the peculiar velocities in linear perturbation theory (eq. 14.6),

$$\begin{aligned} \langle v_{21}{}^{\alpha} v_{21}{}^{\beta} \rangle &\simeq \tfrac{2}{3} \langle v^2 \rangle \delta^{\alpha\beta} - 2 \langle v_1{}^{\alpha} v_2{}^{\beta} \rangle, \\ \langle v_1{}^{\alpha} v_2{}^{\beta} \rangle &\simeq \langle g_1{}^{\alpha} g_2{}^{\beta} \rangle \dot{D}^2 / (4\pi G \rho_b D)^2, \end{aligned} \tag{72.8}$$

from which we obtain

$$\Sigma = -\frac{a^2 I}{4\pi x}\left(\frac{\dot{D}}{D}\right)^2, \qquad \Pi = 0. \tag{72.9}$$

It is an interesting exercise to see how this result follows from equations (27.20) and (27.22). We have from equations (72.4) and (72.9)

$$\frac{\partial v^{\alpha}}{\partial t} + \frac{\dot{a}}{a} v^{\alpha} = -\left[2G\rho_b a I + \frac{aI}{2\pi}\left(\frac{\dot{D}}{D}\right)^2 \right] \frac{x^{\alpha}}{x^3}. \tag{72.10}$$

It is left as an exercise to check that this agrees with equation (71.12) for $v(x, t)$ with equation (11.1) for D and $I \propto D^2$ (eq. 70.3).

The anisotropy of the relative velocity dispersion in this approximation is

$$\Pi - \Sigma = \frac{a^2 I}{4\pi x}\left(\frac{\dot{D}}{D}\right)^2. \tag{72.11}$$

In the power law model for ξ (§ 57),

$$\begin{aligned} \xi(r) &= (r_0/r)^{\gamma}, \qquad r < r_{mx} = 10 h^{-1}\ \text{Mpc}; \\ &= 0, \qquad r > r_{mx}; \\ r_0 &= 4 h^{-1}\ \text{Mpc}, \qquad \gamma = 1.77, \end{aligned} \tag{72.12}$$

we have

$$\Pi - \Sigma = (1300\ \text{km s}^{-1})^2 f(\Omega)^2/(hr), \tag{72.13}$$

where r is measured in megaparsecs and $f(\Omega)$ is shown in figure 14.1. The component Σ is negative because the peculiar velocities $\mathbf{v}_1$ and $\mathbf{v}_2$ are correlated, making the relative velocity dispersion less than the limiting value $2\langle v^2 \rangle/3$ at $x \to \infty$. The expected anisotropy $(\Pi - \Sigma)^{1/2}$ is large, amounting to $\sim 400 f$ km s^{-1} at $hr = 10$ Mpc, which is comparable to the expected value of $2\langle v^2\rangle/3$ (eq. 74.9 below). We see from equation (72.10) that the velocity dispersion plays an important role in fixing the rate of growth of clustering. More careful estimates of Σ and Π may prove useful in estimating the shape of $v(x, t)$ and hence through equation (71.6) the expected shape of $\xi(x, t)$ near $\xi \sim 1$.

73. Similarity solution

A. Scale-invariant solution

The BBGKY hierarchy admits solutions invariant under a similarity transformation if the expansion of the universe approximates the Einstein-de Sitter model and the matter can be described in the fluid limit. Under these assumptions neither the cosmology nor the matter presents characteristic lengths or times.

A convenient approach to the similarity solution is from Inagaki's (1976) fluid limit equations (§ 69),

$$\frac{\partial}{\partial t} f(\mathbf{x}, \mathbf{p}, t) + \frac{p^\alpha}{ma^2}\frac{\partial f}{\partial x^\alpha} - m\frac{\partial \phi}{\partial x^\alpha}\frac{\partial f}{\partial p^\alpha} = 0, \qquad \nabla^2\phi = 4\pi Gma^{-1}\int f\, d^3p, \qquad a \propto t^{2/3}. \tag{73.1}$$

These equations admit solutions of the form

$$f(\mathbf{x}, \mathbf{p}, t) = t^{-3\beta}\hat{f}(\mathbf{x}/t^\alpha, \mathbf{p}/t^\beta), \tag{73.2}$$

where the two constants satisfy

$$\beta = \alpha + 1/3, \tag{73.3}$$

as may be verified by substituting equation (73.2) into equations (73.1). This works only if $a \propto t^{2/3}$. The results of averaging equation (73.2) are the similarity forms for the correlation functions (eqs. 69.3),

$$b = t^{-3\beta}\hat{b}(p/t^\beta), \qquad c = t^{-6\beta}\hat{c}(\mathbf{x}/t^\alpha, \mathbf{p}_1/t^\beta, \mathbf{p}_2/t^\beta), \qquad d = t^{-9\beta}\hat{d}(\mathbf{x}_{12}/t^\alpha, \mathbf{x}_{13}/t^\alpha, \mathbf{p}_1/t^\beta, \mathbf{p}_2/t^\beta, \mathbf{p}_3/t^\beta), \tag{73.4}$$

and so on. The integrals over momenta are

$$\xi = \hat{\xi}(x/t^\alpha), \qquad \zeta = \hat{\zeta}(x_{12}/t^\alpha, x_{23}/t^\alpha, x_{31}/t^\alpha), \tag{73.5}$$

and so on. These equations describe a situation in which the matter distributions at different times are similar with characteristic lengths scaled as t^α and characteristic momenta scaled as t^β. For another derivation that does not use equations (73.1), see Davis and Peebles (1977).

The similarity solution demands of course that the initial values of the correlation functions traced back to small fluctuations in the early universe are power laws in separation because there can be no fixed characteristic separation. The power spectrum of the density fluctuations is (§ 41)

$$\langle|\delta_{\mathbf{k}}|^2\rangle \propto \int d^3x\hat{\xi}(x/t^\alpha)e^{i\mathbf{k}\cdot\mathbf{x}} = t^{3\alpha}\int d^3y\hat{\xi}(y)e^{it^\alpha\mathbf{k}\cdot\mathbf{y}}, \tag{73.6}$$

in the similarity solution. At long wavelength the linear approximation applies (§ 70); we have in this limit $\delta_{\mathbf{k}} \propto t^{2/3}$ or

$$\langle|\delta_{\mathbf{k}}|^2\rangle = t^{3\alpha}F(t^\alpha k) = t^{4/3}G(k). \tag{73.7}$$

The solution to this functional equation can be seen by replacing the variable t with $\hat{t} = tk^{1/\alpha}$ and bringing all the powers of $\hat{t}$ to the left side of the equation, all the powers of k to the right-hand side (Davis and Peebles 1977):

$$\langle|\delta_{\mathbf{k}}|^2\rangle \propto t^{4/3}k^n, \qquad \alpha = 4/(9+3n). \tag{73.8}$$

A popular speculation is that the matter distribution at high redshift approximated a random Gaussian process where the Fourier components $\delta_{\mathbf{k}}$ have random phases so the statistical properties of the distribution are fully described by the power spectrum $|\delta_{\mathbf{k}}|^2$. The natural initial value of the power spectrum might be a power law,

$$\langle|\delta_{\mathbf{k}}|^2\rangle \propto k^n, \qquad t = t_i. \tag{73.9}$$

If so, if the expansion of the universe adequately approximates the Einstein-de Sitter model, and if nongravitational forces can be neglected, then the similarity solution applies with the index n fixing α through equation (73.8) and β through equation (73.3).

B. Nonlinear limit

At small x where the correlation functions are large, the clustering is in statistical equilibrium or close to it, so the relative velocities of neighboring particles have zero mean. This with the similarity solution fixes the variation of the correlation functions with separation. For the two-point function, equations (71.10) and (73.5) give

$$\hat{\xi}(x/t^{\alpha}) = a^3 g(ax), \qquad a \propto t^{2/3}, \qquad \xi \gg 1. \tag{73.10}$$

As in equation (73.7) the solution to this functional equation must be a power law,

$$\xi \propto x^{-\gamma}, \tag{73.11}$$

and the result of matching powers of x and of t in equation (73.10) is

$$\gamma = \frac{6}{2 + 3\alpha} = \frac{9 + 3n}{5 + n}, \tag{73.12}$$

where n is the index for the initial power spectrum (eqs. 73.8 and 73.9). This derivation is from Davis and Peebles (1977). Equation (71.15) is a special case of equation (73.12) with $n = 0$. It will be noted also that equation (73.12) agrees with equation (26.8).

For the three-point function the stability assumption with the similarity solution gives (eqs. 71.27 and 73.5)

$$\begin{gathered} a^6 g_3(a\mathbf{x}_{21}, a\mathbf{x}_{31}) = \hat{h}_3(\mathbf{x}_{21}/t^{\alpha}, \mathbf{x}_{31}/t^{\alpha}), \\ h_3 = 1 + \xi_{12} + \xi_{23} + \xi_{31} + \zeta. \end{gathered} \tag{73.13}$$

Convenient variables are x, u, v where x measures the size of the triangle and u, v, the shape (eqs. 34.9). Here one sees that the solution must be a power law in the variable x,

$$h_3 \propto x^{-\sigma}, \qquad \sigma = 12/(2 + 3\alpha) = 2\gamma. \tag{73.14}$$

Since $\xi \propto x^{-\gamma}$, at small enough x the reduced part of the three-point function scales with the size of the triangle as

$$\zeta \propto x^{-\sigma}, \qquad \sigma = 2\gamma. \tag{73.15}$$

It was shown in Section 37 that ζ/ξ^2 cannot be much less than unity, so equation (73.15) must apply down to $\zeta \sim 1$.

For the four-point function, equation (71.31) gives

$$a^9 g_4(a\mathbf{x}_{21}, a\mathbf{x}_{31}, a\mathbf{x}_{41}), = \hat{h}_4(\mathbf{x}_{21}/t^{\alpha}, \mathbf{x}_{31}/t^{\alpha}, \mathbf{x}_{41}/t^{\alpha}), \tag{73.16}$$

from which one finds as above

$$\eta \propto x^{-3\gamma}, \tag{73.17}$$

when the ratios of separations of points in the quadruplet are fixed.

The results $\zeta \propto \xi^2$ and $\eta \propto \xi^3$ agree with the scale-invariant clustering hierarchy picture discussed in Sections 26, 61, and 62, and it is remarkable that they are closely reproduced by the observed behavior of the galaxy correlation functions (eqs. 61.12 and 61.15). The main problem with the application of the theory is that the universe may not be a close enough approximation to the Einstein-de Sitter model to have established the similarity solution. This is discussed in Section 79.

Equations (57.3) and (73.12) give

$$n = -0.12 \pm 0.16. \tag{73.18}$$

The wanted initial power spectrum thus is quite close to white noise. Whether this is a reasonable result must be a matter for conjecture until we have a theory for the origin of the big bang. Some aspects of this question are discussed in Section 96.

The stability condition in equations (73.10) and (73.13) neglects relaxation of the clustering on small scales. Numerical N-body model simulations provide examples where relaxation is important, the collapse of a protocluster smoothing out subcondensations (Peebles 1970b, Aarseth and Hills 1972, White 1976). However, it has also been found that if the N particles are placed in a clustering hierarchy by the prescription in Section 61B and the motions of the subclusters on each level adjusted to produce dynamic stability, then the n-point correlation functions are quite stable (Peebles 1978). This shows that the stability assumption is self-consistent. It remains to be seen whether the stability condition is a good approximation for the galaxy distribution produced by the gravitational instability process.

74. Cosmic energy equation

The energy equation (24.7) can be derived from the BBGKY hierarchy. Equation (68.7) multiplied by p_1^2 and integrated over momentum is

$$\frac{d}{dt} a^2\langle v_1^2\rangle = \frac{2G}{na^4}\int d^3x_2 d^6p c(1,2) p_1^{\alpha} x_{21}^{\alpha}/x_{21}^3, \tag{74.1}$$

where the mean square peculiar particle velocity is

$$na^3\langle v_1{}^2\rangle = \int d^3p\, b(p, t)p^2/m^2a^2. \tag{74.2}$$

The integral in equation (74.1) can be simplified by integration by parts:

$$\frac{d}{dt}a^2\langle v_1{}^2\rangle = \frac{2G}{na^4}\int \frac{d^3x_2}{x_{21}}\frac{\partial}{\partial x_2{}^\alpha}\int d^6p\, c(1, 2)p_1{}^\alpha. \tag{74.3}$$

This with the first of equations (71.3) becomes

$$\begin{gathered}\frac{d}{dt}a^2\langle v_1{}^2\rangle = G\rho_b a^4\frac{\partial}{\partial t}\int d^3x\,\xi(x, t)/x,\\ \rho_b = nm \propto a^{-3},\end{gathered} \tag{74.4}$$

which is the energy equation (24.7) with $K = \langle v_1{}^2\rangle/2$. This derivation from the BBGKY hierarchy was independently discovered by Gilbert (1965), Fall and Severne (1976), and Davis and Peebles (1977).

Bounds on $\langle v_1{}^2\rangle$ for given $\xi(x)$ were obtained in Section 24. In the similarity solution (eqs. 73.4 and 73.5) we have

$$\begin{gathered}a^2\langle v_1{}^2\rangle = \int d^3p(p/m)^2t^{-3\beta}\hat{b}(p/t^\beta)/na^3 \propto t^{2\beta},\\ \int d^3x\,\hat{\xi}(x/t^\alpha)/x \propto t^{2\alpha},\end{gathered} \tag{74.5}$$

and this in equation (74.4) with equations (73.3) and (73.8) gives

$$\begin{gathered}\langle v_1{}^2\rangle = (\alpha/\beta)G\rho_b\int d^3r\,\xi(r)/r,\\ \alpha/\beta = \alpha/(\alpha + \tfrac{1}{3}) = 4/(7 + n).\end{gathered} \tag{74.6}$$

If $n \geq 1$, $\gamma \geq 2$ (eq. 73.12), and the integral over $\xi \propto r^{-\gamma}$ diverges at $r \rightarrow 0$; if $n \leq -1$, the integral diverges at large r where $\xi \propto r^{-(3+n)}$ (§ 42). Between these two cases we have

$$\tfrac{1}{2} < \alpha/\beta < \tfrac{2}{3}, \qquad -1 < n < 1, \tag{74.7}$$

which is within the limits of equation (24.13).

With the model for ξ in equation (72.12) equation (74.6) is

$$\langle v_1^2 \rangle = \frac{16\pi G\rho_b r_0^{\gamma} r_{mx}^{2-\gamma}}{(7+n)(2-\gamma)} = \frac{6\Omega_0 H_0^2 r_0^{\gamma} r_{mx}^{2-\gamma}}{(7+n)(2-\gamma)}. \tag{74.8}$$

With $n = 0$ this is

$$\langle v_1^2 \rangle^{1/2} \simeq 850\, \Omega_0^{1/2} \text{ km s}^{-1}. \tag{74.9}$$

This is the three-dimensional rms peculiar velocity.

Equation (74.9) (or more generally eqs. 24.9 and 24.13) can be used to estimate the density parameter Ω (Fall 1975). There are two practical problems. Even a small positive correlation on scales $\gg r_{mx}$ could considerably increase the integral over ξ and hence the predicted $\langle v_1^2 \rangle$; for that matter there could be some anticorrelation, $\xi < 0$ at $r > r_{mx}$, which would decrease the predicted $\langle v_1^2 \rangle$. Coupled with this is the second problem: estimating the contribution to $\langle v_1^2 \rangle$ from large-scale currents like the motions of groups of galaxies around clusters and the motions of clusters as superclusters form. Such currents are not well explored; some speculations are described in Sections 77 and 78. An estimate of Ω based on the relative peculiar motions of galaxies is given in Section 76.

75. Cosmic virial theorem

Knowing the two- and three-point correlation functions, we can write down the probability of finding a particle in dV at fixed position $\mathbf{z}$ relative to a pair of particles at separation $\mathbf{r}$ (eq. 34.4). If the separations are small, so $\zeta \gg \xi \gg 1$, it is

$$\delta P = n\delta V \zeta(r, z, |\mathbf{r} - \mathbf{z}|)/\xi(r). \tag{75.1}$$

A particle in δV accelerates the particle at $\mathbf{z} = 0$ by the amount

$$\mathbf{g} = Gm\mathbf{z}/z^3, \tag{75.2}$$

so the mean gravitational acceleration of one particle in the pair as measured by an observer sitting on the other is

$$\mathbf{g} = [2G\rho_b/\xi(r)] \int d^3z\, \zeta\mathbf{z}/z^3. \tag{75.3}$$

This is the average over a random sample of pairs at separation $\mathbf{r}$. The factor of 2 appears because the mean accelerations of the two particles relative to the background are equal and opposite. If clustering on the scale r is in statistical equilibrium, the typical relative velocity of the particle pair must be large enough to balance $\mathbf{g}$. This equilibrium condition is obtained from equation (72.1). At small r the dominant gravitational terms are the first and last ones, and they must be balanced by the third term which is effectively a pressure gradient force. This gives

$$\frac{\partial}{\partial r^\beta}[\xi\langle v_{21}{}^\alpha v_{21}{}^\beta\rangle] + 2Gm\xi r^\alpha/r^3 + 2G\rho_b \int d^3z\zeta(r, z, |\mathbf{r} - \mathbf{z}|)z^\alpha/z^3 = 0, \tag{75.4}$$

where $\mathbf{r} = a\mathbf{x}$, $\mathbf{z} = a(\mathbf{x}_3 - \mathbf{x}_1)$. The second term is the relative gravitational acceleration of the pair due to their direct gravitational interaction and the third term is the mean effect of all the neighbors.

The particles described by equation (75.4) could be galaxies or the stars in galaxies. In these two descriptions the velocity terms differ because of the motions of the stars within the galaxies and the gravity terms differ because of the different roles of m and ζ, but we know that when the separation r is much larger than the size r_g of a galaxy the two descriptions must be equivalent. We can see how this comes about by using the model in Section 63 for the relation between the galaxy and mass correlation functions ζ and ζ_ρ.

When the particles are mass elements within galaxies, the direct interaction term in equation (75.4) is negligible and the relation is

$$\partial/\partial r^\beta(\xi_\rho\langle v_{21}{}^\alpha v_{21}{}^\beta\rangle) + 2G\rho_b \int d^3z\zeta_\rho z^\alpha/z^3 = 0. \tag{75.5}$$

If r is greater than the size r_g of a galaxy, we can set ξ_ρ equal to the galaxy two-point function ξ, and we can use equation (63.18) to replace ζ_ρ with the galaxy three-point function ζ. This introduces two terms in equation (75.5). The integral at $|\mathbf{z} - \mathbf{r}| < r_g$ with equation (63.7) yields the second term in equation (75.4),

$$G\rho_b \int d^3z[\xi(z) + \xi(r)]f(|\mathbf{r} - \mathbf{z}|)z^\alpha/z^3 \simeq 2G\xi(r)\frac{r^\alpha}{r^3}\frac{\langle m^2\rangle}{\langle m\rangle}. \tag{75.6}$$

The factor $\langle m^2\rangle/\langle m\rangle^2$ appears because the velocity is weighted by the galaxy mass. The part of the integral at $|\mathbf{z}| < r_g$ is

$$G\rho_b \int d^3z f(z)\xi(|\mathbf{r} - \mathbf{z}|)z^\alpha/z^3 = -\frac{G\rho_b}{3}\frac{d\xi(r)}{dr}\frac{r^\alpha}{r}\int d^3z f(z)/z$$

$$= -\frac{G}{3\langle m\rangle}\frac{d\xi(r)}{dr}\frac{r^\alpha}{r}\int d^3z d^3y \langle \rho_\alpha(\mathbf{y} + \mathbf{z})\rho_\alpha(y)\rangle/z \qquad (75.7)$$

$$= -\frac{2}{3}\frac{d\xi}{dr}\frac{r^\alpha}{r}\frac{\langle m v_i^2\rangle}{\langle m\rangle}.$$

The third expression comes from equation (63.7), the fourth from the virial theorem for the motion within a galaxy. This term thus subtracts from the first term in equation (75.5) the internal motions of particles within galaxies. What is left is the mean dispersion of the galaxy systemic velocities weighted by particle number, that is, by the product of the galaxy masses.

Equation (75.5) is simplified by assuming the distribution of $\mathbf{v}_{21}$ is close to isotropic,

$$\langle v_{21}{}^\alpha v_{21}{}^\beta\rangle \simeq \langle v_{21}{}^2\rangle\delta^{\alpha\beta}/3, \qquad (75.8)$$

as seems reasonable, so

$$\partial/\partial r(\xi_\rho(r)\langle v_{21}{}^2\rangle) + 6G\rho_b r^{-1}\int d^3z \zeta_\rho \mathbf{r}\cdot\mathbf{z}/z^3 = 0. \qquad (75.9)$$

The integral of this is

$$\langle v_{21}{}^2(r)\rangle = \frac{6G\rho_b}{\xi_\rho(r)}\int_r^\infty \frac{dr}{r}\int d^3z \frac{\mathbf{r}\cdot\mathbf{z}}{z^3}\zeta_\rho(r, z, |\mathbf{r} - \mathbf{z}|). \qquad (75.10)$$

A convenient model for the density autocorrelation function is

$$\xi_\rho = [r_0/(r + r_c)]^\gamma, \qquad \zeta_\rho = Q[\xi_\rho(a)\xi_\rho(b) + \text{cycl.}], \qquad (75.11)$$

where the cut-off r_c makes the density flat within patches of size $\sim r_c$ that might approximate galaxies. In this model equation (75.10) is

$$\langle v_{21}{}^2(r)\rangle = 6G\rho_b Q r_0{}^\gamma r^{2-\gamma} I(r_c/r), \qquad (75.12)$$

$$I(y) = (1 + y)^\gamma \int \frac{d^3t}{t^3}\int_1^\infty \frac{ds}{s}\frac{\mathbf{s}\cdot\mathbf{t}}{(|\mathbf{s} - \mathbf{t}| + y)^\gamma}\left[\frac{1}{(s + y)^\gamma} + \frac{1}{(t + y)^\gamma}\right].$$

When $r \gg r_c$, the dimensionless integral simplifies to

$$(\gamma - 1)(2 - \gamma)(4 - \gamma)I(0) = \pi J, \quad J = \int_0^\infty dy y^{-2}(1 + y^{-\gamma})N,$$
$$N = (1 + y)^{4-\gamma} - |1 - y|^{4-\gamma} - (4 - \gamma)y[(1 + y)^{2-\gamma} + |1 - y|^{2-\gamma}]. \tag{75.13}$$

This brings equation (75.12) to

$$\langle v_{21}{}^2(r)\rangle = \frac{6\pi G\rho_b Q r_0{}^\gamma r^{2-\gamma} J}{(\gamma - 1)(2 - \gamma)(4 - \gamma)}, \qquad J(\gamma = 1.8) = 3.70. \tag{75.14}$$

Equation (75.14) can be interpreted as the usual virial theorem in the clustering hierarchy picture (§ 61). On the scale r matter appears in lumps of typical size r and mass

$$M \sim \rho_b \xi(r) r^3 = \rho_b r_0{}^\gamma r^{3-\gamma}. \tag{75.15}$$

The currents on the scale r are therefore

$$v^2 \sim GM/r \sim G\rho_b r_0{}^\gamma r^{2-\gamma}. \tag{75.16}$$

The statistic $\langle v_{21}{}^2(r)\rangle$ is the sum of contributions from motions within subclusters on all levels $\lesssim r$. Since $\gamma < 2$, the main contribution to this sum is from the currents on the scale r, so equation (75.16) agrees with equation (75.14).

Figure 75.1 shows the relative velocity dispersion given by equations (75.12) with $\gamma = 1.8$, $Q = 1.29$, $r_0 = 4h^{-1}$ kpc (§ 57), and cut-off radius

$$r_c = 20\, h^{-1} \text{ kpc}. \tag{75.17}$$

The dispersion scales with the parameters r_c and Ω_0 (eq. 97.12) as

$$\langle v_{21}{}^2(r)\rangle = \Omega_0 r_c{}^{0.2} F(r/r_c), \tag{75.18}$$

and so is quite insensitive to the choice of r_c. The limiting value of the dispersion at $r \ll r_c$ is

$$\langle v_{21}{}^2(0)\rangle = 2.07\, \Omega_0 Q (H r_c)^2 (r_0/r_c)^{1.8}. \tag{75.19}$$

Since this is mainly the relative velocity of particles both in the same patch, the rms particle velocity within a patch is

$$v \cong (\langle v_{21}{}^2(0)\rangle/2)^{1/2} \cong 270\, \Omega_0{}^{1/2} \text{ km s}^{-1}, \qquad (75.20$$

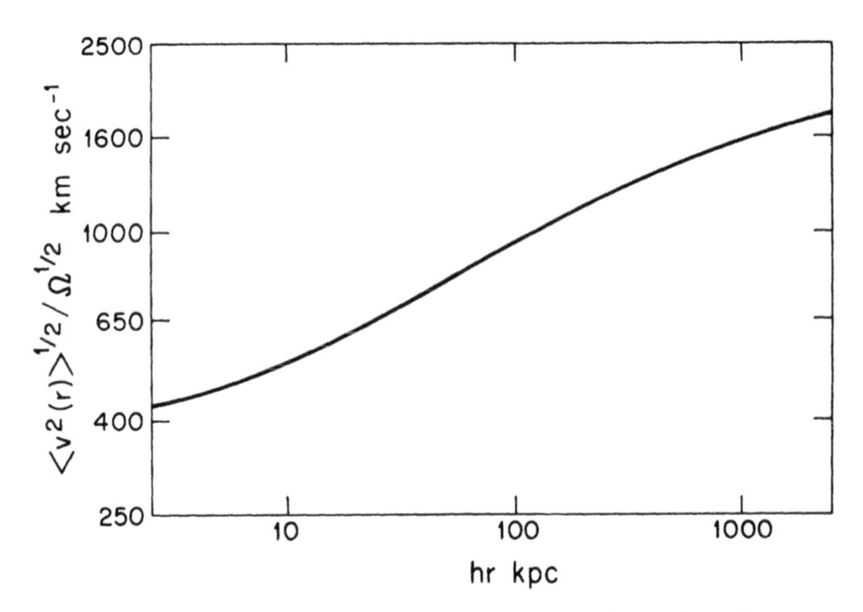

FIG. 75.1. The rms difference of velocities of randomly selected pairs of galaxies as a function of separation. This is computed from equation (75.12) under the assumption of statistical equilibrium (after Peebles 1976c).

interestingly close to the typical velocities in a large galaxy. Of course, this just reflects the apparent continuity of clustering discussed in Section 64.

Equation (75.12) offers a way to estimate the mean mass density. There is the practical advantage over the energy equation (eq. 74.8) that equation (75.12) requires the relative velocity of neighboring galaxies. As is discussed in Section 76, that is much easier to find than the absolute peculiar velocity. Related to this is the fact that the integral over ζ is not sensitive to the poorly known behavior of ζ on large scales because the motion within a cluster of size r is little affected by matter at distance $\gg r$. The relation between $\langle v_{21}{}^2 \rangle$ and the mean mass density is sensitive to the form of ζ_ρ on small scales. The model for this function in equation (75.11) is discussed in Section 64.

The statistic $\langle v_{21}{}^2 \rangle$ was introduced by Geller and Peebles (1973), who compared it to an estimate of the total potential energy of the groups. Equation (75.4) relating $\langle v^2 \rangle$ to the three-point correlation function was derived by Peebles (1976b, c) and Davis and Peebles (1977).

76. JOINT DISTRIBUTION IN REDSHIFT AND POSITION

The precise measures we can hope to have for a fair sample of galaxies are the angular positions and redshifts. These data are conveniently represented in a three-dimensional map with redshift proportional to distance. If peculiar velocities and redshift measuring errors are negligible, this truly represents the galaxy positions, and such maps have proved useful in the study of the large-scale galaxy distribution (Chincarini and Rood 1975, Tifft and Gregory 1976). The pattern of galaxy motions should be revealed in two effects. Compact groups of galaxies should tend to appear elongated along the line of sight because the random motions within the groups add to the apparent spread of distances (Jackson 1972). If well separated galaxies tend to be moving together because new levels of clustering are forming, their apparent separation is reduced, tending to make the large-scale galaxy clustering appear flattened along the line of sight (Sargent and Turner 1977). Some statistical measures of the two effects are discussed in this section. The statistical methods were introduced by Geller and Peebles (1973), Peebles (1976b, c, 1979b) and Davis, Geller, and Huchra (1978). In a related approach Soneira (1978a) examined the joint distribution in redshift and apparent magnitude in model galaxy distributions designed to match the known statistics of galaxy positions and the condition that the clustering be statistically stable.

A. Two-point correlation function in redshift space

We can measure the apparent distortion of the clustering pattern due to peculiar motions by finding the mean distribution of neighbors around a fair sample of galaxies. One way to display the distribution is shown in Figure 76.1. A Cartesian coordinate system with z axis along the line of sight has been centered on each galaxy in the sample and the coordinates $|x|$, $|z|$ of all neighbors with $|y| < 0.5\ h^{-1}$ Mpc recorded. This last restriction is introduced to reduce the distortion in the map due to projection along the y axes. Figure 76.1 is a scatter plot of the positions $|x|$, $|z|$ of the neighbors of all galaxies in the sample. Not all close neighbors are included because of the cutoff on $|y|$, but a fair sample is assured by assigning a different randomly chosen orientation of the x axis for the coordinate system centered on each galaxy. If galaxy peculiar motions and redshift measuring errors were negligibly small, the density of points in this scatter plot would be circularly symmetric and proportional to $1 + \xi(r)$ at $r \gtrsim 0.5\ h^{-1}$ Mpc. Random galaxy motions and redshift measuring errors make the distribution elongated along the z axis. The mean tendency of well-separated galaxies to approach each other would appear as a flattening of the distribution on large scales.

Redshift samples large and accurate enough for statistical analyses are only now becoming available. The data shown in Figures 76.1 and 76.2 are meant only as illustrations of how the statistical analyses might go because the measuring errors in these data are not reliably known. Figure 76.1 is

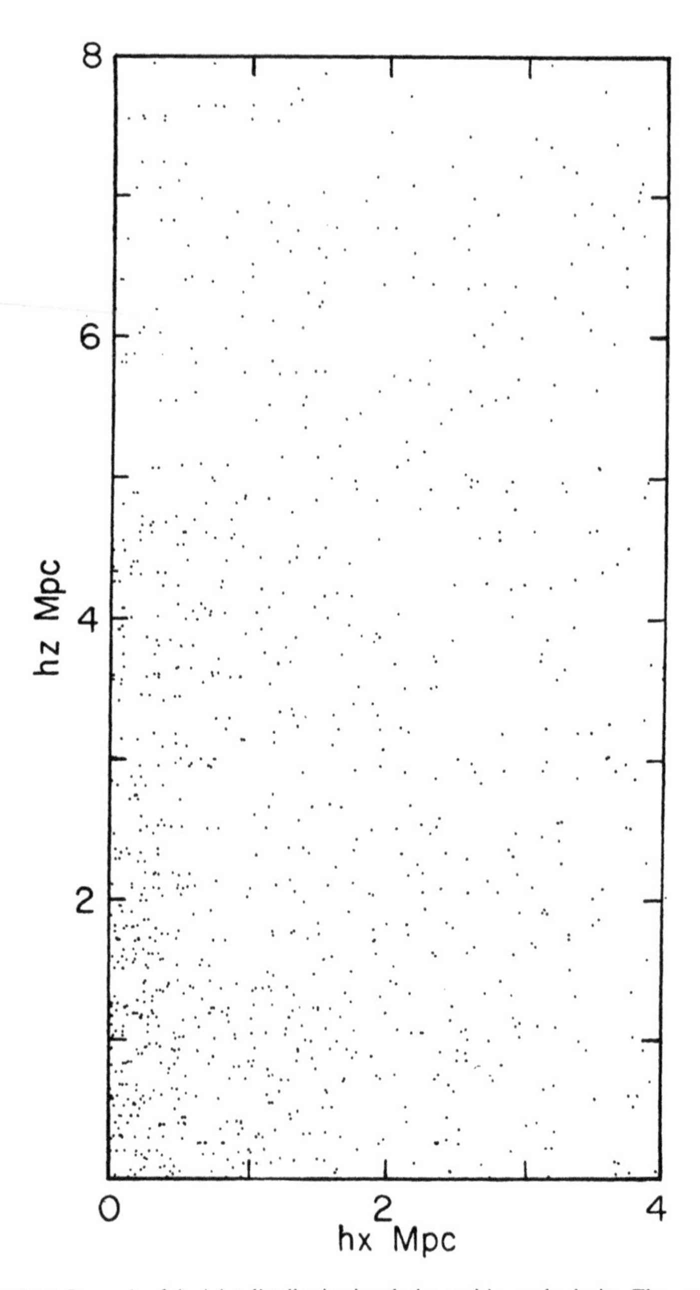

FIG. 76.1. Example of the joint distribution in relative position and velocity. The x axis is normal to the line of sight and the z axis is along the line of sight. Plotted are the positions $|x|$, $|z|$ of all neighbors at $|y| < 0.5\ h^{-1}$ Mpc for all galaxies in the sample.

based on the sample of galaxies brighter than apparent nagnitude $B = 13.2$ at galactic latitude $b > 30°$ and with redshifts in the range $1500 < cz < 4000$ km s^{-1}. The redshifts for these galaxies are summarized by Huchra, Davis, and Geller (1979). The lower bound on v eliminates the major effect of the galaxy concentration around the Virgo cluster. The elongation of the distribution along the line of sight is prominent. What is not yet clear is the contribution to this effect by redshift measuring errors.

The shape of the distribution is measured by the two-point correlation function ξ_v defined like $\xi(r)$ (§ 31) except that since the clustering in the redshift space is anisotropic, ξ_v is a function of two variables, the components π and σ of the separation parallel and perpendicular to the line of sight. The probability that a galaxy is found in the volume element $\delta\pi\,\delta A$ at distance (π, σ) from a randomly chosen galaxy is (Peebles 1979b)

$$\delta P = n[1 + \xi_v(\sigma, \pi)]\delta\pi\delta A. \tag{76.1}$$

It will be assumed that π and σ are small compared to the distance of the sample from us. That makes ξ_v independent of distance, which corresponds to the scaling relation for the angular correlation function $w(\theta)$ (§ 50). It will be assumed that the probability ϕ that a galaxy is included in the catalog is a function only of the galaxy distance r (§ 51). Then the expected number of neighbors at distance π, σ in the range $\delta\pi$, $\delta\sigma$ from galaxy i in the sample is

$$\langle \delta n_i(\sigma, \pi) \rangle = n\phi_i[1 + \xi_v(\sigma, \pi)]\delta\pi\delta A_i, \tag{76.2}$$

where δA_i is the area subtended by the ring $\delta\sigma$ around galaxy i and within the catalog survey area. The expected number of pairs at separation π, σ in the sample is then

$$DD(\sigma, \pi) = [1 + \xi_v(\sigma, \pi)] \sum_i (n\delta\pi\delta A_i\phi_i). \tag{76.3}$$

A convenient way to estimate the second factor on the right-hand side is to replace each galaxy j save the ith with m points all at the same redshift v_j and randomly placed in the survey area, and then count pairs formed between galaxy i and the random points. The expected value of the sum of the pairs over all galaxies i is

$$DR(\sigma, \pi) = \sum_i mn\delta\pi\delta A_i\phi_i. \tag{76.4}$$

The estimate of the correlation function is then (Peebles 1979b)

$$1 + \xi_v(\sigma, \pi) = mDD/DR. \tag{76.5}$$

Figure 76.2 shows estimates of ξ_v based on two samples, the one used in Figure 76.1 and the sample of Kirshner, Oemler, and Schechter (1979) mentioned in Section 57. In the two graphs on the left-hand side of the figure the component π is fixed; in the right-hand graphs, σ is fixed. The asymmetry of ξ_v is evident here as the elongation in the π direction. The consistency of the results from the two samples is encouraging, but still the ξ_v should only be taken as illustrations of the method until we have larger samples with well controlled redshift errors.

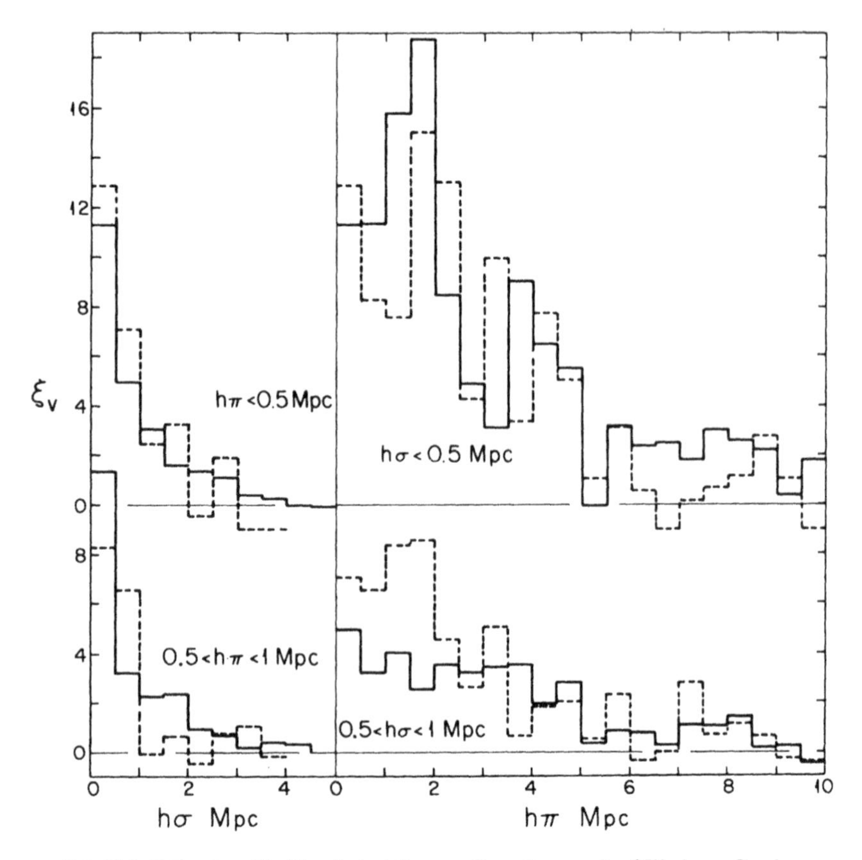

FIG. 76.2. Estimates of ξ_v. The dashed lines are from the sample of Kirshner, Oemler, and Schechter, the solid lines from the bright northern galaxies at $B < 13.2$, $v > 1500$ km s^{-1}.

B. Models for ξ_v

Let $\mathbf{w} = \mathbf{v}_2 - \mathbf{v}_1$ be the relative peculiar velocity of a galaxy pair at separation $\mathbf{r}$ and let $g(\mathbf{r}, \mathbf{w})$ be the distribution of $\mathbf{w}$. Then the relation between the true spatial correlation function $\xi(r)$ and the correlation function in the redshift space is

$$\begin{gathered} 1 + \xi_v(\sigma, \pi) = \int d^3w g(\mathbf{r}, \mathbf{w})[1 + \xi(r)], \\ r_1 = \sigma, \qquad r_3 = \pi - w_3/H, \qquad r^2 = \sigma^2 + (\pi - w_3/H)^2. \end{gathered} \tag{76.6}$$

Two limiting cases of this equation will be discussed.

When Hr is larger than the dispersion of w and $\xi(r) \ll 1$, we can write the equation as

$$1 + \xi_v(\sigma, \pi) \simeq \int g d^3w + \xi((\sigma^2 + \pi^2)^{1/2}). \tag{76.7}$$

The first term on the right-hand side differs from unity because of the mean streaming motion discussed in Section 71. To model this, let us take the distribution of $\mathbf{w}$ to be

$$g = g(\mathbf{w} - \mathbf{r}v(r)/r), \tag{76.8}$$

where $v(r)$ is the mean relative peculiar velocity of galaxy pairs at separation $\mathbf{r}$. The result of changing the variable of integration to $\mathbf{y} = \mathbf{w} - \mathbf{r}v(r)/r$ and expanding the integral to the lowest nontrivial order in v/Hr is

$$\xi_v(\sigma, \pi) \simeq \xi(r) - \frac{v(r)}{Hr}\left[1 + \frac{\pi^2}{r}\frac{d}{dr}\ln\frac{|v|}{r}\right], \qquad r^2 = \sigma^2 + \pi^2. \tag{76.9}$$

With equation (71.12) for $v(r)$ this becomes

$$\xi_v(\sigma, \pi) \simeq \xi(r) - K(\pi^2 - \sigma^2/2), \qquad K = 4f(\Omega)\int s^2 ds \xi(s)/r^5, \tag{76.10}$$

so $\xi_v < \xi$ at $\pi^2 > \sigma^2/2$, $\xi_v > \xi$ at $\pi^2 < \sigma^2/2$. This is the flattening effect mentioned above.

At small r the rms dispersion of w is expected to vary as $\sim r^{1-\gamma/2} \sim r^{0.1}$ (eq. 75.14), which is nearly independent of r. A reasonable approximation here is to write

$$\int dw_1 dw_2 g = f(w_3), \tag{76.11}$$

where f is a function only of the component $\mathbf{w}$ along the line of sight. In this approximation equation (76.6) is

$$\xi_v(\sigma, \pi) = \int dwf(w)\xi((\sigma^2 + (\pi - w/H)^2)^{1/2}). \qquad (76.12)$$

The relative velocity dispersion in equation (75.10) is

$$\langle {v_{21}}^2 \rangle = 3\langle w^2 \rangle = 3 \int dww^2 f(w). \qquad (76.13)$$

Three methods for estimating $\langle w^2 \rangle$ from ξ_v might be noted. First, one can guess at a reasonable functional form for f and then adjust the parameters to make the integral in equation (76.12) agree with ξ_v. In the available data the exponential model (Peebles 1976b, 1979b)

$$f \propto \exp - [2^{1/2}|w|/\langle w^2 \rangle^{1/2}] \qquad (76.14)$$

yields an adequate approximation to ξ_v with (Peebles 1980a)

$$\langle w^2 \rangle^{1/2} = 450 \pm 100 \text{ km s}^{-1}, \qquad \sigma \lesssim 1\ h^{-1}\text{ Mpc}. \qquad (76.15)$$

The correction for measuring error is highly uncertain. The second method of estimating $\langle w^2 \rangle$ uses the equations

$$\begin{gathered} \langle w^2 \rangle = H^2 \int d\pi\pi^2[\xi_v(\sigma, \pi) - \xi(r)] / \int d\pi\xi_v(\sigma, \pi), \\ \int \xi_v(\sigma, \pi)d\pi = \int \xi(r)d\pi, \qquad r^2 = \sigma^2 + \pi^2, \end{gathered} \qquad (76.16)$$

which follow from equation (76.12). Here one can adopt a model for $\xi(r)$, adjust the parameters to fit the second of these equations, and then use the results in the first equation. An example of this method given in Peebles (1979b) yielded the value of r_0 in equation (57.4). Another relation from equation (76.12) is

$$\langle w^2 \rangle = \frac{H^2 \int d\pi d\sigma(\pi^2 - \sigma^2)\xi_v(\sigma, \pi)}{\int d\pi d\sigma\xi_v(\sigma, \pi)}. \qquad (76.17)$$

This requires only the estimates of ξ_v but it averages over the expected slow variation of the dispersion with σ. With the integrals cut off at π, $\sigma = 10\ h^{-1}$ Mpc the sample used for Figure 76.1 gives $\langle w^2 \rangle^{1/2} \sim 400$ km s^{-1} (Peebles 1980).

C. Peculiar velocities and the mean mass density

If the galaxy correlation functions are good approximations to the mass functions, then the mean square relative peculiar velocity of galaxies varies with separation on small scales as (eq. 75.14)

$$\langle {v_{21}}^2 \rangle \propto r^{0.2}, \qquad r \lesssim 4\, h^{-1}\,\mathrm{Mpc}. \tag{76.18}$$

This means the projected dispersion derived from equation (76.16) should vary with projected separation as $\langle w^2 \rangle \propto \sigma^{0.2}$. An adequate test awaits larger redshift samples.

At very large separations the peculiar velocities are statistically independent and $\langle {v_{21}}^2 \rangle$ is twice the galaxy mean square peculiar velocity. The first order correction for the correlation of the velocities was derived in Section 72. With $f(\Omega) \sim \Omega^{0.6}$ (§ 14) we have from equations (72.3), (72.9), and (74.9)

$$\langle {v_{21}}^2 \rangle \simeq 2\langle v^2 \rangle\, (1 - 2.2\, {\Omega_0}^{0.2}/hr), \tag{76.19}$$

with r measured in megaparsecs. If for the purposes of a crude estimate, we extrapolate the two limiting cases in equations (76.18) and (76.19) and join them at $hr = 10$ Mpc, we find

$$\langle v^2 \rangle \sim 3\langle w^2 \rangle\, (\sigma = 1\, h^{-1}\,\mathrm{Mpc}). \tag{76.20}$$

This with equations (74.9) and (76.15) yields $\Omega_0 \sim 0.8$. Equation (75.12) with (76.15) yields a similar result for Ω_0; here there is no correction for correlated motions on large scales. It should be emphasized that these numerical results are not very interesting except as illustrations of the analysis because they are based on sets of data that are too small now but should become adequate in the next several years.

77. Behavior of the halo around a cluster of galaxies

Measurements of galaxy redshifts in the fields of nearby rich clusters of galaxies show that the general enhancement of density extends for some tens of megaparsecs around the cluster (Chincarini and Rood 1975, Tifft and Gregory 1977, Gregory and Thomson 1978). The galaxy distributions in these outer cluster halos are far from spherically symmetric, as might be expected since the crossing time is larger than the Hubble time, so the prospects for analyzing dynamics in any individual halo are doubtful. However, it may be easy to deal with the distribution and motions averaged over a fair sample of clusters. As was discussed in the last section, there are two interesting things to consider. At small enough distances from the cluster center the halos should have reached statistical equilibrium, so we can write down a relation between the galaxy velocity dispersion and the mass in the halo. At large r one would look for a mean streaming toward the cluster, like the relative motion of well-separated galaxies but perhaps more prominent because a cluster is a good deal more prominent.

A. The mean halo of an Abell cluster

The mean density of galaxies around the rich compact clusters picked out by Abell (1958) has been found from the cross correlation of the angular positions of the cluster centers with the Lick galaxy counts: the counts around all clusters in a distance class are stacked to get the mean surface density as a function of angular distance from the cluster center (Peebles 1974e, Seldner and Peebles 1977a). Of course, this average could obscure features peculiar to an individual cluster that might be revealed by the more detailed study of counts around the object, but it is capable of tracing the density enhancement to much greater distances from the cluster center because the background correction is under much better statistical control.

Translation from the stacked counts to the mean space density run $n_1(r)$ around a cluster depends on the galaxy luminosity function, but there is not much freedom of adjustment because it must match the observed variation of the cross correlation function with cluster distance class, that is, with the limiting absolute magnitude down to which the galaxies were counted in the Lick sample. A reasonable fit to the data is obtained with the expression

$$n_1(r) = n[1 + \xi_{cg}(r)] = n(1 + Br^{-\epsilon} + Cr^{-\mu}),$$
$$\epsilon = 2.5, \qquad \mu = 1.7, \qquad B = 130h^{-2.5}, \qquad C = 70h^{-1.7}. \tag{77.1}$$

Here ξ_{cg} is the cluster-galaxy cross correlation function (eq. 44.4); n is the overall mean space density and r is measured in megaparsecs. At $r \lesssim 0.5h^{-1}$ Mpc the best power law index seems to be closer to $\epsilon \sim 2$ than $\epsilon = 2.5$. This model works fairly well to $r \sim 40h^{-1}$ Mpc, where the cross correlation is lost in the noise. The coefficients B and C are evaluated for clusters in Abell's richness class $R = 1$, but the observed cross correlation does not change very much with R at $R \geq 1$, so the coefficients are not much different in the richer clusters.

It should be noted that equation (77.1) takes account of all the galaxies around a cluster including those in other Abell clusters clustered around it.

A model separately treating the halo around each cluster and the correlation among positions of halos is discussed by Seldner and Peebles (1977a)

Fluctuations in the distributions of galaxies around the clusters are measured by the cluster-galaxy-galaxy correlation function defined in equation (44.7). For the cross correlation of the Lick sample with Abell richness class $R = 1$ clusters, Fry and Peebles (1980a) find that the data are well approximated by the model

$$\zeta_{cgg} = Q_1 \xi_{cg}(r_1)\xi_{cg}(r_2) + Q_2 \xi(r_{12})[\xi_{cg}(r_1) + \xi_{cg}(r_2)]. \tag{77.2}$$

The first term in this model is obtained under the assumption that the galaxy number density run around a cluster is $nf(r)M$ where f is the same for all clusters and M is the cluster mass. This gives

$$\xi_{cg} = f\langle M\rangle, \qquad Q_1 = \langle M^2\rangle/\langle M\rangle^2. \tag{77.3}$$

The second term in equation (77.2) represents subclustering around the cluster. For the Lick and Abell samples Fry and Peebles (1980a) find $Q_2 \sim 0.7$.

B. Statistical dynamics

The dynamic behavior of the halo is conveniently analyzed by using the first equation of the BBGKY hierarchy. As in Section 68 the matter in the small patch Γ of six-dimensional phase space is described by the single particle distribution function $f_a(\mathbf{x}, \mathbf{p}, t)$ (eq. 68.1). The average value of f_a is (eq. 68.4)

$$\langle f_a\rangle = b(\mathbf{x}, \mathbf{p}, t). \tag{77.4}$$

Γ is placed not at random but at position $\mathbf{x}$ relative to a cluster center, so b is a function of position and of the direction of $\mathbf{p}$ relative to $\mathbf{x}$. The integral of equation (77.4) over momentum is

$$\int b d^3p = n_1(x, t)a^3 = na^3[1 + \xi_{cg}(x, t)]. \tag{77.5}$$

The Liouville equation for f_a is equation (68.2), and the average of this equation across the ensemble is

$$\frac{\partial}{\partial t} b(\mathbf{x}_1, \mathbf{p}_1, t) + \frac{p_1^{\alpha}}{ma^2}\frac{\partial b}{\partial x_1^{\alpha}} + \frac{Gm^2}{a}\frac{\partial}{\partial p_1^{\alpha}}\int d^3x_2 d^3p_2 \rho_2(1, 2) x_{21}^{\alpha}/x_{21}^3 = 0. \tag{77.6}$$

This differs from equation (68.7) only in the derivative with respect to **x**. The two-point distribution $\rho_2(1, 2)$ is defined as in equation (67.4). The integral of ρ_2 over momenta is

$$\int \rho_2 d^6p = a^6 n_2(\mathbf{x}_1, \mathbf{x}_2, t), \tag{77.7}$$

where n_2 is given by equation (44.7).

Equation (77.6) is simplified by taking momentum moments. The integral over momentum is

$$\frac{\partial}{\partial t} a^3 n_1(x, t) + a^2 \nabla \cdot n_1 \mathbf{v} = 0, \tag{77.8}$$

which is the usual particle conservation law with **v** the peculiar streaming velocity,

$$\mathbf{v} = \int d^3pb\mathbf{p}/(mn_1 a^4). \tag{77.9}$$

The result of multiplying equation (77.6) by p^α and integrating over momentum is

$$\begin{aligned} \frac{\partial}{\partial t} n_1 a^4 v^\alpha &+ a^3 \frac{\partial}{\partial x^\beta} n_1 \langle v^\alpha v^\beta \rangle \\ &= Gma^5 \int d^3x_1 n_2(\mathbf{x}, \mathbf{x}_1)(x_1^\alpha - x^\alpha)/|\mathbf{x}_1 - \mathbf{x}|^3. \end{aligned} \tag{77.10}$$

This is the momentum equation with the pressure term fixed by the velocity moment

$$\langle v^\alpha v^\beta \rangle = \int d^3pbp^\alpha p^\beta/(m^2 n_1(x) a^5). \tag{77.11}$$

The model in equation (77.2) with equation (44.7) for n_2 gives

$$\begin{aligned} \frac{\partial}{\partial t} n_1 a^4 v^\alpha &+ a^3 \frac{\partial}{\partial x^\beta} n_1 \langle v^\alpha v^\beta \rangle \\ &= -GM(<x) na^2 (x^\alpha/x^3)[1 + Q_1 \xi_{cg}(x)] \\ &\quad + Q_2 Gmn^2 a^5 \int d^3x_1 \xi_{cg}(x_1) \xi(|\mathbf{x} - \mathbf{x}_1|)(x_1^\alpha - x^\alpha)/|\mathbf{x}_1 - \mathbf{x}|^3, \end{aligned} \tag{77.12}$$

where $M(<x)$ is the mean mass in excess of a uniform distribution within distance x of a cluster,

$$M(<x) = mna^3 \int_0^x d^3x \xi_{cg}(x). \tag{77.13}$$

Since $Q_2 \sim Q_1 \sim 1$ and $\xi \ll \xi_{cg}$, the last term in equation (77.12) is small and will be dropped. The two limiting cases of equation (77.12) are discussed next.

C. Stability condition

At small x the clusters are in statistical equilibrium and $\xi_{cg} \gg 1$, so we can approximate equation (77.12) as

$$\begin{aligned} \frac{\partial}{\partial r^\beta} n_1(r)\langle v^\alpha v^\beta \rangle &= -GM(<r)n_1(r)Q_1 r^\alpha/r^3, \\ n_1(r) &= n\xi_{cg}(r). \end{aligned} \tag{77.14}$$

This is the usual equation of hydrostatic equilibrium with (in general anisotropic) pressure $mn_1\langle v^\alpha v^\beta \rangle$. The mean mass density at distance r from a cluster is $mn_1(r)$, and the mass excess $M(<r)$ is the integral of $mn_1 - \rho_b$ to r. The mean $\langle v^\alpha v^\beta \rangle$ is weighted by particles, that is by the cluster mass M. Since v^2 is proportional to M, $\langle v^\alpha v^\beta \rangle$ varies as $\langle M^2 \rangle \propto Q_1$ (eq. 77.3). In the following each cluster is given equal weight in the average over the velocity dispersions, so Q_1 is set equal to unity.

If the velocity distribution is isotropic,

$$\langle v^\alpha v^\beta \rangle \cong \langle v^2 \rangle \delta^{\alpha\beta}/3, \tag{77.15}$$

the equilibrium equation becomes

$$\frac{d}{dr} n_1\sigma^2 = -GM(<r)n_1(r)/r^2, \tag{77.16}$$

where $\sigma^2 = \langle v^2 \rangle/3$ is the line of sight velocity dispersion. In the power law model $\xi_{cg} = Br^{-\epsilon}$ the solution to this equation is

$$\sigma^2 = \frac{3\Omega H^2 B r^{2-\epsilon}}{4(3-\epsilon)(\epsilon-1)}, \tag{77.17}$$

where Ω is the density parameter (eq. 97.12). In this single power law model Seldner and Peebles (1977b) found $\epsilon = 2.4$, $B = 165\ h^{-\epsilon}$. These numbers yield

$$\sigma = 1200\,\Omega^{1/2}\,(hr)^{-0.2}\ \mathrm{km\ s^{-1}}, \tag{77.18}$$

with r in megaparsecs. Equation (77.1) for $\xi_{cg}(r)$ in equation (77.16) yields the numbers

$$\begin{aligned} \sigma &= 1390\,\Omega^{1/2}\ \mathrm{km\ s^{-1}}, \quad && hr = 1\ \mathrm{Mpc}; \\ &= 1300\,\Omega^{1/2}\ \mathrm{km\ s^{-1}}, \quad && hr = 2\ \mathrm{Mpc}; \\ &= 1260\,\Omega^{1/2}\ \mathrm{km\ s^{-1}}, \quad && hr = 4\ \mathrm{Mpc}. \end{aligned} \tag{77.19}$$

The difference between equations (77.18) and (77.19) at $r \sim 1\ h^{-1}$ Mpc is a reasonable indication of the uncertainty in σ due to ξ_{cg}.

Table 77.1 lists measurements of the line of sight velocity dispersion in rich clusters. The numbers are collected from discussions by Faber and Dressler (1977), Dressler (1978), Havlen and Quintana (1978), Stauffer, Spinrad, and Sargent (1979), and Hintzen and Scott (1979). Where there is an uncertain correction to σ for neighboring clusters, the lower estimate

TABLE 77.1
VELOCITY DISPERSIONS IN RICH CLUSTERS

Cluster	Richness	σ	Cluster	Richness	σ
A262	0	404	A1656	2	888
A1314	0	678	A1795	2	783
A2589	0	568	A2029	2	778
A2666	0	261	A2065	2	1070
Virgo E + SO	—	550	A2142	2	1241
A154	1	829	A2151	2	628
A576	1	1081	A2199	2	843
A1060	1	771	A2255	2	1222
A2147	1	1079	A2256	2	1274
A2319	1	873	A98	3	786
A168	2	576	A1940	3	715
A401	2	1294	A2670	3	890
A426	2	1396	Centaurus	—	860
A754	2	915	0340-538	—	1011
A1367	2	847			

of σ has been listed. These data show no appreciable correlation with the cluster richness class, consistent with the lack of correlation of ξ_{cg} with richness. The mean of σ^2 for richness $R \geq 1$ is

$$\langle \sigma^2 \rangle = (970 \pm 50 \text{ km s}^{-1})^2. \tag{77.20}$$

This in the first of equations (77.19) yields $\Omega_0 \sim 0.5$.

The main uncertainty in this number for Ω occurs in the translation from the distribution of large galaxies around clusters to the distribution of mass. The calculation assumes the mean value of the mass in and around a bright galaxy is the same whether the galaxy is 1 to 2 h^{-1} Mpc from a rich cluster or well away from any cluster. If, as is sometimes assumed, mass is much more strongly concentrated toward cluster centers than are galaxies, then equations (77.19) greatly overestimate Ω and the velocity dispersion around the cluster ought to vary as $\sigma^2 \propto 1/r$, rather than the slow variation predicted in equations (77.18) and (77.19). An observational test is possible but would require a considerable program of redshift sampling around a fair sample of clusters. Equations (77.18) and (77.19) also assume stability, but it will be noted that that need apply only in the sense of an average across a fair sample of clusters, not to individual ones. With velocity dispersion $\sigma \sim 1000$ km s^{-1} the crossing time is 20 percent of the Hubble time at $r = 2\ h^{-1}$ Mpc, so if ξ_{cg} were not nearly independent of time at $r \lesssim 2\ h^{-1}$ Mpc, we would be seeing the clusters at a very special epoch, which seems unreasonable.

D. Streaming motion at large r

The mean density around an Abell cluster (eq. 77.1) is twice the background, $n_1 = 2n$, at distance

$$r_l = 14\ h^{-1} \text{ Mpc} \tag{77.21}$$

from the cluster center. This marks the transition between the inner nonlinear region where a statistically steady state might apply and the outer parts where the interesting effect is the perturbation of the mean galaxy flow. The flow is measurable in principle in the way discussed for the relative motions of galaxies (§ 76).

At $r \gg r_l$ where $\xi_{cg} \ll 1$, we have from equations (77.8) and (77.12)

$$\frac{\partial M}{\partial t} + 4\pi\rho_b a^2 x^2 v = 0, \qquad \frac{\partial}{\partial t} av + \frac{GM}{ax^2} = 0. \tag{77.22}$$

As before M is the mass excess (eq. 77.13). The result of eliminating the velocity is

$$\frac{\partial^2 M}{\partial t^2} + 2\frac{\dot{a}}{a}\frac{\partial M}{\partial t} = 4\pi G\rho_b M, \tag{77.23}$$

which is the linear perturbation equation (10.3). The growing solution is $D_1(t)$, and if this is much greater than the decaying solution, the first of equations (77.22) gives

$$v = -\frac{M}{4\pi\rho_b(ax)^2}\frac{\dot{D}}{D} = -\frac{2}{3}\frac{GM}{r^2}\frac{f(\Omega)}{\Omega H}, \qquad f = \frac{a}{D_1}\frac{dD_1}{da}, \tag{77.24}$$

which agrees with equation (14.8).

In the model (77.1) for $n_1(r)$ the mean mass excess is

$$M(<r) = 4\pi\rho_b\left(\frac{Br^{3-\epsilon}}{3-\epsilon} + \frac{Cr^{3-\mu}}{3-\mu}\right), \tag{77.25}$$

so the mean peculiar streaming velocity is

$$\begin{aligned} v &= -Hf(\Omega)\left(\frac{Br^{1-\epsilon}}{3-\epsilon} + \frac{Cr^{1-\mu}}{3-\mu}\right) \\ &= -f(\Omega)\left(\frac{2.6\times 10^4}{(hr)^{1.5}} + \frac{5.6\times 10^3}{(hr)^{0.7}}\right)\text{km s}^{-1}. \end{aligned} \tag{77.26}$$

At $r = 20\ h^{-1}$ Mpc from the cluster center, this amounts to $980\,f$ kms s^{-1} peculiar velocity: the motion relative to the cluster is $2000–980\,f$ km s^{-1}.

This current reduces the apparent mean density of galaxies in the background and foreground as measured by the distribution of galaxy redshifts. Consider, for example, the frequency distribution of the redshifts of galaxies seen in projection close to the center of a cluster. The interesting quantity is the velocity difference

$$\Delta = c\,|Z - Z_{cl}|, \tag{77.27}$$

where Z is the galaxy redshift and Z_{cl} the mean for the cluster (with $Z_{cl} \ll 1$). The number of galaxies as a function of distance u from the cluster along the line of sight is (eq. 77.1)

$$\frac{d\mathcal{N}}{du} \propto 1 + B/u^{2.5} + C/u^{1.7}. \tag{77.28}$$

The mean velocity difference for a galaxy at u is

$$\Delta = Hu + v, \tag{77.29}$$

where v is given by equation (77.26), so the distribution in Δ is

$$d\mathcal{N}/d\Delta = (d\mathcal{N}/du)/(d\Delta/du), \tag{77.30}$$

which gives

$$\begin{aligned}\frac{\partial\mathcal{N}}{\partial\Delta} \simeq 1 &+ B\left(\frac{H}{\Delta}\right)^{\epsilon}\left(1 - f(\Omega)\frac{\epsilon - 1}{3 - \epsilon}\right) \\ &+ C\left(\frac{H}{\Delta}\right)^{\mu}\left(1 - f(\Omega)\frac{\mu - 1}{3 - \mu}\right).\end{aligned} \tag{77.31}$$

For example, at $\Delta = 2000\ \mathrm{km\ s^{-1}}$ one finds

$$\partial\mathcal{N}/\partial\Delta \simeq 1 + 0.52(1 - 0.88f(\Omega)). \tag{77.32}$$

As for equation (76.10), one could think of an observational program that ought to reveal this effect if f is close to unity, though again the program would be heroic.

78. Superclusters

Table 78.1 lists estimates (Hauser and Peebles 1973) of the mean number of Abell cluster centers in excess of random within distance r of a

TABLE 78.1
SUPERCLUSTERS

hr (Mpc)	Clusters in excess of random N_n	Clusters in a random sphere N_r	Cluster density contrast N_n/N_r	Galaxy density contrast $N_n(g)/N_r(g)$
10	0.18 ± 0.02	0.02	9 ± 1	6 ± 1
20	0.85 ± 0.07	0.16	5.3 ± 0.4	1.5 ± 0.3
30	1.7 ± 0.3	0.54	3.2 ± 0.6	0.7 ± 0.2
40	2.4 ± 0.5	1.3	1.9 ± 0.4	0.4 ± 0.2

randomly chosen cluster,

$$N_n = n_c \int_0^r \xi_{cc} dV. \tag{78.1}$$

All clusters with richness $R \geq 1$ are counted. The correlation is fairly clearly detected and positive to $r \sim 40\ h^{-1}$ Mpc; the errors listed in the table are rough estimates of one standard deviation based on the scatter in the estimates and the scatter of results from different model fits to the data.

The third column in the table is the mean number of clusters in a randomly placed sphere,

$$N_r = 4/3 \pi n_c r^3, \tag{78.2}$$

where the mean space density of Abell clusters is

$$n_c = 4.8 \times 10^{-6} h^{-3}\ \text{Mpc}^{-3}. \tag{78.3}$$

Another interesting statistic is the mean number of galaxies in excess of random around a cluster (eq. 77.1),

$$N_n(g) = n \int_0^r \xi_{cg} dV. \tag{78.4}$$

This number includes the galaxies in the chosen cluster, galaxies in neighboring correlated clusters, and any general excess density in the intercluster field. The ratio of $N_n(g)$ to $N_r(g) = nV$, the mean number of galaxies found in a randomly placed sphere of the same radius, is listed in the last column of the table.

One might imagine that in the distant past protoclusters were distributed uniformly at random through space and that their positions are correlated now because gravity pulled these massive objects toward each other. However, it cannot be that simple because gravity would pull in galaxies as well as clusters (unless field galaxies had very high random velocities making the Jeans length large). Thus the contrast in the density of galaxies ought to agree with the contrast in the density of clusters (including the chosen one),

$$N_n(g)/N_r(g) \sim (1 + N_n)/N_r. \tag{78.5}$$

This is not observed. For example, by the count of galaxies the mass within $40\ h^{-1}$ Mpc of a cluster is 40 percent higher than expected for a uniform distribution, but by the count of clusters it is larger by the factor $(1 + N_n)/N_r \sim 2.6$. Thus galaxies and clusters cannot both be good tracers of the large-scale mass distribution. The direction of the discrepancy agrees with the idea that the chance that a protocluster evolves into a rich compact cluster is enhanced if it finds itself in a region of generally high density (nurture as well as nature).

To analyze the dynamics of superclustering, we need to know the degree of mass concentration, and as we have just noticed, galaxies and clusters are not equivalent tracers. The galaxies would seem to be the more reliable. If so, the expected perturbation to the mean galaxy flow due to the mean mass concentration around a cluster is (eq. 77.24)

$$v = -\tfrac{1}{3} Hrf(\Omega)N_n(g)/N_r(g). \tag{78.6}$$

This says observers centered on randomly chosen Abell clusters would see that the recession speed of clusters at distance 40 h^{-1} Mpc averages

$$cZ \simeq [4000 - 500 f(\Omega)] \text{ km s}^{-1}, \qquad r = 40\, h^{-1} \text{ Mpc}. \tag{78.7}$$

As was discussed in Sections 76 and 77, it is possible to test for this effect by sampling redshift differences of pairs of clusters close together in the sky. Equally important as a diagnostic of the mass distribution is the test for random relative motions of clusters at smaller space separations (Noonan 1977).

79. Problems and prospects

Much of the discussion in this chapter assumes galaxies are useful tracers of mass. This is by no means obvious: at $r \lesssim 1\, h^{-1}$ Mpc there is the contrary indication from the correlation of galaxy type with the abundance of neighbors, and on this and larger scales there is ample room for mass in other forms, like hot gas, that might be expected to be distributed in a very different way from the galaxies. There is some evidence for the assumption. In the similarity solution discussed in Section 73 the predicted relations between the mass autocorrelation functions ξ_ρ, ζ_ρ, and η_ρ are quite accurately satisfied by the galaxy correlation functions ξ, ζ, and η, a coincidence that would seem surprising if the two were not related. Under the assumption $\xi_\rho = \xi$ and so on, the relative velocity dispersion $\langle v_{21}{}^2(r)\rangle$ at small r should vary as $r^{0.2}$ (eq. 75.14; Fig. 75.1). The present crude observations are at least not inconsistent with this: the observed dispersion does seem to change quite slowly with r. If one assumed galaxies and clusters of galaxies formed by the same physical process, one might expect that $\xi_\rho(r)$ derived from the clustering of galaxies joins smoothly to ξ_ρ derived from the mass distribution within galaxies. It was seen in Section 64 that this is so if $\Omega \sim 1$.

The first point, the relations among ξ, ζ, and η, depends on two more assumptions. The initial mass distribution is taken to approximate a random Gaussian process with power spectrum $\propto k^n$, n an index adjusted to fit ξ at small r. As discussed in the next chapter, it may be that the spectrum was truncated at short wavelength during decoupling of matter and radiation at redshift $Z \sim 1000$. There are initial conditions that avoid this (chapter VI), but if it is present, one must look for some other way to account for the power law shape of $\xi(r)$, as relaxation processes of some sort (Rees 1977, Press and Lightman 1978). Of course, such alternative processes may or may not require $\xi_\rho = \xi$ and so on.

The similarity solution also assumes the density parameter Ω is close to unity and the cosmological constant Λ is negligibly small. When $\Omega \ll 1$, small density fluctuations do not grow (§§ 11 and 13), and so the largest bound clusters in the hierarchy are much more dense than the mean for the background. The expected result is that $\xi(r)$ develops a shoulder where it varies more rapidly than r^{-3} (§ 71), much steeper than is observed. More detailed estimates of the shape of ξ in a low density cosmological model are given by Davis, Groth, and Peebles (1977); it appears from the argument given there that the shoulder in ξ disagrees with the observed shape of the galaxy correlation function unless $\Omega_0 \gtrsim 0.3$. If the density parameter were much less than that, it is doubtful that the universe would have approximated the Einstein-de Sitter model for a large enough span of redshift to have established the power law form of $\xi_\rho(r)$ by the scaling argument of Section 73 for any range of r. For example, if $\Omega_0 = 0.03$ and $H = 50$ km s^{-1} Mpc^{-1}, the Einstein-de Sitter model is a useful approximation only in the range of redshifts

$$\Omega_0^{-1} - 1 \sim 30 \lesssim Z \lesssim 300, \tag{79.1}$$

the upper limit being set by the radiation mass density (§ 12). In this case it is doubtful that the derivation in Section 73 is relevant despite the good fit to the data.

The typical mass within the bright easily seen parts of a large galaxy is given in equation (64.5); this multiplied by the abundance of large galaxies yields $\Omega_0 \sim 0.01$. The discrepancy between this number and what is suggested by the above arguments is stimulating but hardly perplexing. The flat rotation curves of large galaxies show that the mass density varies as r^{-2}, so the standard mass estimate misses an unknown fraction in the faint halo (Rubin 1979). In the great clusters the mass per galaxy greatly exceeds equation (64.5), and if this were applied to all galaxies, it would make $\Omega_0 \sim 0.5$ (§ 77). Thus there is some indication of and considerable room for mass outside the bright parts of galaxies. However, if that is where most of the mass is, it is fair to ask whether the distribution of galaxies is a fair measure of the large-scale distribution of mass. That is the first point mentioned above.

Another way the mass problem could be resolved is by an improvement in the theory of the evolution of $\xi(r)$ to allow for the development of a power law form in a low density cosmological model. An interesting challenge to the present theory has come from the results of numerical N-body model simulations of the motions of matter in an expanding world model. The attractive thing about the N-body model approach is that it should automatically deal with subtle effects like the destruction of the clustering hierarchy by relaxation processes and the collapse of newly forming clusters (§ 71). The N-body models certainly can produce particle distributions that are good visual approximations to maps of galaxy positions (for examples see Peebles 1973d, Groth, Peebles, Seldner, and Soneira 1977). Numerical results from the models agree with equation (70.3) (Peebles and Groth 1976, Efstathiou 1979) and match the power law shapes of the galaxy correlation functions (Miyoshi and Kihara 1975, Gott, Turner, and Aarseth 1979), even when $\Omega_0 \sim 0.1$, though the success of the fit to the data in the case $\Omega_0 \sim 0.1$ has been questioned (Fall 1978, 1979; Efstathiou 1979, Fry and Peebles 1980b). The main problem with N-body models is the limited dynamic range. There are two characteristic lengths, the initial size R of the distribution of N particles and the initial interparticle separation $R_i = R/N^{1/3}$. The initial density fluctuations on scales $r \lesssim R_i$ are large, and we do not know how to arrange these fluctuations to simulate the nonlinear mass clustering and motions that would have developed when a real universe evolved to the starting epoch for the model. We only know how to do that on scales $r \gg R_i$ where the linear perturbation approximation applies. The problem is that it is difficult to integrate the motions of more than $N \sim 1000$ particles, where $R_i \sim 0.1\, R$. Thus the condition $r \gg R_i$ violates the condition that r must be much less than the size R of the system. This means that it is doubtful that conventional N-body models will be capable of providing convincing results on the expected shape of $\xi_\rho(r)$.

The main theme of this chapter has been that the n-point correlation functions that have proved useful as descriptive statistics for galaxy clustering are useful also in the theory of the dynamics of the clustering. As has been summarized here, it is not clear whether the approaches can be married by setting $\xi_\rho = \xi$ and so on, but there is considerable room for observational and theoretical exploration.

V. RELATIVISTIC THEORY OF THE BEHAVIOR OF IRREGULARITIES IN AN EXPANDING WORLD MODEL

80. Role of the relativistic theory

The full relativistic theory rather than the Newtonian approximation of Chapter II is needed to deal with three important aspects of density irregularities in the early universe. First, when the pressure is high the relativistic active gravitational mass and inertial mass associated with pressure affect the dynamics. Second, when the mean density is high, a fluctuation of even modest fractional amount $\delta\rho/\rho$ containing a modest mass can have a large effect on the space curvature. We are thus led to deal with the interaction of speculations on the nature of the mass distribution and of the geometry in the early universe. Third, the horizon ct shrinks to zero at the time of the big bang, $t \rightarrow 0$: the seed fluctuations out of which galaxies might form were larger than the horizon and so were not in causal connection reckoned from the time of the big bang. Of course, this curious point applies as well to the homogeneous background: it was somehow contrived that all parts of the universe now visible were set expanding with quite precise uniformity even though an observer could not have discovered this much before the present epoch.

The situation is further beclouded by the expectation that at redshifts $Z > 1000$ matter was thermally ionized, the free electrons making the mean free path for scattering electromagnetic radiation very short and the thermal relaxation time short for important parts of the spectrum. The opportunities for observational input thus are strongly limited. Because of this practical problem added to the very deep theoretical ones, we must expect that debate on the proper outlines of the picture of the early universe will not soon be resolved. Thus I have not attempted to review all the topics now under discussion but rather have selected those results that seem to me simple and clear enough to be likely to continue to figure in the debate.

It is useful to begin with a review of some main features of the problem. The cosmological equations for the background homogeneous world model are (§ 97)

$$\frac{\dot{a}^2}{a^2} = \frac{8}{3}\pi G\rho - \frac{1}{a^2R^2}; \qquad \frac{d\rho}{dt} = -3\frac{\dot{a}}{a}(\rho + p/c^2), \tag{80.1}$$

where the constant R^{-2} (which can be negative) is related to the density parameter $\Omega(t)$ by

$$\frac{8}{3}\pi G\rho(t) \equiv \Omega(t)\frac{\dot{a}^2}{a^2}, \quad a|R| = 1/(H|1 - \Omega|^{1/2}), \quad H(t) = \dot{a}/a. \tag{80.2}$$

The second of equations (80.1) indicates that if the pressure p is positive, which seems reasonable, the mass density ρ varies more rapidly than a^{-3}. Thus at small enough a the first of equations (80.1) is well approximated by

$$\frac{\dot{a}^2}{a^2} = \frac{8}{3}\pi G\rho. \tag{80.3}$$

The present value of the density parameter is thought to be in the range

$$1 \gtrsim \Omega_0 \gtrsim 0.03, \tag{80.4}$$

which means equation (80.3) is a good approximation at redshift

$$1 + Z = a_0/a(t) \gg |1 - \Omega_0{}^{-1}| \lesssim 30. \tag{80.5}$$

A density irregularity of small amplitude follows the general expansion of the universe so its proper size λ varies as the expansion parameter $a(t)$. The ratio of its size to the distance to the horizon is then

$$\frac{\lambda}{ct} \propto \frac{a(t)}{t}. \tag{80.6}$$

If $\rho \propto a^{-3}$, equation (80.3) yields $a \propto t^{2/3}$, so this ratio diverges as $t^{-1/3}$ at $t \rightarrow 0$. If $p > 0$, the divergence with $t \rightarrow 0$ is more rapid. It is apparent then that at small enough t the pressure gradient cannot prevent the growth of irregularities because there cannot be appreciable response to the pressure gradient in an expansion time.

The geometry in the homogeneous background model is described by the line element (97.11). The present value of the radius of curvature of space at fixed world time is (eqs. 80.2 and 80.4)

$$ca_0 |R| \gtrsim cH_0^{-1} = 3000\, h^{-1}\ \mathrm{Mpc} \sim ct_0. \tag{80.7}$$

That is, the radius $ca_0|R|$ now is not much smaller than the horizon ct_0, and, as in equation (80.6), in the early universe it would have been much larger than the horizon.

The clustering of matter is reliably detected to scales λ as large as perhaps 40 h^{-1} Mpc (compare table 78.1), not more than one percent of the radius of curvature $ca_0|R|$. Since the sizes of these large-scale fluctuations would trace back as $a(t)$, the same ratio would apply in the early universe. Thus in cases of practical interest we can suppose $x \ll c|R|$, so the line element (97.11) is well approximated as

$$\begin{aligned} ds^2 &= c^2dt^2 - a^2(dx^2 + x^2d\theta^2 + x^2 \sin^2\theta\, d\phi^2) \\ &= c^2dt^2 - a^2\delta_{\alpha\beta}dx^\alpha dx^\beta. \end{aligned} \tag{80.8}$$

This fortunate circumstance greatly simplifies the problem: we must deal with perturbations to a geometry that is in effect cosmologically flat.

The theoretical expressions in the following sections have been simplified by choosing units so the velocity of light is equal to unity.

81. Time-orthogonal coordinates

To simplify the analysis I adopt a specific and particularly convenient prescription for the assignment of coordinates. The construction commences with a spacelike hypersurface that is assigned time coordinate t_i. Fundamental observers are placed in the hypersurface: each is provided with a physical clock set to read t_i, and each is assigned a set of spatial coordinates, $(x^1, x^2, x^3) = x^\alpha$. The observers move freely, under no force save gravity. Their initial velocities are normal to the hypersurface,

$$g_{ij}u^i dx^j = 0, \tag{81.1}$$

where u^i is the observer's four-velocity and dx^i is in the hypersurface at the observer. Now the four coordinates assigned to an event in space-time are the three x^α belonging to the fundamental observer whose world line passes through the event and the observer's clock reading t at the event.

To see how this construction affects the components of the metric tensor recall first the interpretation of the line element

$$ds^2 = g_{ij}(x)dx^i dx^j. \tag{81.2}$$

If the two points x^i and $x^i + dx^i$ are separated by a timelike interval, ds is the proper time interval recorded by a clock moving from one point to the other. If the separation is spacelike, $|ds|$ is the proper distance between the points measured on a rod that touches the two points simultaneously as judged by an observer sitting on the rod. In the above coordinate construction, if $dx^\alpha = 0$, then $ds = \mathrm{dt}$, for the coordinate interval dt was marked out by a physical clock. Thus

$$g_{00} = 1. \tag{81.3}$$

The four-velocity u^i of each of the observers in the construction satisfies

$$u^\alpha = dx^\alpha/ds = 0, \tag{81.4}$$

because each observer is assigned fixed x^α. In equation (81.1) the interval dx^i has $dx^0 = 0$ because it is in the hypersurface $x^0 = t_i$, so equation (81.1) yields

$$g_{0\alpha}(t_i) = 0. \tag{81.5}$$

The components $g_{0\alpha}$ off the hypersurface follow from the condition that each observer is moving along a geodesic described by the equation

$$\frac{d}{ds} g_{ij}u^j = \frac{1}{2} g_{jk,i}u^j u^k, \qquad u^i = dx^i/ds. \tag{81.6}$$

With $i = \alpha$, $u^\alpha = 0$, and $g_{00} = 1$, this equation becomes

$$dg_{0\alpha}/ds = 0, \tag{81.7}$$

so $g_{0\alpha} = 0$ everywhere.

Time-orthogonal coordinates defined by the equations $g_{00} = 1$ and $g_{0\alpha} = 0$ always exist (for reasonable g_{ij}) by the above construction, but in general one coordinate system cannot cover all space-time because orbits of the observers tend to intersect in the manner discussed in Section 21. When this happens, one must repeat the construction using a new hypersurface $t' =$ constant.

These coordinates offer a useful way to define a density fluctuation in the early universe. Milne (1935) emphasized the curious point that the hypersurface t_i can be chosen to run through the spots where the mass density is some chosen value (if the mass distribution is smooth enough) thereby defining away the irregularity. This is sometimes taken to say that it is meaningless to think of density irregularities on scales larger than the horizon. However, we can imagine that observers spread through the universe and moving with the matter keep a record of the local density as a function of proper time, $\rho(x^\alpha, t)$. As the observers come within the horizon, their records can be acquired and compared. If they cannot be made to agree by adjusting the starting times, it means the universe has been irregular on scales larger than the horizon. This is the definition that is used in all the following discussion.

One should bear in mind that if the time t is allowed to be a general coordinate, the record $\rho(x^\alpha, t)$ is meaningless because it can be adjusted by a coordinate transformation. This has led to some of the curious results to be found in the literature. For some examples of how the coordinate choice affects $\delta\rho/\rho$, see Sakai (1969). The meaning of irregularities with scales larger than the horizon is further discussed by Press and Vishniac (1980a) and Bardeen (1980).

The large-scale structure of the universe appears to be well approximated by a homogeneous cosmologically flat world model (eq. 80.8). It is supposed for the most part that the irregularities in the mass distribution cause irregularities in the geometry that can be treated in linear perturbation theory. Then in the time-orthogonal coordinates the elements of the perturbed metric tensor can, as we will now see, be taken to have the form

$$g_{00} = 1, \qquad g_{0\alpha} = 0, \qquad g_{\alpha\beta} = -a^2[\delta_{\alpha\beta} - h_{\alpha\beta}(\mathbf{x}, t)], \tag{81.8}$$

where the six fields $h_{\alpha\beta}$ that describe the perturbation all are small, $|h_{\alpha\beta}| \ll 1$, so terms $\sim h^2$ can be dropped. In this approximation the reciprocal tensor is

$$g^{00} = 1, \qquad g^{0\alpha} = 0, \qquad g^{\alpha\beta} = -a^{-2}(\delta_{\alpha\beta} + h_{\alpha\beta}). \tag{81.9}$$

It satisfies the usual equation

$$g_{ij}g^{jk} = \delta_i^{\,k}, \tag{81.10}$$

to order h.

The fields $h_{\alpha\beta}$ can be changed without altering the physical situation by changing the coordinate labels. A general coordinate transformation is

$$g_{ij}(x) = \frac{\partial \hat{x}^k}{\partial x^i}\frac{\partial \hat{x}^l}{\partial x^j}\hat{g}_{kl}(\hat{x}), \qquad \hat{x}^i = \hat{x}^i(x^j). \tag{81.11}$$

An infinitesimal transformation between time-orthogonal coordinates,

$$\hat{t} = t + \psi, \qquad \hat{x}^\alpha = x^\alpha + d^\alpha, \tag{81.12}$$

gives

$$\begin{aligned} g_{00} &= 1 + 2\dot{\psi} = 1, \\ g_{0\alpha} &= \psi_{,\alpha} - a^2\dot{d}^\alpha = 0, \\ g_{\alpha\beta} &= -a^2(\delta_{\alpha\beta} - h_{\alpha\beta}) \\ &= -a^2(\delta_{\alpha\beta} - \hat{h}_{\alpha\beta} + d^\alpha{}_{,\beta} + d^\beta{}_{,\alpha} + 2\psi\delta_{\alpha\beta}\dot{a}/a), \end{aligned} \tag{81.13}$$

where only terms linear in h, ψ and d^α have been kept. The first two of these equations maintain the metric in the form of equation (81.8). The first equation says that ψ is independent of t, the second, that d^α is of the form

$$d^\alpha = \psi(\mathbf{x})_{,\alpha}\int^t dt/a^2 + \chi^\alpha(\mathbf{x}). \tag{81.14}$$

Then the last equation gives the general infinitesimal transformation relation,

$$\hat{h}_{\alpha\beta} = h_{\alpha\beta} + 2\psi_{,\alpha\beta}\int^t dt/a^2 + \chi^\alpha{}_{,\beta} + \chi^\beta{}_{,\alpha} + 2\psi\delta_{\alpha\beta}\dot{a}/a. \tag{81.15}$$

The function $\psi(x)$ describes an infinitesimal shift in the shape of the starting hypersurface, and the first term on the right-hand side of equation (81.14) shows how this shift affects the coordinates x^α off the hypersurface. The term $\chi^\alpha(x^\beta)$ describes a shift of coordinate assignments within the starting hypersurface.

A convenient form in what follows is

$$(-g)^{1/2} = a^3(1 - h/2), \tag{81.16}$$

where g is the determinant of g_{ij} and h is the trace of $h_{\alpha\beta}$. The transformation equation (81.15) for h is

$$\hat{h} = h + 2\nabla^2\psi\int^t dt/a^2 + 2\chi^\alpha{}_{,\alpha} + 6\psi a/a. \tag{81.17}$$

82. The field equations for $h_{\alpha\beta}$

The convenient form for Einstein's gravitational field equations is

$$R_{ij} = 8\pi G(T_{ij} - \tfrac{1}{2}g_{ij}T), \tag{82.1}$$

where the stress-energy tensor for an ideal fluid with proper mass density ρ and pressure p is

$$T^{ij} = (\rho + p)u^i u^j - pg^{ij}, \qquad u^i = dx^i/ds; \tag{82.2}$$

the Ricci tensor is

$$R_{ij} = \Gamma^k{}_{ij,k} - \Gamma^k{}_{ik,j} + \Gamma^k{}_{ij}\Gamma^l{}_{kl} - \Gamma^k{}_{il}\Gamma^l{}_{kj}, \tag{82.3}$$

and Γ is given in terms of the metric tensor by the equation

$$\Gamma^k{}_{ij} = \tfrac{1}{2}g^{kl}(g_{il,j} + g_{jl,i} - g_{ij,l}). \tag{82.4}$$

On using equation (81.8) for g_{ij} in this last expression and keeping only the terms of first order in $h_{\alpha\beta}$, one finds

$$\begin{aligned}
\Gamma^0{}_{0i} &= 0, \qquad \Gamma^\alpha{}_{00} = 0, \\
\Gamma^0{}_{\alpha\beta} &= a\dot{a}(\delta_{\alpha\beta} - h_{\alpha\beta}) - \tfrac{1}{2}a^2\dot{h}_{\alpha\beta}, \\
\Gamma^\alpha{}_{0\beta} &= \delta_{\alpha\beta}\,\dot{a}/a - \tfrac{1}{2}\dot{h}_{\alpha\beta}, \\
\Gamma^\gamma{}_{\alpha\beta} &= \tfrac{1}{2}(h_{\alpha\gamma,\beta} + h_{\beta\gamma,\alpha} - h_{\alpha\beta,\gamma}), \\
\Gamma^j{}_{0j} &= 3\,\dot{a}/a - \tfrac{1}{2}\,\dot{h}, \qquad \Gamma^j{}_{\alpha j} = -\tfrac{1}{2}h_{,\alpha},\ h = \Sigma h_{\alpha\alpha}.
\end{aligned} \tag{82.5}$$

These equations in (82.3) yield the Ricci tensor to first order in $h_{\alpha\beta}$:

$$\begin{aligned}
R_{00} &= -3\frac{\ddot{a}}{a} + \frac{1}{2}\ddot{h} + \frac{\dot{a}}{a}\dot{h}, \\
R_{0\alpha} &= -\tfrac{1}{2}\dot{h}_{\alpha\beta,\beta} + \tfrac{1}{2}\dot{h}_{,\alpha}, \\
R_{\alpha\beta} &= a^2\left[\left(\frac{\ddot{a}}{a} + 2\frac{\dot{a}^2}{a^2}\right)(\delta_{\alpha\beta} - h_{\alpha\beta})\right. \\
&\qquad \left. - \frac{3}{2}\frac{\dot{a}}{a}\dot{h}_{\alpha\beta} - \frac{1}{2}\frac{\dot{a}}{a}\dot{h}\delta_{\alpha\beta} - \frac{1}{2}\ddot{h}_{\alpha\beta}\right] + R_{\alpha\beta}{}^{(3)}, \\
R_{\alpha\beta}{}^{(3)} &= \tfrac{1}{2}\,(h_{\alpha\beta,\gamma\gamma} + h_{,\alpha\beta} - h_{\alpha\gamma,\gamma\beta} - h_{\beta\gamma,\gamma\alpha}).
\end{aligned} \tag{82.6}$$

In the unperturbed background cosmological model the fluid is at rest in the coordinates of (80.8), so $u^i = \delta_0^{\ i}$ and the stress-energy tensor (82.2) is diagonal,

$$T^{00}(b) = \rho_b, \qquad T^{\alpha\beta}(b) = \delta_{\alpha\beta} p_b / a^2, \qquad T(b) = \rho_b - 3p_b. \tag{82.7}$$

This with the unperturbed part of R_{ij} from equations (82.6) in the field equations (82.1) gives

$$\begin{aligned} \frac{\ddot{a}}{a} &= -\frac{4}{3}\pi G(\rho_b + 3p_b), \\ \frac{\ddot{a}}{a} + 2\frac{\dot{a}^2}{a^2} &= 4\pi G(\rho_b - p_b). \end{aligned} \tag{82.8}$$

These are equivalent to equations (80.1) with $R^{-2} = 0$.

When equations (82.8) are substituted for the zeroth order parts in equations (82.6) and these parts are moved over to the source side in the field equations (82.1), one finds

$$\begin{aligned} \frac{1}{2}\ddot{h} + \frac{\dot{a}}{a}\dot{h} &= 4\pi G(2T_{00} - T - \rho_b - 3p_b), \\ \frac{1}{2}\dot{h}_{,\alpha} - \frac{1}{2}\dot{h}_{\alpha\beta,\beta} &= 8\pi G T_{0\alpha}, \end{aligned} \tag{82.9}$$

$$R_{\alpha\beta}^{(3)} - a^2\left[\frac{3}{2}\frac{\dot{a}}{a}\dot{h}_{\alpha\beta} + \frac{1}{2}\frac{\dot{a}}{a}\dot{h}\delta_{\alpha\beta} + \frac{1}{2}\ddot{h}_{\alpha\beta}\right] = 4\pi G[2T_{\alpha\beta} + a^2\delta_{\alpha\beta}(T - \rho_b + p_b)].$$

These are the general linear perturbation equations for the fields $h_{\alpha\beta}$. If the matter is approximated by the ideal fluid model (82.2) the field equations become

$$\begin{aligned} \frac{1}{2}\ddot{h} + \frac{\dot{a}}{a}\dot{h} &= 4\pi G[2(\rho + p)((u^0)^2 - 1) + \rho - \rho_b + 3(p - p_b)], \\ \tfrac{1}{2}\dot{h}_{,\alpha} - \tfrac{1}{2}\dot{h}_{\alpha\beta,\beta} &= -8\pi G a^2(\rho + p)u^0u^\alpha, \\ R_{\alpha\beta}^{(3)} - a^2\left[\frac{3}{2}\frac{\dot{a}}{a}\dot{h}_{\alpha\beta} + \frac{1}{2}\frac{\dot{a}}{a}\dot{h}\delta_{\alpha\beta} + \frac{1}{2}\ddot{h}_{\alpha\beta}\right] &= 4\pi G[2(\rho + p)a^4u^\alpha u^\beta + a^2\delta_{\alpha\beta}(\rho - \rho_b - p + p_b)]. \end{aligned} \tag{82.10}$$

Various solutions and limiting cases are discussed in the following sections.

83. Gravitational waves

We can add to any particular solution of equations (82.9) a homogeneous part $h_{\alpha\beta}$ with no source terms on the right-hand sides, representing a free gravitational wave. The first of the equations is

$$\frac{1}{2}\ddot{h} + \frac{\dot{a}}{a}\dot{h} = 0, \tag{83.1}$$

with the solution

$$\dot{h} = f(x^\alpha)/a^2. \tag{83.2}$$

This can be eliminated by a coordinate transformation: equation (81.17) gives

$$\partial\hat{h}/\partial t = \partial h/\partial t + 2\nabla^2\psi(x^\alpha)/a^2, \tag{83.3}$$

so we can choose ψ to make

$$\dot{h} = 0. \tag{83.4}$$

Since the proper spatial volume measured by an observer at fixed x^α is related to the coordinate volume element by

$$dV = (-g)^{1/2}d^3x = a^3(1 - h/2)d^3x, \tag{83.5}$$

where g is the determinant of $g_{\alpha\beta}$, equation (83.4) means the wave does not affect the volume occupied by a cloud of freely moving particles. This is to be expected since the wave is a tidal field. The second of equations (82.9) gives

$$\dot{h}_{\alpha\beta,\beta} = 0, \qquad h_{\alpha\beta,\beta} = 0, \tag{83.6}$$

for we can choose χ^α (eq. 81.15) to make $h_{\alpha\beta,\beta}$ vanish in the starting hypersurface. This means the wave is transverse. In the last equation we are left with $\nabla^2 h(x^\alpha) = 0$, with h = constant the only allowed solution, and

$$\frac{1}{a^2}\nabla^2 h_{\alpha\beta} = \ddot{h}_{\alpha\beta} + 3\frac{\dot{a}}{a}\dot{h}_{\alpha\beta}. \tag{83.7}$$

This is the gravitational wave equation.

For the Fourier component with spatial wave number $\mathbf{k}$, proper wavelength

$$\lambda = 2\pi a(t)/k, \tag{83.8}$$

the equation for the amplitude of $h_{\alpha\beta}$ is

$$\frac{d^2A}{dt^2} + 3\frac{\dot{a}}{a}\frac{dA}{dt} + \frac{k^2}{a^2}A = 0. \tag{83.9}$$

If the wavelength is small compared to the expansion time $t \sim a/\dot{a}$, we can write down the solution in the adiabatic approximation used in Section 16: set

$$A = B(t)e^{-ik\int^t dt/a}, \tag{83.10}$$

and drop the terms of order t^{-2}. The result is (Hawking 1966)

$$h_{\alpha\beta} \propto \frac{1}{a(t)} \exp i\left(\mathbf{k}\cdot\mathbf{x} - k\int^t dt/a\right). \tag{83.11}$$

The fractional perturbation to the proper distance between neighboring freely moving particles varies as $h_{\alpha\beta}$, and this perturbation decreases with the expansion as $a(t)^{-1}$.

84. Newtonian approximation

The Newtonian approximation used in Chapter II follows from the relativistic equations in the limit that all peculiar velocities are small. To select the dominant terms from the linear perturbation equations (82.9) in this limit, suppose the size of structures to be treated is on the order of λ, the typical time τ for structures to change is less than or comparable to the expansion time t, and the characteristic velocity is $v \sim \lambda/\tau \ll 1$. Then one sees from the first of equations (82.9) that orders of magnitudes for the $h_{\alpha\beta}$ are

$$G\rho \sim \ddot{h} \sim v\dot{h}_{,\alpha}/a \sim v^2 h_{,\alpha\beta}/a^2. \tag{84.1}$$

The sources in equations (82.9) are of the following order:

$$T_{00} \sim \rho, \qquad T_{0\alpha} \sim \rho v a, \qquad T_{\alpha\beta} \sim \rho v^2 a^2, \qquad p \lesssim \rho v^2. \tag{84.2}$$

In the limit $v \ll 1$ the dominant parts of the equations are then

$$\begin{gathered} \ddot{h} + 2\frac{\dot{a}}{a}\dot{h} = 8\pi G(\rho - \rho_b), \qquad \dot{h}_{,\alpha} = \dot{h}_{\alpha\beta,\beta}, \\ h_{\alpha\beta,\gamma\gamma} + h_{,\alpha\beta} - h_{\alpha\gamma,\gamma\beta} - h_{\beta\gamma,\gamma\alpha} = 0. \end{gathered} \tag{84.3}$$

The solution to the second of these equations is

$$\frac{\partial h_{\alpha\beta}}{\partial t} = -\frac{1}{4\pi}\frac{\partial^2}{\partial x^\alpha \partial x^\beta}\int \frac{d^3x'}{|\mathbf{x} - \mathbf{x}'|}\frac{\partial h}{\partial t}. \tag{84.4}$$

Thus $h_{\alpha\beta}$ is of the form

$$\partial h_{\alpha\beta}/\partial t = \Psi_{,\alpha\beta}. \tag{84.5}$$

This agrees with the time derivative of the last of equations (84.3).

The usual Newtonian equations are found by changing coordinates to eliminate $\dot{h}_{\alpha\beta}$ (to first order). The coordinate transformation

$$\hat{t} = t - \psi, \qquad \hat{x}^\alpha = x^\alpha - d^\alpha(\mathbf{x}, t) \tag{84.6}$$

gives, to the first order in ψ, d^α and $h_{\alpha\beta}$ (compare eqs. 81.13),

$$\begin{gathered} \hat{g}_{00} = 1 + 2\dot{\psi}, \qquad \hat{g}_{0\alpha} = \psi_{,\alpha} - a^2\dot{d}^\alpha, \\ \hat{h}_{\alpha\beta} = h_{\alpha\beta} - d^\alpha{}_{,\beta} - d^\beta{}_{,\alpha} - 2\psi\delta_{\alpha\beta}\dot{a}/a. \end{gathered} \tag{84.7}$$

To eliminate $\hat{g}_{0\alpha}$, we take

$$\dot{d}^\alpha = \psi_{,\alpha}/a^2. \tag{84.8}$$

Then the time derivative of the last of equations (84.7) is

$$\frac{\partial}{\partial t}\hat{h}_{\alpha\beta} = \dot{h}_{\alpha\beta} - 2\psi_{,\alpha\beta}/a^2 - 2\frac{\partial}{\partial t}\psi\delta_{\alpha\beta}\,\dot{a}/a,$$

which vanishes if

$$\Psi_{,\alpha\beta} = 2\psi_{,\alpha\beta}/a^2 + 2\frac{\partial}{\partial t}\psi\delta_{\alpha\beta}\dot{a}/a. \qquad (84.9)$$

For the Newtonian limit to apply, the spatial scale of the perturbations that appears in the first term on the right-hand side must be much smaller than the expansion time τ that appears in the second term. In this case, the second term on the right-hand side is negligibly small and we find

$$\psi = \Psi a^2/2. \qquad (84.10)$$

This fixes the time transformation ψ in terms of Ψ (eq. 84.5). The first of equations (84.3) is

$$\frac{1}{2}\frac{\partial}{\partial t}a^2\frac{\partial h}{\partial t} = 4\pi G a^2(\rho - \rho_b), \qquad (84.11)$$

and since

$$\dot{h} = \nabla^2\Psi, \qquad (84.12)$$

this is

$$\nabla^2\dot{\psi} = 4\pi G a^2(\rho - \rho_b) = \nabla^2\phi, \qquad (84.13)$$

where the Newtonian potential is

$$\phi = \dot{\psi} = \frac{1}{2}\frac{\partial}{\partial t}a^2\Psi. \qquad (84.14)$$

The time part of the metric tensor (eq. 84.7) is now

$$\hat{g}_{00} = 1 + 2\phi. \qquad (84.15)$$

This is the standard form in the Newtonian approximation (eq. 6.12).

In the new coordinates the geodesic equations of motion in (81.6) for a slowly moving particle are

$$\frac{d}{dt} a^2 \frac{d\hat{x}^\alpha}{dt} = -\frac{\partial \phi}{\partial x^\alpha}. \tag{84.16}$$

Since the proper velocity in the Newtonian coordinates is, to lowest order,

$$v^\alpha = a\, d\hat{x}^\alpha/dt, \tag{84.17}$$

the equations of motion are

$$\frac{dv^\alpha}{dt} + \frac{\dot{a}}{a} v^\alpha = -\frac{1}{a}\frac{\partial \phi}{\partial x^\alpha}. \tag{84.18}$$

Another way to arrive at this is to note that in the new coordinates in equations (84.6), the proper velocity of a particle at fixed x^α is

$$v^\alpha = a\, d\hat{x}^\alpha/dt = -a\partial d^\alpha/\partial t = -\psi_{,\alpha}/a, \tag{84.19}$$

to first order. With equation (84.14) the acceleration can be reduced to equation (84.18).

Equation (84.18) agrees with equation (7.12), and the field equation (84.13) agrees with equation (7.9). It might be noted also that equations (84.4) and (84.5) indicate

$$\Psi = -\frac{1}{4\pi}\int d^3x' \dot{h}/|\mathbf{x} - \mathbf{x}'|, \tag{84.20}$$

and this with equation (84.14) and the first of equations (84.3) can be reduced to the Green's function solution to Poisson's equation (eq. 8.1).

It was noted in Section 81 that the time-orthogonal coordinates typically can be applied only for a limited time interval before $h_{\alpha\beta}$ diverges because orbits intersect. The derivation given here assumes $|h_{\alpha\beta}| \ll 1$, which can be arranged by using a sequence of time-orthogonal coordinate systems in each of which $|h_{\alpha\beta}| \ll 1$ for some span of time.

In Section 12 models were discussed where there is in addition to the nonrelativistic matter a homogeneous high pressure component. As long as this background is homogeneous, it cancels out of the source term for h (eqs. 82.9), so the results, equations (84.13) and (84.18), still apply (of course with the time variation of $a(t)$ altered by the background density and pressure).

85. Linear perturbation equations for the matter

A standard assumption is that the matter distribution in the early universe was nearly uniform—well approximated by an ideal fluid with pressure and density only slightly perturbed from the mean. It is convenient to set

$$\begin{aligned} \rho &= \rho_b(1+\delta), \qquad p = p_b + c_s^2\rho_b\delta, \\ c_s^2 &= dp/d\rho, \qquad p/\rho = \nu. \end{aligned} \tag{85.1}$$

Here $\rho_b(t)$ and $p_b(t)$ are the mean density and pressure; $\delta \ll 1$ is the fractional density perturbation, and c_s is the velocity of sound. If the ratio ν is constant, then $\nu = c_s^2$. The equations of motion are most simply derived from the conservation equation

$$T_{i;j}^{j} = 0 = ((-g)^{1/2}T_i^j)_{,j} - \tfrac{1}{2}(-g)^{1/2}g_{jk,i}T^{jk}, \tag{85.2}$$

where, in the time-orthogonal coordinates (§ 81),

$$g_{\alpha\beta} = -a^2(\delta_{\alpha\beta} - h_{\alpha\beta}), \qquad (-g)^{1/2} = a^3(1-h/2), \tag{85.3}$$

and, in the ideal fluid model,

$$T^{ij} = (\rho+p)u^iu^j - g^{ij}p, \qquad T = \rho - 3p. \tag{85.4}$$

The result of substituting equations (85.3) and (85.4) in equation (85.2) and keeping only terms of first order in $h_{\alpha\beta}$, δ, and u^α is, for the component $i = \alpha$,

$$\frac{\partial}{\partial t}[a^4\rho_b(1+\nu)v^\alpha] + a^3c_s^2\rho_b\delta_{,\alpha} = 0, \qquad v^\alpha = au^\alpha, \tag{85.5}$$

where v^α is the proper velocity in the time-orthogonal coordinates, and, for the component $i = 0$,

$$\begin{aligned} &\frac{\partial}{\partial t}[a^3\rho_b(1+\delta-h/2)] + a^2\rho_b(1+\nu)v^\alpha{}_{,\alpha} \\ &\qquad = \frac{1}{2}a^3\dot{h}p_b - 3a^2\dot{a}(p_b + c_s^2\rho_b\delta - hp_b/2). \end{aligned} \tag{85.6}$$

The unperturbed part of this last equation is

$$\frac{\partial}{\partial t} a^3 \rho_b + 3a^2 \dot{a} p_b = 0, \tag{85.7}$$

which agrees with the second of equations (80.1). The first order part of equation (85.6) with equation (85.7) becomes

$$\dot{\delta} + (1 + \nu)(\theta - \dot{h}/2) = 3\frac{\dot{a}}{a}(\nu - c_s^2)\delta, \qquad \theta = v^\alpha{}_{,\alpha}/a, \tag{85.8}$$

where θ is the expansion. The divergence of equation (85.5) with equation (85.7) gives

$$\frac{\partial}{\partial t}[(1 + \nu)\theta] + (2 - 3\nu)(1 + \nu)\theta \dot{a}/a + c_s^2 \nabla^2 \delta / a^2 = 0. \tag{85.9}$$

Equations (85.5), (85.8), and (85.9) are the wanted linear perturbation equations. It will be noted that in this ideal fluid approximation the gravitational field enters only as h in equations (85.8). Gravity does not appear in the velocity equation (85.5) because the velocity is measured relative to freely moving observers. In equation (85.8) the term $\dot{h}/2$ is subtracted from the expansion because the proper volume of the fixed coordinate volume d^3x varies as $(1 - h/2)$ (eq. 85.3).

The linear perturbation equations for the matter are completed by the first of equations (82.10),

$$\frac{1}{2}\ddot{h} + \frac{\dot{a}}{a}\dot{h} = 4\pi G \rho_b (1 + 3c_s^2)\delta. \tag{85.10}$$

The other two of the field equations are

$$\begin{gathered} \dot{h}_{,\alpha} - \dot{h}_{\alpha\beta,\beta} = -16\pi G \rho_b (1 + \nu) a v^\alpha, \\ R_{\alpha\beta}{}^{(3)} - a^2\left[\frac{3}{2}\frac{\dot{a}}{a}\dot{h}_{\alpha\beta} + \frac{1}{2}\frac{\dot{a}}{a}\dot{h}\delta_{\alpha\beta} + \frac{1}{2}\ddot{h}_{\alpha\beta}\right] = 4\pi G a^2 \delta_{\alpha\beta} \rho_b (1 - c_s^2)\delta. \end{gathered} \tag{85.11}$$

It might be noted that since equations (85.10) and (85.11) agree with the equation $T^j{}_{i;j} = 0$ from which equations (85.5) and (85.8) were derived, these latter results can equally well be derived by manipulating equations (85.10) and (85.11): for this approach see Weinberg (1972).

86. Behavior of density perturbations at wavelength $\gg ct$

At small enough t, near enough to the time of the big bang, an irregularity with fixed comoving size x has proper size $\lambda = a(t)x$ much greater than the horizon ct. In this limit the pressure gradient term in equation (85.9) can be dropped and the solutions $\delta(\mathbf{x}, t)$ become free functions of x^α multiplied by solutions $D_n(t)$ of the perturbation equations in the one variable t. The solutions are most simply found if, as will be assumed here, the ratio $\nu = p/\rho$ is constant: the interesting cases are $\nu = 0$ for nonrelativistic matter with λ much larger than the Jeans length, $\nu = 1/3$ for a model dominated by electromagnetic radiation or free relativistic particles, and $\nu = 1$ for the possible behavior at extremely high redshifts (Zel'dovich 1961).

When $\lambda \gg c_s t$ and $\nu =$ constant, the perturbation equations from the last section become

$$\begin{aligned}
&\dot{\delta} + (1+\nu)(\theta - \dot{h}/2) = 0,\\
&\dot{\theta} + (2 - 3\nu)\theta\dot{a}/a = 0,\\
&\ddot{h} + 2\frac{\dot{a}}{a}\dot{h} = 8\pi G\rho_b(1+3\nu)\delta,\\
&\dot{h}_{,\alpha} - \dot{h}_{\alpha\beta,\beta} = -16\pi G\rho_b(1+\nu)av^\alpha,\\
&\frac{1}{a^2}[\nabla^2 h_{\alpha\beta} + h_{,\alpha\beta} - h_{\alpha\gamma,\gamma\beta} - h_{\beta\gamma,\gamma\alpha}] - 3\frac{\dot{a}}{a}\dot{h}_{\alpha\beta}\\
&\qquad - \frac{\dot{a}}{a}\dot{h}\delta_{\alpha\beta} - \ddot{h}_{\alpha\beta} = 8\pi G\delta_{\alpha\beta}\rho_b(1-\nu)\delta.
\end{aligned} \tag{86.1}$$

The mean density varies as (eq. 80.1)

$$\dot{\rho}_b = -3\frac{\dot{a}}{a}(1+\nu)\rho_b, \qquad \rho_b \propto a^{-3(1+\nu)}, \tag{86.2}$$

and the expansion rate (eq. 80.3) is

$$\dot{a}^2 = (\tfrac{8}{3}\pi G\rho a^{3+3\nu})a^{-1-3\nu}, \tag{86.3}$$

which gives

$$t^{-2} = 6(1+\nu)^2\pi G\rho_b, \qquad a \propto t^{2/(3+3\nu)}. \tag{86.4}$$

Suppose first that $\theta = \nabla \cdot \mathbf{v}/a$ is not zero. Then the second of equations (86.1) indicates

$$\theta \propto a^{3\nu-2}, \qquad \theta = Tt^m, \qquad m = (6\nu - 4)/(3 + 3\nu), \tag{86.5}$$

where T is a function of x^α. This form in the first and third of equations (86.1) yields the particular solution

$$\begin{gathered} \delta = Dt^{m+1}, \qquad \dot{h} = (m+1)Ht^m, \\ \frac{D}{H} = \frac{\nu(9\nu - 1)}{2(1 + 3\nu)}, \qquad \frac{D}{T} = \frac{3\nu(1+\nu)^2}{(6\nu+1)(1-\nu)}. \end{gathered} \tag{86.6}$$

The fourth of equations (86.1) fixes the sizes of the components of $h_{\alpha\beta}$. Suppose the perturbation has been broken up into plane waves. For a wave with propagation vector along the x^3 axis we have

$$\dot{h} - \dot{h}_{33} = 16\pi G\rho_b(1+\nu)\theta a^2/k^2 \propto a^{-3}, \tag{86.7}$$

with the particular solution

$$h_{11} = -8\pi G\rho_b\theta t(a/k)^2(1+\nu)^2/(1-\nu) \propto t^{-(1-\nu)/(1+\nu)}. \tag{86.8}$$

One sees from equations (86.5) and (86.6) that this is on the order of

$$h_{11} \sim \delta(\lambda/t)^2, \qquad \lambda = 2\pi a/k. \tag{86.9}$$

Since $\lambda \gg ct$, $\dot{h}_{11}$ is much larger than $\dot{h} \sim \delta/t$ (eqs. 86.6), and therefore

$$\dot{h}_{33} = -2\dot{h}_{11}. \tag{86.10}$$

If $\nu = 0$, equations (86.6) yield $D = 0$: there is no density perturbation in the mode and of course it did not appear in Chapter II. In a radiation-dominated universe, $\nu = 1/3$, $m = -1/2$, and $\delta \propto t^{1/2}$.

If $\theta = 0$, equations (86.1) give

$$\ddot{\delta} + \frac{4}{3+3\nu}\frac{\dot{\delta}}{t} = \frac{2(1+3\nu)}{3(1+\nu)}\frac{\delta}{t^2}. \tag{86.11}$$

The two solutions are powers of time:

$$\begin{gathered} \delta \propto D_1 = t^n, \qquad n = (2+6\nu)/(3+3\nu), \\ \delta \propto D_2 = t^{-1}. \end{gathered} \tag{86.12}$$

The first of equations (86.1) gives

$$\dot{h} = 2\dot{\delta}/(1 + \nu). \qquad (86.13)$$

The sizes of the components $h_{\alpha\beta}$ are fixed by the last of equations (86.1). Taking again a plane wave with propagation vector along the x^3 axis, we have from the components of this equation

$$\begin{aligned} \frac{k^2}{a^2} h_{11} + 3\frac{\dot{a}}{a}\dot{h}_{11} + \frac{\dot{a}}{a}\dot{h} + \ddot{h}_{11} &= -8\pi G\rho_b(1 - \nu)\delta, \\ 2\frac{k^2}{a^2} h_{11} + 3\frac{\dot{a}}{a}\dot{h}_{33} + \frac{\dot{a}}{a}\dot{h} + \ddot{h}_{33} &= -8\pi G\rho_b(1 - \nu)\delta. \end{aligned} \qquad (86.14)$$

With the third of equations (86.1) these can be combined to

$$\ddot{h}_{11} + 3\frac{\dot{a}}{a}\dot{h}_{11} = 8\pi G\rho_b\nu\delta, \qquad \frac{k^2}{a^2} h_{11} + \frac{\dot{a}}{a}\dot{h} + 8\pi G\rho_b\delta = 0. \qquad (86.15)$$

The first equation gives, for the growing solution in equations (86.12),

$$\dot{h}_{11} = \frac{4\nu}{(3\nu + 5)(\nu + 1)}\frac{\delta}{t}, \qquad \delta \propto D_1, \qquad (86.16)$$

and, for the decaying solution,

$$\dot{h}_{11} = -\frac{2}{3 + 3\nu}\frac{\delta}{t}, \qquad \delta \propto D_2. \qquad (86.17)$$

The second of equations (86.15) with equation (86.13) is

$$(k/a)^2 h_{11} = 0, \qquad \delta \propto D_2. \qquad (86.18)$$

This means that, for the decaying solution, h_{11} is bounded as the wavelength $\lambda = 2\pi\, a/k \to \infty$. For the growing solution we have

$$h_{11} = -\frac{4(5 + 9\nu)}{9(1 + \nu)^3}\left(\frac{a}{kt}\right)^2 \delta, \qquad \delta \propto D_1. \qquad (86.19)$$

One sees from equations (86.4) and (86.12) that this number for h_{11} is independent of time. There is no conflict with equations (86.16) because that indicates $t\dot{h}_{11} \sim \delta$, down from equations (86.19) by the factor $(\lambda/t)^2 \gg 1$.

The index n for the growing solution $D_1 \propto t^n$ (eq. 12) is $n = 2/3$ if $\nu = 0$, which agrees with equation (11.7). In a radiation-dominated universe where $\nu = 1/3$, the three modes in equations (86.5), (86.6), and (86.12) are

$$D_1 = t, \qquad D_2 = t^{-1}, \qquad D_3 = t^{1/2}. \tag{86.20}$$

These solutions were derived by Lifshitz (1946), Hawking (1966), Yu (1968), Adams and Canuto (1975), and Olson (1976).

For the two growing solutions, $|h_{11}| \sim |\delta| (\lambda/t)^2$ (eqs. 86.9 and 86.19). This is simply interpreted: the size of the perturbation is $\lambda \sim a/k$, the density excess is $\rho_b \delta$, so the mass excess is

$$\delta M \sim \rho_b \lambda^3 \delta. \tag{86.21}$$

Since $G\rho_b \sim t^{-2}$, equation (86.19) is

$$|h_{11}| \sim G\rho_b \lambda^2 |\delta| \sim G|\delta M|/\lambda, \tag{86.22}$$

the ordinary Newtonian potential. If we imagine that the epoch t and the amplitude δ of the perturbation both are fixed, then we see that the size of the perturbation to g_{ij} varies with the assumed extent of the perturbed patch as λ^2. If λ is so large as to make $G\delta M/\lambda \sim 1$, space in the perturbation is forced to curve into a knob that is in danger of pinching off into a relativistic singularity.

It was noted that h_{11} for the most rapidly growing mode D_1 is nearly constant when $\lambda \gg ct$. Thus if $|h_{11}| \ll 1$ at very high redshift, equations (86.19) say that when the horizon becomes comparable to the extent of the perturbation, that is, $\lambda \sim ct$, the amplitude of the density perturbation is $\delta \ll 1$: the fluctuations appearing on the horizon are small. This certainly agrees with what is seen now, and it means that when a density fluctuation reaches $\delta \sim 1$ and stops expanding, it is within the horizon and so nonrelativistic, which also agrees with what is observed. This restriction on δ was first pointed out by Novikov (1964a).

The time derivative of h_{33} is determined (eqs. 86.10, 86.13, 86.16, and 86.17) but not the value of h_{33}. That is because we are always free to adjust h_{33} by adjusting the way the coordinate x^3 is assigned along the plane wave in the initial hypersurface.

The decaying mode D_2 in equation (86.12) can be removed by adjusting the initial hypersurface of fixed cosmic time. The coordinate transformation in equations (81.12) and (81.15) with equations (86.4) yields

$$\hat{\delta} = \delta + 2\psi(\mathbf{x})/t, \qquad \frac{d}{dt}\hat{h}_{\alpha\beta} = \dot{h}_{\alpha\beta} + 2\psi_{,\alpha\beta}/a^2 - \frac{4}{3+3\nu}\frac{\psi}{t^2}\delta_{\alpha\beta}, \qquad (86.23)$$

where $\hat{\delta}$ is the density perturbation in the new time-orthogonal coordinates. We see from the first equation that we can eliminate the part δ_2 of δ that varies as t^{-1} by setting $\psi = -t\delta_2/2$. The second term in the right-hand side of the second equation is negligible compared to the last term because we are assuming that the spatial scale of the perturbation is much greater than t. With $\psi = -t\delta_2/2$ we see from equations (86.13) and (86.17) that the last term in equation (86.23) cancels the part of $\dot{h}_{\alpha\beta}$ associated with D_2. Since the decaying mode thus can be eliminated, it has no physical significance: it is simply the result of the freedom of choice of how the initial hypersurface of fixed cosmic time is assigned. For detailed discussion of this point see Press and Vishniac (1980a).

The solutions D_1 and D_2 in equations (86.12) are obtained from equations (10.12) by setting $\Lambda = 0 = R^{-2}$ in equation (10.10) and using equations (86.2) and (86.4). In these solutions different parts of the universe evolve like Friedman-Lemaître models with different values of the starting time and curvature constant R^{-2}. This means that irregularities as measured by R^{-2} or by h_{11} (eq. 86.19) are built into the cosmological model. Since Einstein's field equations say the curvature fluctuations are fixed, we cannot ascribe the origin of the irregularities described by this solution to any process operating at epochs when general relativity theory is an adequate approximation.

A way to avoid this conclusion by broadening the class of solutions was pointed out by Press and Vishniac (1980a) and Bardeen (1980). We start with the assumption that in the limit $t \rightarrow 0$, at the moment of the big bang, the universe was exactly homogeneous and isotropic and that at some epoch t_i there were spontaneous fluctuations in the pressure. It will be supposed that in the background model the ratio $p_b/\rho_b = \nu$ is constant. The fluctuating part of the pressure is

$$\delta p = p_b \epsilon, \qquad (86.24)$$

where p_b is the mean value and $\epsilon \ll 1$, the fractional perturbation to the pressure. With this change of variable equation (85.9) is changed to

$$(1+\nu)\dot{\theta} + \tfrac{2}{3}(2-3\nu)\theta/t + \nu\nabla^2\epsilon/a^2 = 0. \qquad (86.25)$$

In the solutions obtained above the last term is negligibly small because $\epsilon \lesssim \delta$ and the scale $a\lambda$ over which ϵ varies is much larger than t. Within the framework of Einstein's gravitational field equations we are allowed to speculate that at some epoch t_i there were spontaneous pressure fluctuations coherent over scales $a(t_i)\lambda \gg t_i$. (By ascribing the fluctuations to some quantum process, we avoid having to explain how the coherence length could be much larger than the horizon.) If we suppose the fluctuation lasted for about an expansion time t_i, then with $\nabla^2\epsilon \sim \epsilon/\lambda^2$ we see from equations (85.8), (85.10), and (86.25) that the pressure fluctuations generate the perturbations in the growing mode

$$\theta_i \sim t_i\epsilon_i/(\lambda a_i)^2, \qquad \delta_i \sim \epsilon_i(t_i/\lambda a_i)^2,$$
$$h \sim \epsilon_i \sim \delta_i(\lambda a_i/t_i)^2. \tag{86.26}$$

If at $t > t_i$ the spontaneous pressure fluctuations vanish and $\nu = 1/3$ so equations (86.20) apply, then at the epoch when the mass densities of matter and radiation are equal (eq. 92.42), we have

$$\delta_{eq} \sim \epsilon_i(t_{eq}/\lambda a_{eq})^2. \tag{86.27}$$

Since the observed galaxy clustering has scales smaller than t_{eq} (eq. 92.47), we see that pressure fluctuations in the early universe with $\epsilon \lesssim 1$ in principle can produce density irregularities of the observed size.

87. Spherical model

A particularly simple and useful model assumes that a density fluctuation can be approximated as spherically symmetric about one point, so that all quantities can be written as functions of just two independent variables x and t, and also that the pressure is negligible, $p \ll \rho c^2$. Under these conditions it is easy to see how the assumed density irregularity behaves, and it is easy to understand some of the results from the last section.

The spherical symmetry allows us to write the line element as

$$ds^2 = dt^2 - e^{2\alpha}dx^2 - e^{2\beta}(d\theta^2 + \sin^2\theta\, d\phi^2), \tag{87.1}$$

where α and β are functions of x and t. The gravitational field equations for a diagonal g_{ij} were writen out by Dingle (1933c). For the present line element Dingle's equations become

$$\begin{aligned} 8\pi G T_0{}^0 &= \dot\beta^2 + 2\dot\alpha\dot\beta + e^{-2\beta} - e^{-2\alpha}(2\beta'' + 3\beta'^2 - 2\alpha'\beta'),\\ 8\pi G T_1{}^1 &= 2\ddot\beta + 3\dot\beta^2 + e^{-2\beta} - \beta'^2 e^{-2\alpha},\\ 8\pi G T_2{}^2 &= \ddot\alpha + \dot\alpha^2 + \ddot\beta + \dot\beta^2 + \dot\alpha\dot\beta - e^{-2\alpha}(\beta'' + \beta'^2 - \alpha'\beta'),\\ 8\pi G e^{2\alpha} T_0{}^1 &= 2\dot\beta' + 2\dot\beta\beta' - 2\dot\alpha\beta'. \end{aligned} \tag{87.2}$$

The prime means a partial derivative with respect to the radial coordinate x and the dot, a partial derivative with respect to time.

The assumption that pressure is negligible means the matter orbits approximate geodesics. Thus given the spherical symmetry, we can always choose the coordinates to begin with so that $u^\alpha = 0$. Then the only nonzero source term in equation (87.2) is

$$T_0{}^0 = \rho(x, t). \tag{87.3}$$

The last of equations (87.2) with $T_0{}^1 = 0$ is

$$\begin{gathered}\dot{\beta}'/\beta' = \frac{\partial}{\partial t}\log\beta' = \dot{\alpha} - \dot{\beta}, \qquad \beta' \propto e^{\alpha-\beta}, \\ e^\beta\beta' = \frac{\partial}{\partial x}e^\beta = g(x)e^\alpha,\end{gathered} \tag{87.4}$$

where g is some function of x alone. Now it is convenient to introduce the expansion parameter $a(x, t)$ by the equation

$$e^\beta = xa(x, t), \tag{87.5}$$

and then rewrite the free function g in equations (87.4) in terms of a new function $R(x)$,

$$e^\alpha = (ax)'/(1 - x^2/R(x)^2)^{1/2}. \tag{87.6}$$

The results of substituting equations (87.5) and (87.6) into the first and second of equations (87.2) are

$$\begin{gathered}\frac{8}{3}\pi G\rho\frac{\partial}{\partial x}(ax)^3 = \frac{\partial}{\partial x}(\dot{a}^2ax^3 + ax^3/R^2), \\ 0 = 2\frac{\ddot{a}}{a} + \frac{\dot{a}^2}{a^2} + \frac{1}{a^2R^2}.\end{gathered} \tag{87.7}$$

These equations describe the evolution of the model.

The second of equations (87.7) multiplied by $a^2\dot{a}$ is

$$0 = \frac{\partial}{\partial t}(\dot{a}^2a + a/R^2), \qquad \dot{a}^2a + a/R^2 = F(x). \tag{87.8}$$

The function F can be adjusted by adjusting the coordinate assignments x on the initial hypersurface. In particular if space is not too strongly perturbed, we can make F independent of x, so equations (87.7) and (87.8) become

$$\dot{a}^2 a + a/R(x)^2 = A = \text{constant}, \qquad {}^8\!/\!_3\,\pi G\rho a^2 (ax)' = A. \qquad (87.9)$$

Lemaître (1933a, b) was the first (of many) to notice that the spherical model is a pleasantly simple generalization of the usual homogeneous world model. The first of equations (80.1) with $\rho \propto a^{-3}$ agrees with equations (87.9), where now R is a function of x. The line element is a generalization of the Robertson-Walker form (eq. 97.11):

$$ds^2 = dt^2 - \frac{((ax)'dx)^2}{1 - x^2/R(x)^2} - (ax)^2(d\theta^2 + \sin^2\theta\, d\phi^2). \qquad (87.10)$$

Figure 87.1 shows a geometrical interpretation in the case $R^{-2} \geq 0$. The hypersurface t = constant is represented by the surface of the two-dimensional axially symmetric curved sheet: proper lengths measured at fixed t are supposed to be measured on the sheet. The two open circles on the sheet represent points at the same t and polar angles and at radial

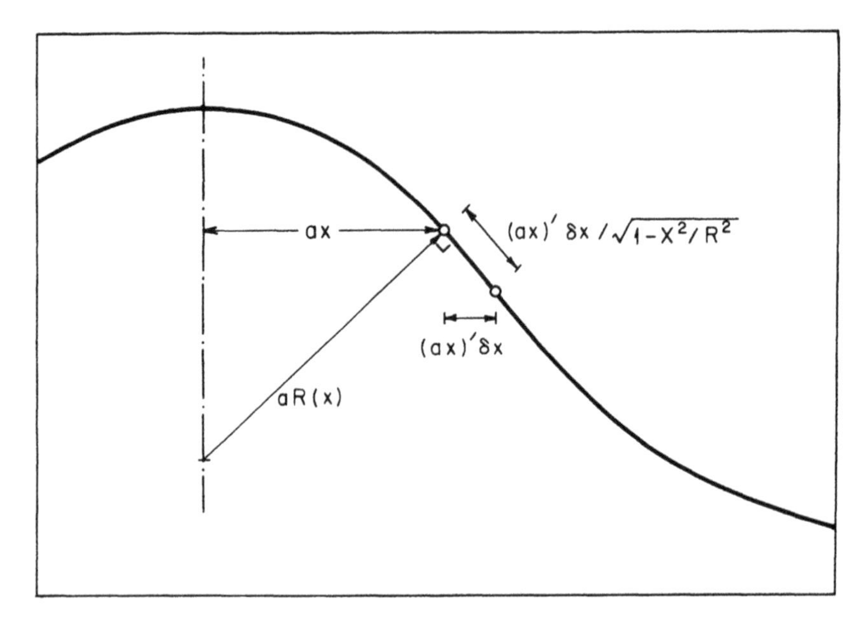

FIG. 87.1. The geometry around a growing mass concentration.

coordinate positions x and $x + \delta x$. The perpendicular distance from the axis is $xa(x, t)$, so the circumference of the circle x = constant is $2\pi ax$, as in equation (87.10). The radial distance between the two points, measured on the surface, is

$$\delta r = (ax)'\delta x/(1 - x^2/R^2)^{1/2}, \tag{87.11}$$

again in agreement with the proper distance given by equation (87.10). The slope of the surface in the radial direction is then

$$m = (x/R)/(1 - x^2/R^2)^{1/2}, \tag{87.12}$$

and so the normal to the surface intersects the axis at distance aR from the axis. If the curvature of the sheet is constant, aR is the curvature. Where $x = R(x)$, the surface is vertical so there is a knob; if $x \ll R$ everywhere, space is only slightly wrinkled.

A useful measure of the effective distance r_e of the mass shell x from the center is the quantity

$$r_e = ax. \tag{87.13}$$

The proper length of a great circle at x is $2\pi r_e$, so if $R(x) \gg x$, this agrees with what we would normally call the proper distance even if the matter well within x suffers a relativistic collapse. The gravitational acceleration at r_e, for a freely moving observer at x = constant, is (eqs. 87.7 and 87.9)

$$\ddot{r}_e = x\ddot{a} = -Ax^3/2r_e^2. \tag{87.14}$$

A measure of the effective gravitational mass within r_e is then

$$M_e = Ax^3/2G. \tag{87.15}$$

The rate of change of this effective mass with r_e at fixed t is (eq. 87.9)

$$\frac{\partial M_e}{\partial r_e} = \frac{3Ax^2}{2G(ax)'} = 4\pi\rho r_e^2, \tag{87.16}$$

which agrees with the usual Newtonian expression. Another interesting quantity is the number of particles within r_e: if the mass per particle is m, the proper number density is ρ/m and we have from equations (87.9) and (87.10)

$$
\begin{aligned}
mN(<r_e) &= \int_0^x 4\pi(ax)^2\,dx\,\rho(ax)'/(1 - x^2/R^2)^{1/2} \\
&= \frac{3}{2}\frac{A}{G}\int_0^x \frac{x^2\,dx}{(1 - x^2/R(x)^2)^{1/2}},
\end{aligned}
\tag{87.17}
$$

so

$$
\frac{mN}{M_e} = \frac{3}{x^3}\int_0^x \frac{x^2\,dx}{(1 - x^2/R^2)^{1/2}}. \tag{87.18}
$$

If $x \ll R$, the geometry is only slightly wrinkled, and this ratio is close to unity; where $R^{-2} > 0$ and space is strongly curved, M_e may be substantially reduced by the negative gravitational binding energy.

Consider now the evolution of a positive density fluctuation that is strong enough to make $R^{-2} > 0$. The coordinate size of the patch is $x \sim l$. The time variation of the proper radius $r_e = la(l, t)$ is the same as for a homogeneous cosmological model: the parametric solution to the first of equations (87.9) is, for $R^{-2} > 0$,

$$
a = \tfrac{1}{2} R^2 A(1 - \cos\eta), \qquad t = \tfrac{1}{2} R^3 A(\eta - \sin\eta), \tag{87.19}
$$

so the maximum size of the patch is

$$
r_{mx} \sim la_{mx} \sim lAR(l)^2, \tag{87.20}
$$

and this is reached at epoch

$$
t_{mx} \sim AR(l)^3. \tag{87.21}
$$

The ratio is

$$
r_{mx}/t_{mx} \sim l/R. \tag{87.22}
$$

As we noticed in the last section, if $l \ll R$ so the density fluctuation only wrinkles the geometry, the patch does not stop expanding until it is well within the horizon, $r_{mx} \ll ct_{mx}$. Consistent with this,equation (87.15) says

$$
\frac{GM_e(l)}{r_{mx}} \sim \left(\frac{l}{R}\right)^2, \tag{87.23}
$$

which is small if the geometry is only wrinkled.

We saw also that the condition $l \ll R(l)$ places a limit on the amplitude of the growing mode (eq. 86.22). To find this limit in the spherical model, write

$$a = a_b(t)(1 - \epsilon(x, t)), \tag{87.24}$$

where equations (87.9) for a_b are

$$\dot{a}_b^{\,2} a_b = A = \tfrac{8}{3}\pi G \rho_b a_b^{\,3}. \tag{87.25}$$

If terms $\sim\epsilon^2$ are discarded, equations (87.9) for $a(x, t)$ are

$$\dot{\epsilon} = -\frac{3}{2}\frac{\dot{a}_b}{a_b}\epsilon + \frac{\dot{a}_b}{2AR^2}, \qquad \frac{\rho}{\rho_b} = 1 + 3\epsilon + x\epsilon', \tag{87.26}$$

with the solution

$$\epsilon = \frac{B(x)}{a_b^{\,3/2}} + \frac{a_b}{5AR^2}. \tag{87.27}$$

Since $a_b \propto t^{2/3}$, the decaying solution varies as t^{-1} and the growing solution as $t^{2/3}$, as usual. The size of the growing density perturbation is

$$\delta \sim \epsilon \sim a_b(t)/AR^2, \tag{87.28}$$

so if the perturbed patch has size $x \sim l$, we see from equation (87.25) that

$$G\rho(a_b l)^2 \delta \sim (l/R)^2 \ll 1, \tag{87.29}$$

if space is to be only wrinkled. This is equivalent to equation (86.22).

If this limit is violated and there is a shell where $x \sim R(x)$, then the perturbation is relativistic and in danger of collapsing to a black hole. This would happen at $t \sim r_{mx} \sim AR(x)^3$ (eqs. 87.21 and 87.22), so the residual mass of the object is, from equation (87.15),

$$M_e \sim t/G, \tag{87.30}$$

comparable to the mass within the horizon t in the unperturbed model. Of course, the net mass of the baryons in the black hole could be much larger than M_e (eq. 87.18). It will be noted that there is no known way to heal over a massive black hole once it has formed: M_e can only grow through accretion (as in eq. 87.16). Such primeval black holes resulting from overly strong density fluctuations in the early universe may exist (as is discussed by Hawking 1971, Carr 1978, and others) but there must not be too many with masses larger than that of a galaxy, for otherwise they should have been noticed through the disruption of galaxies. The absence of extremely massive black holes is one basis for the argument that the large-scale structure of the universe must have been very accurately regulated at the time of the big bang: otherwise it would be difficult to understand how the universe avoided producing them (Peebles 1972).

88. Evolution of acoustic waves

The solutions in Section 86 assume $\lambda \gg c_s t$ so pressure gradients can be ignored. At later epochs, when the scale λ becomes less than $c_s t$, the irregularity oscillates like an acoustic wave. The first thing to consider is the effect of the general expansion on the amplitude of the wave. Later sections deal with dissipation due to nonlinearities (§ 89) and photon diffusion (§ 92).

When the wavelength λ is much less than $c_s t$, the perturbation δ oscillates with period $\tau \sim \lambda/c_s$, so we have from equation (85.10) (with $G\rho \sim t^{-2}$)

$$\dot{h} \sim \dot{\delta}(\tau/t)^2 \sim \dot{\delta}(\lambda/c_s t)^2. \tag{88.1}$$

Thus $\dot{h}$ can be dropped from equation (85.8) in this limit. If the perturbation is broken up into plane waves, equations (85.8) and (85.9) for the component with propagation vector $\mathbf{k}$ are

$$\begin{aligned} &\dot{\delta} + (1 + \nu)\theta = 3(\nu - c_s^2)(\dot{a}/a)\delta, \\ &\frac{\partial}{\partial t}[(1 + \nu)\theta] + (2 - 3\nu)(1 + \nu)\theta(\dot{a}/a) = (c_s^2 k^2/a^2)\delta. \end{aligned} \tag{88.2}$$

On differentiating the first equation with respect to time and using the second to eliminate $\partial_t(1 + \nu)\theta$, we arrive at a wave equation for δ. The dominant terms are on the order of δ/τ^2, others are of order $\delta/\tau t$ and δ/t^2, where t is the expansion time. Dropping the smallest terms $\sim\delta/t^2$, we have

$$\ddot{\delta} + (2 - 6\nu + 3\,c_s^2)(\dot{a}/a)\dot{\delta} + (c_s^2 k^2/a^2)\delta = 0. \tag{88.3}$$

The solution in the adiabatic approximation is found as in Section 16: write

$$\delta = A(t)e^{-i\phi(t)}, \tag{88.4}$$

and set

$$\phi = \int^t (kc_s/a)\,dt. \tag{88.5}$$

This eliminates the parts of equation (88.3) of order τ^{-2}, and the parts of order $(t\tau)^{-1}$ can be rearranged to

$$\frac{\ddot{\phi}}{\dot{\phi}} + 2\frac{\dot{A}}{A} + \frac{\dot{a}}{a}(2 - 6\nu + 3c_s^2) = 0. \tag{88.6}$$

From

$$\dot{\rho} = -3\frac{\dot{a}}{a}\rho(1 + \nu), \tag{88.7}$$

we have

$$3\frac{\dot{a}}{a}\nu = -3\frac{\dot{a}}{a} - \frac{\dot{\rho}}{\rho}, \tag{88.8}$$

and from

$$c_s^2 = \dot{p}/\dot{\rho}, \tag{88.9}$$

we arrive at

$$3\frac{\dot{a}}{a}c_s^2 = -3\frac{\dot{a}}{a} - \frac{d}{dt}\ln(\rho + p). \tag{88.10}$$

These expressions bring the last term of equation (88.6) to a total time derivative, so the equation can be integrated with the result

$$\delta \propto \left(\frac{\rho + p}{c_s\rho^2 a^4}\right)^{1/2} \exp - i\int^t (kc_s/a)\,dt. \tag{88.11}$$

This is the adiabatic solution. It agrees with equations (16.16) if $p \ll \rho \propto a^{-3}$. If $\nu = \text{constant} = c_s^2$, the amplitude varies as (eqs. 86.2 and 86.4)

$$A \propto a^{(3\nu-1)/2} \propto t^{(3\nu-1)/(3\nu+3)}. \tag{88.12}$$

In a radiation-dominated universe, $\nu = 1/3$, the amplitude is not affected by the expansion. If ν were greater than $1/3$, as it is speculated might be so in the very early universe, the amplitudes at the very short wavelengths then within the horizon would grow. At redshift $Z \sim 10^4$ the mass density is well approximated as a roughtly equal mixture of radiation and zero pressure matter. The mass densities in matter and radiation are μ and $\mathcal{E}$, and the total density and pressure are

$$\rho = \mu + \mathcal{E}, \qquad p = \mathcal{E}/3. \tag{88.13}$$

Assuming matter and radiation are tightly coupled so $\mathcal{E} \propto \mu^{4/3}$, we can relate the fractional perturbations to the densities of each by

$$\delta_m = \delta\mu/\mu = (3/4)\delta\mathcal{E}/\mathcal{E}. \tag{88.14}$$

With equations (88.13) and (88.14), equation (88.11) can be reduced to (Peebles and Yu 1970)

$$\begin{aligned} \delta_m &\propto (1 + R)^{-1/4} \exp - i\phi, \\ R &= 3\mu/4\mathcal{E} = 3.2 \times 10^4 \Omega h^2/(1 + Z). \end{aligned} \tag{88.15}$$

The number $R(t)$ is evaluated assuming the present radiation temperature is 2.7 K. We see that at decoupling of matter and radiation (redshift $Z \sim 1000$), the amplitude is only slightly depressed from its maximum value.

89. Nonlinear acoustic waves

In some scenarios the pressure waves discussed in the last section can reach amplitudes close to unity. If so, nonlinear hydrodynamics tend to shift the energy to short wavelengths where it is dissipated by shocks or viscosity. This can have two interesting consequences: it limits the amplitude of the residual pressure waves and it converts part of the irregular distribution of the original matter-radiation fluid into an irregular distribution of the matter alone. Two models for this nonlinear dissipation of pressure waves are discussed here.

A. Relativistic simple waves

Liang (1977a, b) has pointed out that Taub's (1948) relativistic simple wave solution provides a very convenient way to understand the nonlinear development of acoustic waves in the early universe. The solution applies to the ideal fluid model. It describes a running plane wave, with fluid velocity in the x direction, all fields single-valued functions of the density ρ, and ρ a function of the two variables x and t. It assumes also that the wavelength is much shorter than the Jeans length so gravity and the general expansion can be ignored. Then the equations of fluid dynamics in this one-dimensional problem are found from the two components of $T_{i\ ,j}^{j} = 0$:

$$\begin{aligned} 0 &= T_{0\ ,0}^{0} + T_{0\ ,0}^{1}, \qquad 0 = T_{1\ ,0}^{0} + T_{1\ ,1}^{1}, \\ T_i^j &= (\rho + p) u_i u^j - \delta_i^j p, \qquad g_{ij} = 1, -1, -1, -1, \\ u^1 &= v/(1 - v^2)^{1/2}, \qquad u^0 = 1/(1 - v^2)^{1/2}. \end{aligned} \tag{89.1}$$

Here $v = dr/dt$ is the ordinary fluid speed in proper Cartesian coordinates. The results are

$$\begin{aligned} \frac{\partial}{\partial t} \frac{\rho + pv^2}{1 - v^2} + \frac{\partial}{\partial r} \frac{(\rho + p)v}{1 - v^2} &= 0, \\ \frac{\partial}{\partial t} \frac{(\rho + p)v}{1 - v^2} + \frac{\partial}{\partial r} \frac{\rho v^2 + p}{1 - v^2} &= 0. \end{aligned} \tag{89.2}$$

Taub's running wave solution to equations (89.2) commences with the function

$$\phi(\rho) = \int^{\rho} c_s d\rho/(\rho + p). \tag{89.3}$$

The partial derivative of this with respect to time at fixed r gives

$$\frac{\partial \rho}{\partial t} = \frac{(\rho + p)}{c_s} \frac{\partial \phi}{\partial t}. \tag{89.4}$$

Since

$$c_s^2 = dp/d\rho, \tag{89.5}$$

we have also

$$\frac{\partial p}{\partial t} = c_s(\rho + p) \frac{\partial \phi}{\partial t}. \tag{89.6}$$

In the same way,

$$\frac{\partial \rho}{\partial r} = \frac{(\rho + p)}{c_s} \frac{\partial \phi}{\partial r}, \qquad \frac{\partial p}{\partial r} = c_s(\rho + p) \frac{\partial \phi}{\partial r}. \tag{89.7}$$

On using equations (89.4), (89.6), and (89.7) to eliminate the derivatives of ρ and p in equations (89.2) we find

$$\begin{aligned} v(1 - v^2)(1/c_s + c_s)\frac{\partial\phi}{\partial r} + (1 + v^2)\frac{\partial v}{\partial r} \\ + (1 - v^2)(v^2 c_s + 1/c_s)\frac{\partial\phi}{\partial t} + 2v\frac{\partial v}{\partial t} = 0, \\ (1 - v^2)(v^2/c_s + c_s)\frac{\partial\phi}{\partial r} + 2v\frac{\partial v}{\partial r} \\ + v(1 - v^2)(1/c_s + c_s)\frac{\partial\phi}{\partial t} + (1 + v^2)\frac{\partial v}{\partial t} = 0. \end{aligned} \tag{89.8}$$

These equations are simplified by taking linear combinations: multiply the first by $c_s - v$ and the second by $(1 - c_s v)$ and add, or multiply the first by $(c_s + v)$ and the second by $(1 + c_s v)$ and subtract. The results are

$$\begin{aligned} (1 - v^2)D_+\phi + D_+ v = 0, \\ (1 - v^2)D_-\phi - D_- v = 0, \end{aligned} \tag{89.9}$$

where $D_\pm$ are the differential operators

$$D_\pm = (1 \pm c_s v)\frac{\partial}{\partial t} \pm (c_s \pm v)\frac{\partial}{\partial r}. \tag{89.10}$$

Finally, Taub notes that

$$D_\pm \log\left(\frac{1 + v}{1 - v}\right)^{1/2} = \frac{1}{2}\left(\frac{1}{1 + v} + \frac{1}{1 - v}\right)D_\pm v = \frac{D_\pm v}{1 - v^2}, \tag{89.11}$$

so equations (9) are

$$\begin{aligned} D_+\left[\phi + \log\left(\frac{1 + v}{1 - v}\right)^{1/2}\right] = 0, \\ D_-\left[\phi - \log\left(\frac{1 + v}{1 - v}\right)^{1/2}\right] = 0, \end{aligned} \tag{89.12}$$

or

$$D_+ q = 0, \qquad D_- s = 0,$$
$$q = \phi + \log\left(\frac{1+v}{1-v}\right)^{1/2}, \qquad s = \phi - \log\left(\frac{1+v}{1-v}\right)^{1/2}. \tag{89.13}$$

A plane wave travelling in the direction of increasing r is represented by s = constant. This satisfies the second of equations (89.12) and from the first we have, since q is supposed to be a single-valued function of ρ alone,

$$D_+ \rho = (1 + c_s v)\frac{\partial \rho}{\partial t} + (c_s + v)\frac{\partial \rho}{\partial r} = 0. \tag{89.14}$$

The solution is

$$\rho = f(r - ut), \qquad u = (c_s + v)/(1 + c_s v). \tag{89.15}$$

The fluid speed v is fixed by the condition $s = \phi_0$ = constant,

$$\phi = \phi_0 + \log\left(\frac{1+v}{1-v}\right)^{1/2} = \int^{\rho} c_s d\rho/(\rho + p). \tag{89.16}$$

This miraculously simple result generalizes the nonrelativistic simple wave solution (Landau and Lifshitz 1959). We see from Equation (89.15) that the position r at which the density has some chosen value ρ moves toward increasing r at speed u. This speed is the relativistic sum of the velocity of sound in the rest frame of the fluid and the fluid speed v. Equation (89.16) gives the fluid speed as a function of ρ. Since v increases with increasing ρ, peaks tend to overtake troughs in the wave. Where this happens, the gradient sharpens to a shock and the wave energy dissipates. The rate of damping of the wave once the shock forms is computed by Liang and Baker (1977).

If pressure is directly proportional to density so that $\nu = c_s^2$ = constant, equation (89.16) gives

$$v = \frac{(\rho/\rho_b)^\alpha - 1}{(\rho/\rho_b)^\alpha + 1}, \qquad \alpha = 2c_s/(1 + c_s^2), \tag{89.17}$$

where the constant ϕ_0 has been chosen so that where ρ is equal to the mean ρ_b the fluid speed vanishes. For a peak or trough of amplitude $\delta\rho/\rho = \pm\delta$, with $\delta \ll 1$, this can be reduced to

$$v = \pm\alpha\delta/2, \tag{89.18}$$

and the propagation speed (eq. 15) is

$$u = c_s \pm \delta c_s(1 - c_s^2)/(1 + c_s^2). \tag{89.19}$$

For a radiation-dominated model, $c_s^2 = 1/3$,

$$u = 0.58 \pm 0.29\delta. \tag{89.20}$$

If there is a trough initially at distance λ ahead of a peak with densities at peak and trough $\rho_b\,(1 \pm \delta)$, then the peak is overtaken in time

$$t^* \sim 3^{1/2}\lambda/\delta, \tag{89.21}$$

and the wave starts to dissipate in a shock.

The simple wave solution is not a realistic explicit model: the pressure distribution is thought to be more like that of the water level in a choppy pond. Thus if the coherence length, a typical distance from peak to nearest trough, is λ, the running wave solution is a reasonable approximation for a coherence time $\sim\lambda/c_s$ after which the wave runs off in another direction. In a coherence time the distance between peak and trough changes by the typical amount

$$|\Delta\lambda| \sim c_s|\delta|(\lambda/c_s), \tag{89.22}$$

if $c_s \ll 1$. This grows in successive coherence times as a random walk, so the coherence length has appreciably shortened in time t^* such that

$$|\Delta\lambda|(t^* c_s/\lambda)^{1/2} \sim \lambda, \tag{89.23}$$

or

$$t^* \sim \lambda/(c_s\delta^2). \tag{89.24}$$

With t^* equal to the expansion time t, this sets the limit on the amplitude of the fluctuations δ that can exist without substantial nonlinear dissipation.

B. Second order perturbation theory

A more direct though tedious approach to the choppy water picture starts with the density perturbation approximated as a random Gaussian process with power spectrum cut off at some minimum wavelength $2\pi/k_{mx}$, so the coherence length is $\sim k_{mx}^{-1}$. Then in second-order perturbation theory we can compute the rate of growth of the short wavelength tail of the power spectrum $k > k_{mx}$ (Peebles 1970a, Tomita 1971).

The following calculation treats the case of special interest, a fluid of zero pressure matter tightly coupled to radiation. The mass densities in matter and radiation are μ and $\mathscr{E}$, and the total density and pressure are

$$\rho = \mu + \mathscr{E}, \qquad p = \mathscr{E}/3. \tag{89.25}$$

If the matter and radiation are tightly coupled, the entropy per baryon is fixed, so

$$\mathscr{E}(\mathbf{r}, t) \propto \mu(\mathbf{r}, t)^{4/3}. \tag{89.26}$$

The perturbed total density field is

$$\rho = \rho_b(1 + \delta(\mathbf{r}, t)). \tag{89.27}$$

On differentiating out equation (89.26), one finds that to second order the perturbed pressure is

$$p = p_b + \rho_b(c_s^2\delta + B\delta^2),$$
$$c_s^2 = 1/(3 + 3R), \qquad B = \frac{R(1 + 4R/3)}{24(1 + R)^3}, \tag{89.28}$$

where p_b is the mean pressure and

$$R = 3\mu_b/4\mathscr{E}_b \propto a(t). \tag{89.29}$$

Another useful expression is the relation between δ and the fractional perturbation to the matter density:

$$\mu = \mu_b(1 + \delta_m), \qquad \delta_m = S\delta,$$
$$S = \frac{R + 3/4}{R + 1} = \frac{\rho_b}{\rho_b + p_b}. \tag{89.30}$$

Next, let us write down expressions for the fluid density and velocity in linear perturbation theory. The perturbation δ (eq. 89.27) will be written as a Fourier series (representing a distribution periodic in some large volume V_u),

$$\delta = \sum_{\mathbf{k}} f_{\mathbf{k}}(T_{\mathbf{k}}^{+} + T_{\mathbf{k}}^{-})e^{i\mathbf{k}\cdot\mathbf{r}},$$
$$T_{\mathbf{k}}^{+} = e^{-ikc_s t + i\phi_{\mathbf{k}}}, \qquad T_{\mathbf{k}}^{-} = e^{ikc_s t - i\phi_{-\mathbf{k}}}. \tag{89.31}$$

The phases $\phi_{\mathbf{k}}$ are random, which makes δ a random Gaussian process, $f_{\mathbf{k}}$ is real with $f_{-\mathbf{k}} = f_{\mathbf{k}}$, and $2f_{\mathbf{k}}^2$ is the power spectrum of δ. This expression for δ is a sum of running waves like $\cos(\mathbf{k}\cdot\mathbf{r} - kc_s t + \phi_{\mathbf{k}})$, with randomly assigned phases. For the velocity field we can observe that the matter density satisfies

$$\partial v^\alpha/\partial r^\alpha = -\,\partial\delta_m/\partial t = -\,S\partial\delta/\partial t. \tag{89.32}$$

The wanted solution for v^α is (compare eq. 27.22)

$$v^\alpha = \Sigma v_{\mathbf{k}}^{\alpha} e^{i\mathbf{k}\cdot\mathbf{x}}, \qquad v_{\mathbf{k}}^{\alpha} = (k^\alpha/k)Sc_s(T_{\mathbf{k}}^{+} - T_{\mathbf{k}}^{-})f_{\mathbf{k}}. \tag{89.33}$$

Now let us find the perturbation equation for the evolution of the power spectrum. As before the general expansion and gravity are ignored and the fluid is taken to be ideal, but the motion now is in three dimensions so equation (89.1) is replaced with

$$T_{i\ ;j}^{\ j} = 0, \tag{89.34}$$

and the $i = 0$ and $i = \alpha$ parts are

$$\frac{\partial}{\partial t}\frac{\rho + pv^2}{1 - v^2} + \frac{\partial}{\partial r^\alpha}\frac{(\rho + p)v^\alpha}{1 - v^2} = 0,$$
$$\frac{\partial}{\partial t}\frac{(\rho + p)v^\alpha}{1 - v^2} + \frac{\partial}{\partial r^\beta}\frac{(\rho + p)v^\alpha v^\beta}{1 - v^2} + \frac{\partial p}{\partial r^\alpha} = 0. \tag{89.35}$$

These can be combined to

$$\frac{\partial^2}{\partial t^2}\frac{\rho + pv^2}{1 - v^2} = \nabla^2 p + \frac{\partial^2}{\partial r^\alpha \partial r^\beta}\frac{(\rho + p)v^\alpha v^\beta}{1 - v^2}. \tag{89.36}$$

This is the full nonlinear equation (neglecting gravity). If all terms smaller than $\sim\delta^2$ are dropped, it can be reduced to

$$\ddot{\delta} - c_s^2\nabla^2\delta = B\nabla^2\delta^2 + S^{-1}(\partial_\alpha\partial_\beta - \delta_{\alpha\beta}\partial_t^2)v^\alpha v^\beta. \tag{89.37}$$

The pressure is given by equation (89.28) and S is defined in equation (89.30).

The right-hand side of this equation, with δ and v^α given by equations (89.31) and (89.33), acts as a source for the growth of the short wavelength part of the spectrum. A convenient way to proceed is to write δ on the left-hand side in the form of equation (89.31) with $f_{\mathbf{k}}$ a slowly varying function of time. Then on picking out the Fourier component $\mathbf{k}$ from both sides of the equation, we have

$$\begin{aligned} &2ikc_s\dot{f}_{\mathbf{k}}[-\exp i\phi_{\mathbf{k}} + \exp i(2kc_st - \phi_{-\mathbf{k}})] \\ &\quad = -e^{ikc_st}\sum_{\mathbf{k}'}[k^2B\delta_{\mathbf{k}'}\delta_{\mathbf{k}''} + S^{-1}(k_\alpha k_\beta + \delta_{\alpha\beta}\partial_t^2)v_{\mathbf{k}'}{}^\alpha{}_{\mathbf{k}''}{}^\beta \end{aligned} \tag{89.38}$$

where

$$\mathbf{k}'' = \mathbf{k} - \mathbf{k}', \qquad \delta_{\mathbf{k}} = f_{\mathbf{k}}(T_{\mathbf{k}}^+ + T_{\mathbf{k}}^-). \tag{89.39}$$

We can drop the term on the left side with frequency $2kc_s$ because if $k > k_{mx}$, it cannot resonate with the source terms (where the maximum frequency is 2 $\mathrm{k}_{mx}c_s$). Then the time integral of equation (89.38) gives $f_k(t)$, the short wavelength part that develops through the nonlinear interaction. The second time derivative on the right-hand side can be eliminated by integrating by parts and dropping the oscillating part, leaving us with the expression

$$\begin{aligned} 2ikc_sf_{\mathbf{k}}(t)e^{i\phi_{\mathbf{k}}} &= \sum_{k'} F\frac{e^{i(k-k'-k'')c_st} - 1}{i(k - k' - k'')c_s}, \\ F &= Gf_{\mathbf{k}'}f_{\mathbf{k}''}e^{i\phi_{\mathbf{k}'} + \phi_{\mathbf{k}''}}, \\ G &= k^2B + Sc_s^2(k_\alpha k_\beta - k^2c_s^2\delta_{\alpha\beta})k'^\alpha k''^\beta/(k'k''), \end{aligned} \tag{89.40}$$

where

$$k > k_{mx}, \qquad k' < k_{mx}, \qquad k'' = |\mathbf{k} - \mathbf{k}'| < k_{mx}. \tag{89.41}$$

The power spectrum is found by squaring the first of equations (89.40) and averaging the random phases. This reduces the sum to the squared terms because the phases must cancel in pairs ($\phi_{\mathbf{k}'} = \phi_{\mathbf{q}'}$ and $\phi_{\mathbf{k}''} = \phi_{\mathbf{q}''}$ or $\phi_{\mathbf{k}'} = \phi_{\mathbf{q}''}$ and $\phi_{\mathbf{k}''} = \phi_{\mathbf{q}'}$), and the sum can be changed to an integral:

$$\sum_{\mathbf{k}'} = \frac{V}{8\pi^3} \int d^3k'. \tag{89.42}$$

The main contribution to the integral is where the denominator vanishes,

$$k'' = |\mathbf{k} - \mathbf{k}'| = k - k'. \tag{89.43}$$

This resonance condition indicates $\mathbf{k}'$ and $\mathbf{k}''$ are parallel to $\mathbf{k}$ so G (eq. 89.40) is

$$G = k^2 H, \qquad H = B + Sc_s^2(1 - c_s^2), \tag{89.44}$$

and the first of equations (89.40) becomes

$$\begin{aligned} 2f_k^2 &= \frac{V}{\pi^2} \frac{k^2 H^2}{c_s^2} \int_{k-k_{mx}}^{k_{mx}} (k'^2 dk' f_{k'}^2 f_{k''}^2 I, \\ I &= \int d\mu \frac{\sin^2 (k - k' - k'') c_s t/2}{(k - k' - k'')^2 c_s^2}. \end{aligned} \tag{89.45}$$

Finally, we can evaluate the integral over $\mu = \cos\theta$ when $kc_s t \gg 1$ by writing

$$\begin{aligned} k'' &= (k^2 + k'^2 - 2kk'\mu)^{1/2} \\ &\simeq k - k' + (1 - \mu) kk'/(k - k'), \end{aligned} \tag{89.46}$$

from which

$$I = \frac{\pi}{4} \frac{k - k'}{kk'} \frac{t}{c_s}, \tag{89.47}$$

so the power spectrum is

$$2f_k^2 = \frac{V}{4\pi} \frac{kH^2 t}{c_s^3} \int_{k-k_{mx}}^{k_{mx}} k'(k - k') dk' f_{k'}^2 f_{k''}^2. \tag{89.48}$$

This is the general expression for the growth of the short wavelength tail of the power spectrum. To evaluate it, let us suppose that the initial power spectrum is flat at k less than the cutoff k_{mx}. Then the initial variance of the density is

$$v = \langle(\delta\rho/\rho)^2\rangle = \sum_{k<k_{mx}} 2f_{\mathbf{k}}^2 = Vf^2 k_{mx}^3/(3\pi^2). \qquad (89.49)$$

The contribution to the variance by the short wavelength part is the integral of equation (89.48),

$$\delta v = (V/\pi^2)\int_{k_{mx}}^{2k_{mx}} k^2 dk f^2, \qquad (89.50)$$

which gives

$$\frac{\delta v}{v^2} = \frac{9\pi t}{8}\frac{(B + Sc_s^2(1 - c_s^2))^2}{c_s^3 k_{mx}^6}\int_{k_{mx}}^{2k_{mx}} k^3 dk \int_{k-k_{mx}}^{k_{mx}} k'(k - k')dk'. \qquad (89.51)$$

With equations (89.28) and (89.30) we have finally (Peebles 1970a).

$$\frac{\delta v}{v^2} = \frac{213\pi}{1120}\frac{(1 + 4R/3)^2(1 + 7R/4)^2}{(1 + R)^4} k_{mx} c_s t. \qquad (89.52)$$

This lengthy calculation yields a simple condition: if the variance of the density in the acoustic perturbation exceeds about $\lambda/(c_s t)$, where $\lambda \sim k_{mx}^{-1}$ is the coherence length, then in an expansion time t a substantial part of the energy shifts to shorter wavelength. The noise thus is dissipated until the variance on the scale $\lambda \sim k^{-1}$ satisfies

$$(\delta\rho/\rho)^2 \lesssim (kc_s t)^{-1}. \qquad (89.53)$$

This limit, which agrees with equation (89.24), is not very sensitive to the parameter $R = 3\mu/4\mathcal{E}$. The application of course depends on how irregular the universe is assumed to be at high redshift where matter and radiation behaved like a single fluid. Some scenarios are summarized in Chapter VI.

90. Incompressible flow

As was noted in Section 4D, primeval turbulence scenarios have had a large part in discussions of the origin of galaxies. There are two points to consider here: the effect of the general expansion on the rms turbulence velocity and the manner in which this assumed turbulent flow might trace back to very early times where the eddy size is much larger than the horizon.

For the first problem, the evolution of eddies well within the horizon, the ideal fluid model will be applied and it will be supposed that the flow is subsonic, $v \ll c_s$, so ρ is nearly constant and gravity can be ignored. The equations of motion are derived from $T_{i;j}^{j} = 0$ (eq. 85.2) with $h_{\alpha\beta} = 0$ and T_i^j given by equation (85.4). The component $i = 0$ with $v \ll c$ is the usual energy conservation equation

$$\frac{\partial \rho}{\partial t} + 3\frac{\dot{a}}{a}(\rho + p) + \frac{1}{a}\frac{\partial}{\partial x^\alpha}[(\rho + p)v^\alpha] = 0. \qquad (90.1)$$

The component $i = \alpha$ is

$$\frac{\partial}{\partial t}(a^4(\rho + p)v^\alpha) + a^3\frac{\partial}{\partial x^\beta}[(\rho + p)v^\alpha v^\beta + p\delta^{\alpha\beta}] = 0, \qquad (90.2)$$

if $v \ll c$. On differentiating this out and using equation (90.1) we find

$$(\rho + p)\left(\frac{\partial v^\alpha}{\partial t} + \frac{v^\beta}{a}\frac{\partial v^\alpha}{\partial x^\beta}\right) + \frac{\dot{a}}{a}(\rho + p)v^\alpha + \frac{1}{a}\frac{\partial p}{\partial x^\alpha} + v^\alpha\frac{\partial p}{\partial t} = 0. \qquad (90.3)$$

This can be compared to the usual Euler-Lagrange equations of motion. The inertial mass density is $(\rho + p)$. The second term says v^α tends to decrease as a^{-1}. As discussed in Section 7, this is a purely kinematic effect. The last term corrects the pressure gradient to the rest frame of the fluid.

A kinetic energy equation is found by multiplying equation (90.3) by v^α:

$$(\rho + p)\left(\frac{\partial}{\partial t} + \frac{v^\beta}{a}\frac{\partial}{\partial x^\beta}\right)\frac{v^2}{2} + (\rho + p)v^2\,\dot{a}/a$$

$$+ \mathbf{v}\cdot\nabla p/a + v^2\partial p/\partial t = 0. \qquad (90.4)$$

This equation averaged over x^α gives the time variation of the mean square peculiar velocity. We can use the assumption of subsonic flow to reduce the pressure gradient term: the mean pressure is p_b and the deviation from the mean satisfies

$$\mathbf{v} \cdot \nabla(p - p_b)/a = \nabla \cdot [\mathbf{v}(p - p_b)/a] - (p - p_b)\, \nabla \cdot \mathbf{v}/a. \qquad (90.5)$$

Since $(p - p_b)$ and $\nabla \cdot \mathbf{v}/a$ both are of order v^2, the last term can be dropped leaving a total divergence that vanishes when the equation is averaged over position. Using equation (90.1) we can write the second term in equation (90.4) as

$$\frac{\partial}{\partial x^\beta}[(\rho + p)v^\beta v^2/2a] + (v^2/2)(\dot\rho + 3(\rho + p)\, \dot a/a), \qquad (90.6)$$

and again the first part vanishes in the average over the x^α. These equations bring the space average of equation (90.4) to

$$(\rho_b + p_b)\, \partial\langle v^2\rangle/\partial t + 2(\rho_b + p_b)\, \langle v^2\rangle\, \dot a/a + 2\, \langle v^2\rangle \partial p/\partial t = 0. \qquad (90.7)$$

With the energy conservation equation (80.1) this yields the final result (*PC*, chapter VIIb).

$$\langle v^2\rangle \propto a^{-8}\, (\rho_b + p_b)^{-2}. \qquad (90.8)$$

If $p \ll \rho$, then $\rho \propto a^{-3}$ and $\langle v^2\rangle$ decays as a^{-2}; if $p = \rho/3$, then $\rho \propto a^{-4}$ and $\langle v^2\rangle$ is constant.

A standard heuristic argument for equation (90.8) uses angular momentum conservation (for example, Ozernoy and Chernin 1968). An eddy of coordinate size x has inertial mass $M \sim (\rho + p)(ax)^3$ (eq. 90.3), hence angular momentum $\sim M(ax)v$. Conservation of this angular momentum at fixed x reproduces equation (90.8).

This result is based on an ideal fluid and so of course does not describe the viscous dissipation when the kinetic energy cascades down to small enough scales. At epoch t this energy cascade dissipates eddies of size x if the turnover time is less than the expansion time,

$$t_{to} \sim ax/v(x, t) \lesssim t. \qquad (90.9)$$

When $p = \rho/3$, the eddy turnover time varies as $t_{to} \propto t^{1/2}$ and so would have exceeded t at high enough redshift. Thus one might imagine that primeval currents on scale x become fully turbulent when $t_{to}(x)/t$ approaches unity and that these nascent eddies serve to drive the turbulence on smaller scales (Ozernoy 1974 and references therein). As noted by Barrow (1977), one can even imagine that at extreme redshifts the ratio p/ρ exceeds $1/3$, so ρ varies faster than a^{-4}, and $\langle v^2 \rangle$ in equation (90.8) approaches zero as $Z \rightarrow \infty$ (eq. 86.2).

There is, however, a problem with space curvature in the scenario (as has been noted by Lifshiftz 1946, Ozernoy and Chernin 1968, and Zel'dovich and Novikov 1970). It will be supposed that the primeval mass currents at very high redshift can be described in linear perturbation theory (§ 85). Solutions with $\delta \neq 0$ were discussed in Section 86; we will consider here the case where the peculiar velocity field v^α is not zero but is so arranged that δ and the expansion θ vanish to first order.

With $\lambda \gg ct$ the pressure gradient term in equation (85.5) is negligible and the equation becomes

$$a^4(\rho + p)v^\alpha = F^\alpha(\mathbf{x}). \tag{90.10}$$

This gives the same time variation as equation (90.8). The first of equations (85.11) is

$$\frac{\partial^2}{\partial t \partial x^\beta}(h\delta_{\alpha\beta} - h_{\alpha\beta}) = -16\pi G F^\alpha(\mathbf{x})/a^3. \tag{90.11}$$

In general F^α can have a part with zero vorticity and a part with zero divergence. The former leads to equations (86.5) to (86.8). We are interested here in the case

$$\partial F^\alpha/\partial x^\alpha = 0, \tag{90.12}$$

where the solution to equation (90.11) is

$$h_{\alpha\beta} = -4G\int^t dt/a^3 \int d^3x' \,(F^\alpha{}_{,\beta} + F^\beta{}_{,\alpha})/|\mathbf{x} - \mathbf{x}'|, \qquad h = 0. \tag{90.13}$$

Since $v^\alpha{}_{,\alpha} = 0$ and $h = 0$, the density perturbation δ vanishes so the dominant part of the second of equations (85.11) is

$$\ddot{h}_{\alpha\beta} + 3\frac{\dot{a}}{a}\dot{h}_{\alpha\beta} = 0, \tag{90.14}$$

which agrees with the time variation of $h_{\alpha\beta}$ in equation (90.13).

The problem with this solution is that $h_{\alpha\beta}$ diverges at $t \to 0$, as $t^{-1/2}$ if $p = \rho/3$, more generally as (eq. 86.4)

$$h_{\alpha\beta} \propto t^{-(1-\nu)/(1+\nu)}, \tag{90.15}$$

if $p/\rho = \nu$ is constant. A fairly common assumption is that at $Z \sim 10^4$ there were random currents on scales comparable to the horizon ct and with velocities within an order of magnitude or so of c, which makes $h_{\alpha\beta}$ comparable to unity. At $Z = 10^6$ these currents would be larger than the horizon and $h_{\alpha\beta} \sim 100$ times larger. Thus we are forced to a situation that seems contrived: space at high redshift is strongly and irregularly buckled by the random primeval currents, but the mass distribution has been arranged so it always is homogeneous as it appears on the horizon.

91. Behavior of collisionless particles

Particles with long mean free path, like neutrinos or photons at $Z \lesssim 1000$, are described by a single-particle distribution function rather than a fluid model. It will be supposed here that scattering can be neglected; electromagnetic radiation scattered by free electrons is discussed in the next section. All particles have the same mass m. The results are valid for $m = 0$, but it will be convenient to suppose at first that the mass is greater than zero.

If $m > 0$, the motion satisfies the action principle

$$0 = \delta \int m\,ds = \delta \int m(g_{ij}\dot{x}^i\dot{x}^j)^{1/2}\,dt, \tag{91.1}$$

where the particle is moving along the path $x^\alpha(t)$ and

$$\dot{x}^i = (1, dx^\alpha/dt), \tag{91.2}$$

is the coordinate velocity. The Lagrangian is then

$$\mathcal{L} = m(g_{ij}\dot{x}^i\dot{x}^j)^{1/2}, \tag{91.3}$$

so the canonical momenta are

$$p_\alpha = \frac{\partial\mathcal{L}}{\partial\dot{x}^\alpha} = m\,g_{\alpha i}\,\dot{x}^i/(g_{ij}\dot{x}^i\dot{x}^j)^{1/2}. \tag{91.4}$$

The denominator is ds/dt so

$$p_\alpha = mg_{\alpha i}\, dx^i/ds = mg_{\alpha i}u^i = mu_\alpha, \tag{91.5}$$

where u^i is the particle four-velocity. If $h_{\alpha\beta} = 0$ and $ds/dt \approx 1$, this is

$$p_\alpha = -\, ma^2 dx^\alpha/dt, \tag{91.6}$$

which agrees with equation (7.10) apart from the sign.

The Hamiltonian is

$$H = \Sigma\, p_\alpha \dot{x}^\alpha - \mathcal{L} = mu_\alpha dx^\alpha/dt - mds/dt = -mu_0 = -p_0. \tag{91.7}$$

This is the fourth component of the four-vector p_i.

The Euler-Lagrange equations of motion are (eq. 81.6)

$$\frac{dp_\alpha}{dt} = \frac{\partial \mathcal{L}}{\partial x^\alpha} = \frac{1}{2}\, g_{jk,\alpha} p^j \dot{x}^k, \tag{91.8}$$

where the coordinate velocity in equation (91.2) is

$$\dot{x}^i = p^i/p^0. \tag{91.9}$$

The energy equation is

$$\frac{dH}{dt} = \frac{\partial H}{\partial t}. \tag{91.10}$$

The partial derivative with respect to time of the expression

$$m^2 = g^{ij} p_i p_j, \tag{91.11}$$

with equations (91.7) and (91.10) yields the fourth geodesic equation,

$$\frac{dp_0}{dt} = \frac{1}{2}\, g_{jk,0} p^j \dot{x}^k. \tag{91.12}$$

Equations (91.8) to (91.12) describe the motions of free particles. These equations remain valid in the limit $m = 0$, and so we can use them to describe a gas of neutrinos, or electromagnetic radiation approximated as a gas of photons.

The single particle distribution function $f(x^\alpha, p_\alpha, t)$ gives the number of particles per unit volume in the $\mathbf{x}$, $\mathbf{p}$ space,

$$\begin{aligned} \delta N &= f(\mathbf{x}, \mathbf{p}, t)\,\delta x^1 \delta x^2 \delta x^3 \delta p_1 \delta p_2 \delta p_3, \\ f'(\mathbf{x}', \mathbf{p}', t') &= f(\mathbf{x}, \mathbf{p}, t). \end{aligned} \tag{91.13}$$

Liouville's theorem says f is constant along the path of a particle (in the absence of the collisions not described by $\mathcal{L}$). Since this is true however the coordinates were assigned, f must be a scalar unchanged by coordinate transformations, as indicated in the second equation.

The stress-energy tensor for the gas of particles is

$$T^{ij} = \int \frac{d^4p}{(-g)^{1/2}}\, 2\delta(g^{ij}p_i p_j - m^2) p^i p^j f. \tag{91.14}$$

This is a tensor because the distribution function f and the argument of the delta function are scalars, and $(-g)^{1/2}$ makes $d^4p/(-g)^{1/2}$ a scalar. In locally Minkowski coordinates the argument of the delta function is $p_0{}^2 - \mathbf{p}^2 - m^2$, so the integral can be reduced to

$$T^{ij} = \int d^3p f p^i v^j, \tag{91.15}$$

which is the usual expression for the densities of energy and momentum. In time-orthogonal coordinates it is convenient to use as independent variables the particle energy p_0, the polar angles θ, ϕ of the p_α, and the auxilliary variable e,

$$p_\alpha = -pa(t)e\gamma_\alpha, \qquad p = (p_0{}^2 - m^2)^{1/2}, \qquad \Sigma(\gamma_\alpha)^2 = 1, \tag{91.16}$$

where the γ_α are the direction cosines. The argument of the delta function becomes

$$g^{ij}p_i p_j - m^2 = p^2[1 - e^2(1 + h_{\rho\sigma}\gamma_\rho\gamma_\sigma)], \tag{91.17}$$

so the integral over e reduces equation (91.14) to

$$T^{ij} = (1 + h/2) \int p^2 dp d\Omega (p^i p^j / p_0) f (1 - \tfrac{3}{2} h_{\rho\sigma} \gamma_\rho \gamma_\sigma),$$
$$p_\alpha = -pa\gamma_\alpha (1 - \tfrac{1}{2} h_{\rho\sigma} \gamma_\rho \gamma_\sigma), \tag{91.18}$$

in the weak field approximation, $|h_{\alpha\beta}| \ll 1$.

The mass density measured by an observer at fixed x^α is

$$T_{00} = (1 + h/2) \int p_0 p^2 \, dp d\Omega \, f (1 - \tfrac{3}{2} h_{\rho\sigma} \gamma_\rho \gamma_\sigma). \tag{91.19}$$

In a linear perturbation calculation it is assumed that the distribution function f is only slightly perturbed from the background function $f_b(p, t)$. To first order in the perturbations $h_{\alpha\beta}$ and $f - f_b$, equation (91.19) is

$$T_{00} = \int p_0 p^2 \, dp d\Omega [f + f_b(\tfrac{1}{2} h - \tfrac{3}{2} h_{\rho\sigma} \gamma_\rho \gamma_\sigma)], \tag{91.20}$$

and with

$$\int d\Omega \gamma_\alpha \gamma_\beta = 4\pi \delta_{\alpha\beta}/3, \tag{91.21}$$

we have

$$T_{00} = \int p_0 p^2 \, dp d\Omega \, f \equiv \rho_b(t)[1 + \delta(\mathbf{x}, t)]. \tag{91.22}$$

In the same way, using equation (91.21) and

$$\int \gamma_\alpha \gamma_\beta \gamma_\gamma \gamma_\delta \, d\Omega / 4\pi = (\delta_{\alpha\beta}\delta_{\gamma\delta} + \delta_{\alpha\gamma}\delta_{\beta\delta} + \delta_{\alpha\delta}\delta_{\beta\gamma})/15, \tag{91.23}$$

one arrives at the components

$$T_{0\alpha} = -a \int p^3 dp d\Omega \gamma_\alpha f,$$
$$T_{\alpha\beta} = \frac{4\pi}{3} a^2 (\delta_{\alpha\beta} - h_{\alpha\beta}) \int (p^4/p_0) \, dp f_b \tag{91.24}$$
$$+ a^2 \int d\Omega \gamma_\alpha \gamma_\beta \int (p^4/p_0) \, dp (f - f_b).$$

If the particle rest mass m is zero or negligibly small, the expressions for the T^{ij} simplify because $p_0 = p$ (eq. 91.16). A convenient variable in this case is the fractional perturbation to the brightness integrated over particle energy, defined by the equation

$$\int p^3 dpf = \rho_b[1 + \iota(\theta, \phi)]/4\pi, \qquad \int p^3 dpf_b = \rho_b/4\pi. \quad (91.25)$$

This integral appears in each of the expressions for the T_{ij}:

$$\begin{aligned} T_{00} &= \rho_b(1 + \delta), & \delta &= \int \iota d\Omega/4\pi, \\ T_{0\alpha} &= -a\rho_b f_\alpha, & f_\alpha &= \int \gamma_\alpha \iota d\Omega/4\pi, \\ T_{\alpha\beta} &= \rho_b a^2[{}^1\!/\!_3(\delta_{\alpha\beta} - h_{\alpha\beta}) + \eta_{\alpha\beta}], & \eta_{\alpha\beta} &= \int \gamma_\alpha\gamma_\beta \iota d\Omega/4\pi. \end{aligned} \quad (91.26)$$

Liouville's equation expressed in the variables x^α, p_0, γ_β is

$$\frac{\partial f}{\partial t} + \frac{\partial f}{\partial x^\alpha}\frac{dx^\alpha}{dt} + \frac{\partial f}{\partial p_0}\frac{dp_0}{dt} + \frac{\partial f}{\partial \gamma_\alpha}\frac{d\gamma_\alpha}{dt} = 0. \quad (91.27)$$

Both factors in the last term are of first order in the perturbation, so in a linear calculation it can be dropped. In the second term dx^α/dt can be replaced with the zeroth order expression $p\gamma_\alpha/(p_0 a)$ (eqs. 91.9 and 91.18) because $\partial f/\partial x^\alpha$ is first order in the perturbation. This brings the equation to

$$\frac{\partial f}{\partial t} + \frac{p\gamma_\alpha}{p_0 a}\frac{\partial f}{\partial x^\alpha} + \frac{\partial f}{\partial p_0}\frac{dp_0}{dt} = 0, \quad (91.28)$$

where the energy equation (91.12) in the new variables is

$$\frac{dp_0}{dt} = -\left(\frac{\dot{a}}{a} - \frac{1}{2}\gamma_\alpha\gamma_\beta\dot{h}_{\alpha\beta}\right)\frac{p^2}{p_0}. \quad (91.29)$$

These two equations describe the behavior of the gas of collisionless particles in linear perturbation theory.

If $m \ll p_0$ so $p = p_0$, equations (91.28) and (91.29) are homogeneous in the particle energy p. The evolution of the integrated brightness (eq. 91.25) is found by multiplying equation (91.28) by p^3 and integrating over p:

$$\frac{\partial}{\partial t}\rho_b(1 + \iota) + \rho_b\frac{\gamma_\alpha}{a}\frac{\partial \iota}{\partial x^\alpha} + 4\rho_b(1 + \iota)\left(\frac{\dot{a}}{a} - \frac{1}{2}\dot{h}_{\alpha\beta}\gamma_\alpha\gamma_\beta\right) = 0. \quad (91.30)$$

The unperturbed part is

$$\partial \rho_b / \partial t + 4(\dot{a}/a)\rho_b = 0. \tag{91.31}$$

This is the usual expression for radiation or free relativistic particles (eq. 80.1 with $p = \rho/3$). The first order part is

$$\frac{\partial \iota}{\partial t} + \frac{\gamma_\alpha}{a}\frac{\partial \iota}{\partial x^\alpha} = 2\gamma_\alpha\gamma_\beta \frac{\partial h_{\alpha\beta}}{\partial t}. \tag{91.32}$$

This is the brightness equation in the linear perturbation approximation.
The result of integrating equation (91.32) over direction is

$$\frac{\partial \delta}{\partial t} + \frac{1}{a}\frac{\partial f_\alpha}{\partial x^\alpha} = \frac{2}{3}\frac{\partial h}{\partial t}, \tag{91.33}$$

where δ is the fractional perturbation to the mass density and f_α is proportional to the net energy flux measured by an observer at fixed x^α (eqs. 91.26). With

$$\frac{4}{3}\theta \equiv \frac{1}{a}\frac{\partial f_\alpha}{\partial x^\alpha}, \tag{91.34}$$

the equation becomes

$$\frac{\partial \delta}{\partial t} + \frac{4}{3}\left(\theta - \frac{1}{2}\frac{\partial h}{\partial t}\right) = 0, \tag{91.35}$$

which agrees with equation (85.8) with $\nu = c_s^2 = 1/3$. The result of multiplying equation (91.32) by γ_α and integrating over angles is (eq. 91.26)

$$\frac{\partial f_\alpha}{\partial t} + \frac{1}{a}\frac{\partial \eta_{\alpha\beta}}{\partial x^\beta} = 0, \tag{91.36}$$

and the divergence of this with equation (91.34) is

$$\frac{4}{3}\frac{\partial \theta}{\partial t} + \frac{4}{3}\frac{\dot{a}}{a}\theta + \frac{1}{a^2}\frac{\partial^2}{\partial x^\alpha \partial x^\beta}\int \gamma_\alpha\gamma_\beta\iota \, d\Omega/4\pi = 0, \tag{91.37}$$

which can be compared to equation (85.9) with $\nu = 1/3$.

Let us suppose next that these particles with $m = 0$ are the main contributors to the total stress-energy tensor. Then $T_i^i = 0$ (eq. 91.14), and equations (91.26) and (82.9) give

$$\frac{1}{2}\frac{\partial^2 h}{\partial t^2} + \frac{\dot{a}}{a}\frac{\partial h}{\partial t} = 8\pi G\rho_b\delta,$$
$$\frac{\partial^2}{\partial x^\beta \partial t}(h\delta_{\alpha\beta} - h_{\alpha\beta}) = -16\pi G a\rho_b f_\alpha, \tag{91.38}$$

in agreement with equation (85.10). If in addition the characteristic scale of the density fluctuation is $\lambda \gg ct$, the last term in equation (91.37) can be dropped, and we come to the first three of equations (86.1) with $\nu = 1/3$ and, as was derived in Section 86, the solutions $\delta \propto t^n$, $n = 1$, $1/2$, and -1. Here the gas of collisionless particles behaves like an ideal fluid because the distance a particle moves is much less than a wavelength.

Another simple and interesting limiting case is $\lambda \ll ct$. Here $h_{\alpha\beta}$ is negligible if $m = 0$ and the brightness equation (91.32) becomes

$$\frac{\partial\iota}{\partial t} + \frac{\gamma_\alpha}{a}\frac{\partial\iota}{\partial x^\alpha} = 0, \tag{91.39}$$

with the solution

$$\iota(t, \gamma, \mathbf{x}) = \iota(t_i, \gamma, \mathbf{x} - \gamma\int_{t_i}^{t} dt/a). \tag{91.40}$$

When the time integral in the last argument is large, it makes ι a rapidly fluctuating function of direction and so washes out the fluctuations in the net density δ. As an example of how this goes, suppose that at the starting time t_i the brightness ι is independent of direction but a fluctuating function of position. Then we can write the initial value as

$$\iota(t_i, \mathbf{x}) = \Sigma\, \iota_{\mathbf{k}} e^{i\mathbf{k}\cdot\mathbf{x}}. \tag{91.41}$$

At time t the density perturbation is

$$\delta = \sum \iota_{\mathbf{k}} e^{i\mathbf{k}\cdot\mathbf{x}} \int \frac{d\mu}{2} \exp - ik\mu \int_{t_i}^{t} dt/a, \tag{91.42}$$

where $\mathbf{k} \cdot \boldsymbol{\gamma} = k\mu$. The mean square fluctuation is then (Stewart 1972)

$$\langle \delta^2 \rangle = \sum |\iota_{\mathbf{k}}|^2 (\sin \kappa)^2/\kappa^2, \qquad \kappa = k \int_{t_i}^{t} dt/a. \tag{91.43}$$

In this limit the rms fluctuation in the density decays as $a(t)/t$.

For examples of numerical integration of equations (91.32) and (91.38) see Peebles (1973c).

These results assume the particles are relativistic. It is left as an exercise to verify that in the nonrelativistic limit $p \ll m$ the distribution behaves in the way discussed in Chapter II.

92. Linear dissipation of adiabatic perturbations

At high redshift matter and radiation are strongly coupled and so behave like a single ideal fluid. Any irregularities present in the mass distribution grow until the wavelength is comparable to the horizon and then oscillate like pressure waves (§§ 86, 88) developing a spectrum of acoustic noise. The short wavelength end is damped by photon diffusion leaving residual fluctuations in the matter distribution. The interesting result is that, as will be described, the process yields three characteristic lengths.

A. Description of the matter and radiation

The important variables are the mass density of the matter,

$$\mu = \mu_b(1 + \delta_m), \tag{92.1}$$

the matter velocity v^α relative to the time-orthogonal coordinates, the radiation brightness $\propto \iota$ (eq. 91.25), the mass density of the radiation (eq. 91.26),

$$\mathscr{E} = \mathscr{E}_b(1 + \delta_r), \qquad \delta_r = \int \iota d\Omega/4\pi, \tag{92.2}$$

and the gravitational fields $h_{\alpha\beta}$. Matter pressure can be neglected.

The brightness equation (91.32) must be adjusted to take account of scattering by the matter. The dominant effect is ordinary nonrelativistic Thomson scattering by the free electrons, and it is a reasonable approximation to simplify things by taking the differential scattering cross section to be isotropic in the matter rest frame.

In this frame the scattering changes the photon distribution function f at neighboring points 1 and 2 along a path by the amount

$$f(2)' - f(1)' = \sigma n_e \delta t'(f'_+(p'_0) - f'). \qquad (92.3)$$

The primes refer to the matter rest frame. In the approximation that the scattering is isotropic, the distribution scattered into the beam is just

$$f'_+(p'_0) = \int f'(p'_0, \gamma'_\alpha)\, d\Omega'/4\pi. \qquad (92.4)$$

The Thomson scattering cross section is σ, n_e is the free electron density in this frame, and $\delta t'$ is the time interval between points (1) and (2). The time measured in the time-orthogonal coordinates is δt:

$$\delta t' = \frac{\partial t'}{\partial x^i}\frac{dx^i}{dt}\delta t = \frac{\partial t'}{\partial x^i}\frac{p^i}{p_0}\delta t = \frac{p'_0}{p_0}\delta t. \qquad (92.5)$$

From the invariance of $p_i u^i$, with u^i the matter four-velocity, and equation (91.18), we have

$$p'_0 = u^0(p_0 + p_\alpha dx^\alpha/dt) = p(1 - \gamma_\alpha v^\alpha), \qquad (92.6)$$

to first order.

Since f is a scalar, we can directly write down the collision equation in the time-orthogonal coordinates:

$$\frac{\partial f}{\partial t} + \frac{\partial f}{\partial x^\alpha}\frac{dx^\alpha}{dt} + \frac{\partial f}{\partial p}\frac{dp}{dt} + \frac{\partial f}{\partial \gamma_\alpha}\frac{d\gamma_\alpha}{dt} = \sigma n_e (p'_0/p_0)(f_+ - f). \qquad (92.7)$$

In the linear perturbation approximation the equation is (eqs. 91.28 and 91.29)

$$\frac{\partial f}{\partial t} + \frac{\gamma_\alpha}{a}\frac{\partial f}{\partial x^\alpha} - p\frac{\partial f}{\partial p}\left(\frac{\dot a}{a} - \frac{1}{2}\gamma_\alpha\gamma_\beta\frac{\partial h_{\alpha\beta}}{\partial t}\right) = \sigma n_e (f_+ - f). \qquad (92.8)$$

On multiplying this by p^3 and integrating over p, we find (eq. 91.32)

$$\frac{\partial \iota}{\partial t} + \frac{\gamma_\alpha}{a}\frac{\partial \iota}{\partial x^\alpha} - 2\gamma_\alpha\gamma_\beta\frac{\partial h_{\alpha\beta}}{\partial t} = \sigma n_e (\iota_+ - \iota), \qquad (92.9)$$

where ι is defined in equation (91.25). Finally, the radiation scattered into the beam is computed by noting that (eq. 92.4)

$$f_+(p, \gamma) = \int f'(p', \gamma')\, d\Omega'/4\pi, \qquad (92.10)$$

where p' is given in terms of p and γ_α by equation (92.6), so

$$1 + \iota_+(\gamma) = \int p^3 dp \int d\Omega' f'(p', \gamma')/\mathscr{E}_b = (1 + 4\gamma_\alpha v^\alpha)\mathscr{E}'/\mathscr{E}_b, \quad (92.11)$$

where

$$\mathscr{E}' = T'_{00} = T_{ij}u^i u^j \simeq T_{00} = \mathscr{E}_b(1 + \delta_r), \quad (92.12)$$

to first order (eq. 91.26). Thus the collision equation can be reduced to

$$\frac{\partial\iota}{\partial t} + \frac{\gamma_\alpha}{a}\frac{\partial\iota}{\partial x^\alpha} - 2\gamma_\alpha\gamma_\beta\frac{\partial h_{\alpha\beta}}{\partial t} = \sigma n_e(\delta_r + 4\gamma_\alpha v^\alpha - \iota). \quad (92.13)$$

This is the differential equation for the radiation brightness in the linear perturbation approximation. It was derived by Peebles and Yu (1970). Related cases of the equation were derived by Dautcourt (1969, 1970).

The time variation of the perturbation to the matter mass density is given by equation (85.8) with $p = 0 = \nu$,

$$\frac{\partial\delta_m}{\partial t} = \frac{1}{2}\frac{\partial h}{\partial t} - \frac{1}{a}\frac{\partial v^\alpha}{\partial x^\alpha}. \quad (92.14)$$

The brightness seen by an observer at fixed x^α is

$$I = \mathscr{E}_b(1 + \iota)/4\pi, \quad (92.15)$$

to first order, and the corresponding brightness of the radiation scattered by the matter is (eqs. 92.11 and 92.12)

$$I_+ = \mathscr{E}_b(1 + \delta_r + 4\gamma_\alpha v^\alpha)/4\pi, \quad (92.16)$$

so the net volume force on the matter is

$$\begin{aligned} F_\alpha &= \int d\Omega(\mathscr{E}_b/4\pi)(\iota - 4\gamma_\alpha v^\beta - \delta_r)\gamma_\alpha\sigma n_e \\ &= \sigma n_e \mathscr{E}_b(f_\alpha - \tfrac{4}{3} v^\alpha), \quad (92.17) \\ f_\alpha &= \int \iota\gamma_\alpha \, d\Omega 4\pi. \end{aligned}$$

Thus the equations of motion of the matter are (eq. 7.12)

$$\frac{\partial v^\alpha}{\partial t} + \frac{\dot{a}}{a}v^\alpha = \sigma n_e(\mathscr{E}/\mu)(f_\alpha - \tfrac{4}{3}v^\alpha). \quad (92.18)$$

The equations are completed by the gravitational field equations (82.9) with T_i^j given by equations (91.26). Some properties of the solutions are discussed next.

B. Limit $t_c \rightarrow 0$

The mean free time is

$$t_c = 1/\sigma n_e. \tag{92.19}$$

When the mean free path is much less than the size of the fluctuations, matter and radiation act like a viscous fluid. In this limit it is convenient to rewrite the collision equation (92.13) as

$$\iota = \delta_r + 4\gamma_\alpha v^\alpha - t_c\left(\frac{\partial \iota}{\partial t} + \frac{\gamma_\alpha}{a}\frac{\partial \iota}{\partial x^\alpha} - 2\gamma_\alpha\gamma_\beta\frac{\partial h_{\alpha\beta}}{\partial t}\right), \tag{92.20}$$

and then write down the solution by iteration:

$$\iota = \delta_r + 4\gamma_\alpha v^\alpha, \tag{92.21}$$

to lowest order,

$$\begin{aligned}\iota = \delta_r + 4\gamma_\alpha v^\alpha - t_c\bigg(&\frac{\partial \delta_r}{\partial t} + 4\gamma_\alpha\frac{\partial v^\alpha}{\partial t} + \frac{\gamma_\alpha}{a}\frac{\partial \delta_r}{\partial x^\alpha} \\ &+ 4\frac{\gamma_\alpha\gamma_\beta}{a}\frac{\partial v^\alpha}{\partial x^\beta} - 2\gamma_\alpha\gamma_\beta\frac{\partial h_{\alpha\beta}}{\partial t}\bigg),\end{aligned} \tag{92.22}$$

to order t_c, and so on. With the definition of δ_r this last equation yields

$$\delta_r = \int \iota\, d\Omega/4\pi = \delta_r - t_c\left(\frac{\partial \delta_r}{\partial t} + \frac{4}{3a}\frac{\partial v^\alpha}{\partial x^\alpha} - \frac{2}{3}\frac{\partial h}{\partial t}\right), \tag{92.23}$$

so, to zeroth order in t_c,

$$\frac{\partial \delta_r}{\partial t} = \frac{4}{3}\left(\frac{1}{2}\frac{\partial h}{\partial t} - \frac{1}{a}\frac{\partial v^\alpha}{\partial x^\alpha}\right) = \frac{4}{3}\frac{\partial \delta_m}{\partial t}. \tag{92.24}$$

The first equation agrees with equation (85.8) with $\nu = 1/3$, the second with equation (92.14) agrees with the conservation of entropy per baryon, $\mathscr{E} \propto \mu^{4/3}$. The flux defined in equation (92.17) with equation (92.22) is

$$f_\alpha = \int \iota\gamma_\alpha d\Omega/4\pi = \frac{4}{3}v^\alpha - t_c\left(\frac{4}{3}\frac{\partial v^\alpha}{\partial t} + \frac{1}{3a}\frac{\partial \delta_r}{\partial x^\alpha}\right), \tag{92.25}$$

to first order in t_c, and this in equation (92.18) gives

$$\left(\mu + \frac{4}{3}\mathcal{E}\right)\frac{\partial v^\alpha}{\partial t} + \mu v^\alpha \dot{a}/a = -\frac{\mathcal{E}}{3a}\frac{\partial \delta_r}{\partial x^\alpha}, \tag{92.26}$$

to zeroth order. This agrees with equation (90.3): as expected, in the limit $t_c \to 0$ the matter and radiation behave like an ideal fluid with $\rho = \mu + \mathcal{E}$, $p = \mathcal{E}/3$.

The evolution of a pressure wave in an ideal fluid was discussed in Section 88. For a mixture of matter and radiation the time variation is (eq. 88.15)

$$\begin{gathered} \delta_r \propto e^{-i\phi}/(1 + R)^{1/4}, \\ \phi = \int^t dtkc_s/a, \qquad c_s = [3(1 + R)]^{-1/2}, \\ R = 3\mu_b/4\mathcal{E}_b \propto a(t). \end{gathered} \tag{92.27}$$

Photon diffusion adds an exponential damping term, which is discussed next.

C. Damping by photon diffusion

The simplest way was shown by Field (1971) and Chibisov (1972): since we are interested in the limit $\lambda \ll c_s t \lesssim ct$, we can ignore gravity, and let us ignore also the time variation of the coefficients a and t_c. This brings equations (92.13) and (92.18) to

$$\begin{gathered} \frac{\partial \iota}{\partial t} + \frac{\gamma_\alpha}{a}\frac{\partial \iota}{\partial x^\alpha} = (\delta_r + 4\gamma_\alpha v^\alpha - \iota)/t_c, \\ \frac{\partial v^\alpha}{\partial t} = \frac{\mathcal{E}}{t_c \mu}\left(f_\alpha - \frac{4}{3}v^\alpha\right). \end{gathered} \tag{92.28}$$

With the coefficients independent of position and time we can look for solutions of the form

$$v^\alpha \propto \iota \propto e^{i\mathbf{k}\cdot\mathbf{x} - \omega t}, \tag{92.29}$$

for which the equations become

$$\iota = \frac{\delta_r + 4\mu v}{1 - it_c(\omega - k\mu/a)}, \qquad v = \frac{(3f/4)}{1 - i\omega R t_c}, \tag{92.30}$$

with

$$\delta_r = \int \iota d\mu/2, \qquad f = \int \iota\mu d\mu/2, \qquad \mu = \cos\theta = \boldsymbol{\gamma}\cdot\mathbf{k}/k. \tag{92.31}$$

The vectors **k**, **v**, and **f** all are parallel, and θ is the angle between the wave vector **k** and the direction $\boldsymbol{\gamma}$. The convenient symbol $\mu = \cos\theta$ should not be confused with the mass density.

The result of integrating the first of equations (92.28) over direction $\boldsymbol{\gamma}$ is

$$\omega\delta_r = fk/a. \tag{92.32}$$

With this we can combine equations (92.30) to get

$$\iota = \delta_r \frac{1 + (3a\mu\omega/k)(1 - i\omega t_c R)^{-1}}{1 - it_c(\omega - k\mu/a)}. \tag{92.33}$$

This is valid to all orders in t_c (but of course it assumes the characteristic times for changes in t_c and R are much longer than the times from the real and imaginary parts of ω). The dispersion relation $\omega(k)$ is found by substituting equation (92.33) in the first of equations of (92.31) and eliminating δ_r, but the general result is lengthy and not very illuminating, so it is better to expand equation (92.33) in a series in t_c and then integrate over directions μ. The result to first order in t_c is

$$\omega = \omega_0 - i\gamma, \qquad \omega_0 = kc/a[3(1 + R)]^{1/2}, \qquad \gamma = \frac{k^2c^2t_c}{6a^2}\frac{(R^2 + 4(R + 1)/5)}{(R + 1)^2}. \tag{92.34}$$

The real part agrees with the ideal fluid limit in equation (92.27). The imaginary part γ is the exponential damping rate. Finally, we can take account of a relatively slow time variation of γ by modifying equation (92.27) to

$$\delta_r \propto (1 + R)^{-1/4} \exp\left(-i\phi(t) - \int^t \gamma dt\right). \tag{92.35}$$

Another way to arrive at this result is to continue the expansion in equation (92.22) to order t_c^2, derive the equations of motion in equations (92.24) and (92.26) to first order in t_c, and then solve these equations in the adiabatic approximation. This derivation of equations (92.34) and (92.35) is worked out by Peebles and Yu (1970). Field (1971) and Chibisov (1972) derived γ by the method presented here. One can also find γ from the transport coefficients for the matter and radiation described as a fluid with viscosity and conductivity. This was pioneered by Silk (1968a) and worked out in detail by Weinberg (1971).

D. Linear transfer function for adiabatic perturbations: characteristic masses

Suppose the most rapidly growing mode of the density irregularity at some very high redshift Z_i is broken up into plane waves like in equation (92.29). The amplitude belonging to the propagation vector $\mathbf{k}$ is $\delta_{\mathbf{k}}(i)$ at this starting time, and the amplitude at this same comoving $\mathbf{k}$ has grown to $\delta_{\mathbf{k}}(f)$ at epoch $Z_f = Z_{dec} \sim 1300$, just after matter and radiation have decoupled. In linear perturbation theory we can write

$$\delta_{\mathbf{k}}(f) = T(k)\delta_{\mathbf{k}}(i), \tag{92.36}$$

where $T(k)$ is the linear transfer function for the cosmological model. As will be discussed here, there are three interesting and possibly important characteristic quantities in the function $T(k)$: a short wavelength cutoff M_s that may set the scale of the first generation of bound systems, more or less regularly spaced zeros of $T(k)$, and the Jeans mass M_x longward of which T is independent of k.

The proper wavelength belonging to wave number k is

$$\lambda = 2\pi a(t)/k. \tag{92.37}$$

It is convenient and standard to express λ in terms of the mass in baryons within a sphere with diameter equal to the wavelength:

$$M = \pi\mu_b\lambda^3/6, \qquad \mu_b = 1.88 \times 10^{-29}\Omega_0 h^2 \text{ g cm}^{-3}. \tag{92.38}$$

This gives

$$\lambda = \frac{5.87 \times 10^{20}\,(M/\mathrm{M}_\odot)^{1/3}}{(1 + Z)(\Omega_0 h^2)^{1/3}} \text{ cm}, \tag{92.39}$$

at redshift Z. To set the scale we might note that the correlation functions $\xi_{cg}(r)$ and $\xi_{gg}(r)$ are positive to r at least as large as 40 h^{-1} Mpc (table 78.1). At $\lambda = 40\ h^{-1}$ Mpc the characteristic mass is

$$M(40h^{-1}\ \mathrm{Mpc}) \sim 1 \times 10^{16} h^{-1}\Omega_0\ \mathrm{M}_{\odot}. \tag{92.40}$$

The ratio of the mean mass densities in matter and radiation is

$$\frac{\mathscr{E}_b}{\mu_b} = \frac{aT^4}{\mu_b c^2} = 2.38 \times 10^{-5}(1 + Z)/(\Omega_0 h^2), \tag{92.41}$$

where the present radiation temperature is taken to be $T_0 = 2.7$ K. The densities of matter and radiation are equal at redshift

$$1 + Z_{eq} = 4.2 \times 10^4 \Omega_0 h^2. \tag{92.42}$$

The cosmic time t at epoch Z is found by integrating the cosmological equations (80.1). Two convenient limiting cases will be used here. At $Z \gtrsim Z_{eq}$ radiation energy dominates and the R^{-2} term in equation (80.1) is negligible. This leaves

$$t \simeq (32\pi G\mathscr{E}_b/3)^{-1/2} = 3.17 \times 10^{19}\,(1 + Z)^{-2}\,\mathrm{sec}, \qquad Z > Z_{eq}. \tag{92.43}$$

At redshifts less than Z_{eq} but still high enough that the R^{-2} term may be neglected, we have

$$\begin{gathered} t \simeq (6\pi G\mu_b)^{-1/2} = 2.06 \times 10^{17}\,\Omega_0^{-1/2} h^{-1}\,(1 + Z)^{-3/2}\,\mathrm{sec}, \\ |\Omega_0^{-1} - 1| \lesssim 1 + Z \lesssim 1 + Z_{eq}. \end{gathered} \tag{92.44}$$

Now we can write down the phase integral in equations (92.27) and (92.35). At $Z > Z_{eq}$, where $a \propto t^{1/2}$, it is

$$\begin{aligned} \phi &= \int_0^t dt\, kc_s/a(t) \simeq (kc/3^{1/2}) \int dt/a \\ &\simeq 1.18 \times 10^{10}\,(\Omega_0 h^2)^{1/3}\,(M/\mathrm{M}_{\odot})^{-1/3}\,(1 + Z)^{-1}\ \mathrm{radians}. \end{aligned} \tag{92.45}$$

This is a reasonable approximation down to redshift Z_{eq}, where it amounts to

$$\phi_{eq} \simeq 2.8 \times 10^5\,(M/\mathrm{M}_{\odot})^{-1/3}\,(\Omega_0 h^2)^{-2/3}. \tag{92.46}$$

The phase is $\phi_{eq} = 1$ radian at

$$\lambda_x = 53\,(\Omega_0 h^2)^{-1}\,(1+Z)^{-1}\,\text{Mpc},\quad M_x = 2\times 10^{16}\,(\Omega_0 h^2)^{-2}\,\text{M}_\odot. \qquad (92.47)$$

This marks the scale of irregularities that are just starting to oscillate at $Z = Z_{eq}$. At $Z < Z_{eq}$ the phase integral is

$$\phi \cong \phi_{eq} + 3.2\times 10^5\,(M/\text{M}_\odot)^{-1/3}\,(\Omega_0 h^2)^{-2/3}\ln\left[(1+Z_{eq})/(1+Z)\right]. \qquad (92.48)$$

Since the logarithm factor cannot grow very much greater than unity, we see that at $M > M_x$, where $\phi_{eq} < 1$, the phase never does get very large: pressure is unimportant and the transfer function T is nearly independent of wavelength. At $M = M_x$ the acoustic time $\sim\lambda/c_s$ agrees with the expansion time: M_x is the Jeans mass for the matter-radiation fluid. Because $p \propto (1+Z)^4$ and $\rho \propto (1+Z)^3$ at $Z < Z_{eq}$, the Jeans mass is independent of redshift (§ 16). At $M \ll M_x$ the phase advances by many radians each time the radius of the universe doubles, so we have acoustic fluctuations.

Photon diffusion tends to dissipate the acoustic part. At M close to M_x, γt (eq. 34) is fairly small until $Z_{dec} \sim 1300$, where the temperature falls to $\sim$4000 K, the plasma combines to neutral atomic hydrogen, and matter and radiation decouple. During the decoupling process the photon mean free path increases from a small fraction of a wavelength to one wavelength in a time span Δt that is shorter than the expansion time t but of comparable order of magnitude. If the wave makes many oscillations during this interval Δt, it is smoothly damped and the residual irregularity in the matter is quite small. If there is less than one oscillation during Δt, the decoupling is abrupt and the matter is deposited with irregularity close to what it was before decoupling. One finds from numerical integrations that the division between these cases is about where the wave is making five oscillations while $1 + Z$ changes by the factor e or (Peebles and Yu 1970)

$$\left[(1+Z)\frac{d\phi}{dZ}\right]_{Z_{dec}} \sim 30 \text{ radians} \sim \frac{k_s c}{[3(1+R)]^{1/2}}\frac{dt}{da}. \qquad (92.49)$$

This gives the characteristic Silk mass at the short wavelength cutoff of the transfer function:

$$M_s = 1.2\times 10^{12}\,\text{M}_\odot\,(\Omega_0 h^2)^{-2}\,(1 + 0.036/(\Omega_0 h^2))^{-3}. \qquad (92.50)$$

Another way to estimate M_s is to integrate the exponential dissipation factor γ (eq. 94.34) up to the epoch that matter and radiation stop acting like a single fluid (for example, Chibisov 1972, Silk 1974b). Bonometto and Lucchin (1976) and Press and Vishniac (1980b) treated matter and radiation as coupled fluids, each with dissipation. As it happens, direct numerical integration of the transfer equation is fairly easy because in a plane wave perturbation there are only two independent variables to deal with, t and θ: the parameter in equation (92.49) was adjusted to make equation (92.50) about agree with the numerical integrations (Peebles and Yu 1970). It might be noted that for a plane wave with characteristic mass equal to M_s the damping rate at $Z = 1300$ is $\gamma t = 2.3/x_e$, where x_e is the fractional ionization of the hydrogen, if $\Omega_0 h^2 = 1$; and $\gamma t = 1.1/x_e$ if $\Omega_0 h^2 = 0.03$. Damping at M_s prior to recombination thus is appreciable. However, it is partly compensated by the residual matter velocity that tends to enhance the final amplitude. The result is that the transfer function through decoupling is fairly close to flat at $M_s \lesssim M \lesssim M_x$.

Depending on the spectrum of the primeval irregularities, the cutoff in the linear transfer function at M_s may fix the size of the first generation of objects to form after decoupling. If $\Omega_0 h^2 = 1$, M_s amounts to about 10^{12} $M_\odot$; if $\Omega_0 h^2 = 0.03$, $M_s \sim 10^{14}$ $M_\odot$. It is interesting that one can find a reasonable interpretation for either of these numbers: the mass in the visible parts of the largest galaxies is $\sim 10^{12}$ $M_\odot$ and the nominal mass in a moderately rich cluster is $\sim 10^{14}$ $M_\odot$

The boundary condition $\delta \rightarrow 0$ at $t \rightarrow 0$ forces $\delta(\mathbf{x}, t)$ to be a sum of standing waves, and it fixes the phase for the time variation of each Fourier component. This means $\delta_{\mathbf{k}}(f)$ is forced to vanish at discrete values of k where the phase of the standing wave at decoupling is such that the residual effects of the displacement and velocity just cancel each other. If the residual velocity is negligible, the zeros are where the phase at decoupling is an integral multiple of π (Sunyaev and Zel'dovich 1970):

$$\int_{t=0}^{dec} dt\, k_s c_s/a \sim n\pi. \tag{92.51}$$

This with equation (92.48) for ϕ indicates the zeros are at wavelengths

$$\lambda_n \sim \lambda_0/n, \qquad \lambda_0 = \frac{20\text{ Mpc}}{\Omega_0 h^2} \ln(80\,\Omega_0 h^2). \tag{92.52}$$

The zeros are somewhat perturbed from this by the effects of gravity and the residual velocity. This is seen in the numerical examples of Michie (1967) and Peebles and Yu (1970). Since galaxy clustering on the scale of

λ_0 is weak, in the sense that the two-point correlation function $\xi(\lambda_0)$ is much less than unity, the linear approximation should still be valid and the zeros of the power spectrum should still be present if they were at decoupling. When extensive redshift samples are available, it may be possible to test for this: a detection and measurement of the length λ_0 certainly would afford a valuable probe of the model.

Finally, let us consider the shape of the envelope of $T(k)$ longward of M_s. At $M < M_x$ the amplitude $\delta_{\mathbf{k}}(t)$ grows as $(1 + Z)^{-2}$ (if we assume $Z \gtrsim Z_{eq}$) until the wavelength appears within the horizon, and then it oscillates as an acoustic wave, with nearly constant amplitude (eq. 92.27), until γt becomes appreciable. At the first oscillation the proper wavelength is

$$\lambda = 2\pi a(t)/k \sim ct \propto a^2, \tag{92.53}$$

so this happens at redshift

$$1 + Z_k \propto a^{-1} \propto k. \tag{92.54}$$

The growth factor is therefore

$$T(k) \propto (1 + Z_k)^{-2} \propto k^{-2}. \tag{92.55}$$

Thus if the power spectrum of the primeval density irregularities is

$$|\delta_{\mathbf{k}}(i)|^2 \propto k^{\nu}, \tag{92.56}$$

the part that has come within the horizon is tilted to

$$|\delta_{\mathbf{k}}|^2 \propto k^{\nu-4}, \qquad 2\pi a/k \lesssim t. \tag{92.57}$$

Since the Fourier component with characteristic baryon mass M has wave number $k \propto M^{-1/3}$ (eq. 92.39), equation (92.54) says this wave starts to oscillate at epoch

$$1 + Z_k \sim (1 + Z_{eq})(M_x/M)^{1/3}. \tag{92.58}$$

The normalization is fixed by the condition that M_x is just on the edge of oscillation at redshift Z_{eq}. The envelope of the transfer function at $M < M_x$ is then

$$\delta_f \sim \delta_i \left(\frac{1 + Z_i}{1 + Z_{eq}}\right)^2 \left(\frac{M}{M_x}\right)^{2/3} \left(\frac{1 + Z_{dec}}{1 + Z_{eq}}\right)^{1/4}, \qquad M \lesssim M_x. \tag{92.59}$$

The last factor, which applies if $Z_{eq} > Z_{dec}$, is the adiabatic variation of the wave amplitude (eq. 92.27). Longward of the Jeans length the pressure is unimportant so the growth factor to Z_{dec} is

$$\delta_f \sim \delta_i \left(\frac{1 + Z_i}{1 + Z_{eq}}\right)^2 \left(\frac{1 + Z_{eq}}{1 + Z_{dec}}\right), \qquad M \gtrsim M_x, \tag{92.60}$$

if $Z_{dec} \lesssim Z_{eq}$. If $\Omega_0 h^2 \sim 0.03$, so $Z_{eq} \sim Z_{dec}$, these two expressions for δ_f agree at $M \sim M_x$. If Z_{eq} is much larger than Z_{dec}, then the envelope of the short wavelength part is down from the long wavelength part at $M \sim M_x$ by the factor

$$\frac{\delta(M_x+)}{\delta(M_x-)} \sim \left(\frac{1 + Z_{eq}}{1 + Z_{dec}}\right)^{5/4}. \tag{92.61}$$

The power spectrum thus may have a spike at $M \sim M_x$. This effect was pointed out by Field and Shepley (1968). The spike would add to the Fourier transform $\xi(r)$ a term that is roughtly constant at $r < \lambda_x$ and approaches zero at $r > \lambda_x$, where λ_x is given by equation (92.47). This length is somewhat outside the range that can be tested with present data, but again it will be something interesting to look for in better samples.

93. Residual fluctuations in the microwave background

This is important as a possible probe of conditions now and at decoupling. One must bear in mind, however, that the irregularity in the radiation distribution left over from the early universe could be diminished or augmented by scattering and emission along the line of sight: a full application thus awaits a fuller understanding of the galaxy building process.

A. Small angular scales

The radiation detected now along two lines of sight intersecting at angle θ was at epoch t at proper separation (eq. 97.11)

$$d = a(t) x \theta. \tag{93.1}$$

This is the proper length at epoch t subtended by the angle of observation θ. The coordinate distance x is given by equation (56.7). In an open universe $\Omega_0 < 1$, the density parameter was $\Omega_z = 0.5$ at redshift (eq. 97.20)

$$1 + Z_c = \Omega_0^{-1} - 1, \tag{93.2}$$

and at this redshift x amounts to

$$x_c = \frac{2}{H_0 a_0 \Omega_0}\left[3 - \frac{2^{1/2}(2 - \Omega_0)}{(1 - \Omega_0)^{1/2}}\right] \simeq \frac{0.34}{(H_0 a_0 \Omega_0)}, \tag{93.3}$$

for $\Omega_0 \ll 1$. In the limit $Z \gg Z_c$, x approaches

$$x_m = 2(H_0 a_0 \Omega_0)^{-1}. \tag{93.4}$$

Present attempts to find small-scale irregularities in the microwave background reach upper limits $\delta T/T \lesssim 10^{-3}$ on angular scales 3 to 10 minutes of arc (Boynton 1978, Parijski 1978). The angle $\theta = 3'$ subtends at redshift $Z \gg Z_c$ the proper length

$$d = 5.2\,[h\Omega_0(1 + Z)]^{-1}\ \text{Mpc}, \quad M = 2 \times 10^{13}\,(\Omega_0{}^2 h)^{-1}\ M_\odot. \tag{93.5}$$

The second quantity is the mean mass in baryons in a sphere of diameter d (eq. 92.39). The measurements thus are probing the structure on scales larger than that of a galaxy and, for $\Omega_0 = 0.1$, comparable to the largest well-observed scale clustering of galaxies, $d_0 \sim 40\ h^{-1}$ Mpc.

A Jeans wavelength for the matter-radiation mixture is defined in equation (92.47). It subtends the angle

$$\theta_x \sim 30' h^{-1}, \tag{93.6}$$

independent of Ω_0 and Z at $Z_c \lesssim Z \lesssim Z_{eq}$. If $Z_{eq} \gtrsim Z_{dec}$, this angle marks the division between irregularities on large scale that grow unimpeded by the pressure and those on smaller scales that, in the linear theory, have been oscillating as acoustic waves prior to decoupling. This latter oscillation is damped out when the photon mean free path becomes comparable to the wavelength for the matter distribution. The matter is then deposited in a more or less irregular fashion, but since the radiation mean free path is less than the horizon for some time after this the radiation is smoothed out. The strongest residual irregularities in the radiation come from the first one or two (longest wavelength) peaks of the transfer function and from the primeval shape of the fluctuations longward of that. Thus one would look for a cutoff in the spectrum of the angular fluctuations at θ somewhat smaller than θ_x, and, if the plasma recombines on schedule, one expects the amplitude $\delta T/T$ at this shoulder is comparable to $\delta\rho/\rho$ at the first peak at the epoch of decoupling. For $Z_{eq} \gtrsim Z_{dec}$ the growing mode of a linear perturbation grows by the total factor (eq. 11.21)

$$\frac{\delta_\infty}{\delta_{dec}} = \frac{5}{2}\frac{1 + Z_{dec}}{\Omega_0^{-1} - 1}, \tag{93.7}$$

so if the first peak is to develop into nonlinear clumps, we need

$$\delta T/T \sim \tfrac{1}{3}\,\delta\rho/\rho \gtrsim 10^{-4}\,\Omega_0^{-1}, \tag{93.8}$$

close to the observational limits. More detailed estimates require numerical integration of the brightness equation through decoupling (Peebles and Yu 1970, Silk and Wilson 1980). Discussions of the orders of magnitudes of this and competing effects are given by Longair and Sunyaev (1969), Chibisov and Ozernoy (1969), Sunyaev and Zel'dovich (1970), Silk (1974c), Partridge (1980), and Boynton (1980).

B. Large angular scales

Sachs and Wolfe (1967) showed how the residual fluctuations in the background at $\theta \gg \theta_x$ can be computed in a useful approximation based on the following consideration. The proper length $d(Z)$ subtended by angle θ (eq. 93.1) is equal to the horizon ct at epoch Z if

$$\theta \simeq \frac{1}{3}\left(\frac{\Omega_0}{1 + Z}\right)^{1/2}, \qquad Z_c \lesssim Z \lesssim Z_{eq}. \tag{93.9}$$

This fixes the epoch Z at which fluctuations observed now at angular scale θ appeared on the horizon. The angle amounts to

$$\begin{aligned} \theta_{\rm dec} &\sim 30'\,\Omega_0^{1/2}, \qquad & Z &= Z_{dec}; \\ \theta_{eq} &\sim 6'\,h^{-1}, \qquad & Z &= Z_{eq}. \end{aligned} \tag{93.10}$$

For irregularities on scales $\theta \gg \theta_{dec}$ scattering never had much effect (provided the plasma recombines on schedule) because by the time the photons could have responded to an irregularity the mean free path was longer than ct. Thus we can greatly simplify the computation by dropping the scattering term in the brightness equation.

Neglecting scattering, we find the brightness equation in the linear perturbation approximation is (eq. 92.8)

$$\frac{\partial f}{\partial t} + \frac{\gamma_\alpha}{a}\frac{\partial f}{\partial x^\alpha} = p\frac{\partial f}{\partial p}\left(\frac{\dot a}{a} - \frac{1}{2}\gamma_\alpha\gamma_\beta\frac{\partial h_{\alpha\beta}}{\partial t}\right). \tag{93.11}$$

The unperturbed part is

$$\frac{\partial f_b}{\partial t} = p\frac{\dot{a}}{a}\frac{\partial f_b}{\partial p}, \qquad f_b = F(ap) \propto (e^{ap/kTa} - 1)^{-1}. \tag{93.12}$$

The solution is a function of ap alone, and, if the spectrum is blackbody, the temperature varies as $T \propto a^{-1}$, as usual. For the perturbation we can write

$$\begin{aligned} f &= f_b(1+\epsilon), \qquad \epsilon = \epsilon(ap, \gamma_\alpha, \mathbf{x}, t), \\ \frac{\partial \epsilon}{\partial t} &+ \frac{\gamma_\alpha}{a}\frac{\partial \epsilon}{\partial x^\alpha} = -\frac{1}{2}\frac{p}{f_b}\frac{\partial f_b}{\partial p}\frac{\partial h_{\alpha\beta}}{\partial t}\gamma_\alpha\gamma_\beta. \end{aligned} \tag{93.13}$$

The time derivative is at fixed ap. To simplify the following expressions, it will be supposed that the photon energy p is in the long wavelength Rayleigh-Jeans part of the spectrum, so $f_b \propto p^{-1}$ (as is the case of practical interest). Here the last of equations (93.13) becomes

$$\frac{\partial \epsilon}{\partial t} + \frac{\gamma_\alpha}{a}\frac{\partial \epsilon}{\partial x^\alpha} = \frac{1}{2}\gamma_\alpha\gamma_\beta\frac{\partial h_{\alpha\beta}}{\partial t}. \tag{93.14}$$

To simplify things some more, let us reverse the sign of γ_α so it points back, along the direction of observation. The observer can be placed at $x^\alpha = 0$. The radiation received now, at epoch t_0, was at epoch t at coordinate position

$$x^\alpha = \gamma_\alpha x(t), \qquad x(t) = \int_t^{t_0} dt'/a(t'). \tag{93.15}$$

With these variables we can write the integral of equation (93.14) as

$$\epsilon = \frac{1}{2}\gamma_\alpha\gamma_\beta\int_0^{t_0} dt\,\frac{\partial}{\partial t}h_{\alpha\beta}\,(t, \gamma_\alpha x(t)). \tag{93.16}$$

This shows how the radiation brightness is perturbed by the gravitational fields along the line of sight. The result is simply interpreted by considering the observers, each at fixed x^α, that define the time-orthogonal coordinates. A packet of radiation passes an observer at epoch t, and it passes a second at $t + \delta t$. The coordinate separation of the observers is (eqs. 91.9 and 91.18)

$$x^\alpha = \gamma_\alpha \delta t/a. \tag{93.17}$$

The proper separation is $r = \delta t$, and this is increasing due to the general expansion plus the perturbation $h_{\alpha\beta}$. To compute the rate of increase of the separation of the two observers, let us use equation (93.17) with equations (81.8) to write

$$r = (-g_{\alpha\beta}\delta x^\alpha \delta x^\beta)^{1/2} \propto a(1 - \tfrac{1}{2} h_{\alpha\beta}\gamma_\alpha\gamma_\beta), \tag{93.18}$$

which gives

$$\frac{dr}{dt} = \left(\frac{\dot{a}}{a} - \frac{1}{2}\frac{\partial h_{\alpha\beta}}{\partial t}\gamma_\alpha\gamma_\beta\right) r. \tag{93.19}$$

Therefore, the first-order Doppler effect makes the frequency of the radiation in the packet as measured by the two observers differ by the fractional amount

$$\frac{\partial \nu}{\nu} = -\frac{dr}{dt}, \qquad \frac{\delta(a\nu)}{a\nu} = \frac{\delta(aT)}{aT} = \frac{1}{2}\frac{\partial h_{\alpha\beta}}{\partial t}\gamma_\alpha\gamma_\beta\delta t. \tag{93.20}$$

This is an ordinary Lorentz transformation, valid because the two observers are close to each other. As indicated, the Doppler shift $\delta\nu/\nu$ causes a like perturbation $\delta T/T$ to the measured brightness temperature (for example, Peebles and Wilkinson 1968). Equation (93.16) is the integral of equations (93.20) over a sequence of observers along the path of the radiation.

It might be noted that if $\epsilon = \delta T/T$ with T the equivalent blackbody brightness temperature, then equation (93.20) and hence equation (93.16) are valid at any photon energy p, not just in the Rayleigh-Jeans limit. It is left as an exercise to see how this comes about in the first derivation.

For the final simplification we note that if $\theta \gg \theta_{dec}$, then $\theta \gg \theta_{eq}$ (eq. 93.10), so pressure gradient forces are negligible: we can choose coordinates so $T_0^{\ \alpha} = 0$ and hence, as discussed in Section 84, the field $h_{\alpha\beta}$ can be derived from the potential Ψ (eqs. 84.5 and 84.20). Assuming the most rapidly growing mode $\propto D_1(t)$ dominates, we can write $h_{\alpha\beta}$ as

$$\frac{\partial h_{\alpha\beta}}{\partial t} = \frac{dD_1(t)}{dt}\frac{\partial^2 k(\mathbf{x})}{\partial x^\alpha \partial x^\beta}, \tag{93.21}$$

where, since the matter velocity in the time-orthogonal coordinates is negligible (eq. 92.14),

$$2\frac{\partial \delta_m}{\partial t} = \frac{\partial h}{\partial t} = \frac{dD_1}{dt}\nabla^2 k. \tag{93.22}$$

With $D_1 \propto t^{2/3}$, equation (84.20) gives

$$D_1(t)\, k(x^\alpha) = \phi/(2\pi G\rho_b a^2),$$
$$\phi = -G\rho_b a^2 \int d^3x' \delta(\mathbf{x}', t)/|\mathbf{x} - \mathbf{x}'|, \tag{93.23}$$

where $\phi(\mathbf{x})$ is the Newtonian potential (computed on the hypersurface of fixed time t in the time-orthogonal coordinates). The function k evaluated along the path of the radiation is $k(x(t)\gamma_\alpha)$, so

$$\frac{d^2k}{dx^2} = \frac{\partial^2 k}{\partial x^\alpha \partial x^\beta}\gamma_\alpha \gamma_\beta, \tag{93.24}$$

and equation (93.16) is therefore

$$\epsilon = \frac{1}{2}\int_0^{t_0} dt \frac{dD_1}{dt}\frac{d^2k}{dx^2}, \qquad dt = -a\,dx. \tag{93.25}$$

When this is integrated by parts twice, it becomes

$$\epsilon = -\frac{a_0}{2}\frac{dD_1}{dt_0}\frac{\partial k}{\partial x^\alpha}\gamma_\alpha - \left[\frac{a}{2}\frac{d}{dt}\left(a\frac{dD_1}{dt}\right)k\right]_0^{t_0} + \frac{1}{2}\int_0^{t_0} dt\, k \frac{d}{dt}\, a \frac{d}{dt}\, a \frac{dD_1}{dt}. \tag{93.26}$$

This equation was derived by Sachs and Wolfe (1967). They computed from the redshift of the wave packet rather than the brightness equation.

The first term in equation (93.26) with equations (93.23) is

$$\epsilon = -t_0 \gamma_\alpha \phi_{,\alpha}/a_0 = t_0 g_\alpha \gamma_\alpha. \tag{93.27}$$

The peculiar gravitational acceleration is g_α, and, in the Einstein-de Sitter model, the peculiar velocity in the growing mode is $v^\alpha = g_\alpha t$ (eq. 14.7). This term thus is just the 24-hour anisotropy due the peculiar motion of the freely moving observer at $x^\alpha = 0$ (in the time-orthogonal coordinates). Since $D_1 \propto t^{2/3}$ in this model, the third term in equation (93.26) vanishes and the second can be reduced to

$$\epsilon = -\tfrac{1}{3}\,[\phi(0) - \phi(x\gamma_\alpha)], \tag{93.28}$$

the final Sachs-Wolfe result. The local gravitational redshift at the observer is $-\phi(0)$. One third of the effect appears in this term. The rest comes from the fact that ϵ is defined as the perturbation from the mean of the brightness along a hypersurface of fixed time t kept by the observers at fixed x^α. The correction to Newtonian time is $\hat{t} = t(1 - \phi)$ (eq. 84.15 with ϕ independent of t), which brings the perturbation at fixed $\hat{t}$ to

$$\frac{\delta T}{T} = -\frac{\phi(0)}{3} - \frac{\delta a}{a} = -\phi(0), \tag{93.29}$$

the full effect.[1]

Equation (93.28) says ϵ is unaffected by what lies along the path from the position at high redshift, $x^\alpha = x\gamma_\alpha$, to the observer at $x^\alpha = 0$. To see how this comes about, let us write down, in the uniformly expanding Newtonian coordinates used in Chapter II, the integral expressing the effect of the peculiar gravitational acceleration along the path on the photon energy (and brightness temperature):

$$\frac{\delta(a\nu)}{a\nu} = -\int \frac{\partial\phi(\mathbf{x}, t)}{\partial x^\alpha}\, dx^\alpha. \tag{93.30}$$

In the Einstein-de Sitter model ϕ for a linear, growing perturbation is a function of x^α alone (because $\delta \propto a$ in eq. 93.23), so the integral depends only on the end points as in equation (93.28). Rees and Sciama (1968) have noted that in other cosmological models, or in nonlinear mass concentrations, ϕ at fixed $\mathbf{x}$ may vary with t, so the integral may depend on what lies along the path: it can happen that in a growing potential well the photons lose more energy moving out of the well than they gain while moving in. However, if the clustering length $\ll cH^{-1}$, the change in ϕ during the time taken for the radiation to cross the irregularity generally is a small fraction of ϕ so this is a small correction to equation (93.28)

Equation (93.28) has been derived in an Einstein-de Sitter model but it applies equally well in the open or closed case at small enough angular scales where the length subtended by θ at high redshift is a small fraction of $|R|a(t)$. In an open model this condition is (eq. 80.8)

$$d = \frac{2a\theta}{H_0 a_0 \Omega_0} \ll a|R|, \qquad a_0|R| \sim H_0^{-1}, \qquad \theta \ll \Omega_0 \text{ radians}. \tag{93.31}$$

[1]The renormalization of the time from t to $\hat{t}$ is not applied at the other end of the integration: under the assumed boundary conditions, $\epsilon \rightarrow 0$ at $t \rightarrow 0$, equations (93.16) and (93.28) give the full difference in brightness measured in different directions.

Generalization of the analysis to an open or closed model is discussed by Anile and Motta (1976).

We can write the fluctuating part of equation (93.28) as

$$\delta T/T = \epsilon = \phi(x^\alpha)/3 = -\tfrac{1}{3}\, G\rho_b a^2 \int d^3x' \delta_m(\mathbf{x}', t)/|\mathbf{x}' - \mathbf{x}|,$$
$$x^\alpha = \gamma_\alpha \int_0^{t_0} dt'/a(t'). \tag{93.32}$$

It will be noted that since ϕ is independent of t in the Einstein-de Sitter model, this can be evaluated at any convenient t. To see the order of magnitude, consider a patch with proper size l and density contrast δ at epoch t. Here equation (93.32) amounts to

$$\delta T/T \sim -G\rho_b l^2 \delta \sim -\delta(l/t)^2. \tag{93.33}$$

If the epoch is chosen so $l = t$, then $|\delta T/T| \sim |\delta|$: the residual perturbation to the radiation is about equal to the density contrast in the matter when it appears on the horizon, as seems reasonable. It is interesting, however, that the signs of δ and ϵ are opposite. An account of this is at best only heuristic because it involves $l \gtrsim ct$, but we can note that there are two effects to consider. At high redshift where $l(t) \gg ct$ the matter and radiation at each spot behave as if they were in a homogeneous world model: where the matter density is high so is the radiation density $\epsilon = \delta/3$. When t approaches $l(t)$, the excess radiation within a dense patch leaves moving out of a potential well $\phi \sim \delta$ (eq. 93.33) and suffering a like decrease in brightness temperature. The two effects are of the same order but opposite sign, and, as it appears, the latter fixes the sign of the residual ϵ.

To make the relation between ϵ and δ_m more precise, consider the power spectrum of ϵ. With

$$\delta = \Sigma \delta_{\mathbf{k}} e^{i\mathbf{k}\cdot\mathbf{x}}, \tag{93.34}$$

equation (93.32) is (compare eq. 27.6)

$$\epsilon = -\frac{2}{9}\frac{a^2}{t^2} \sum_{\mathbf{k}} (\delta_{\mathbf{k}}/k^2)\, e^{i\mathbf{k}\cdot\mathbf{x}}. \tag{93.35}$$

Suppose ϵ is measured in a square patch of the sky, size Θ by Θ with $\Theta \ll 1$ radian. Then we can write in equation (93.35)

$$e^{i\mathbf{k}\cdot\mathbf{x}} \simeq e^{ix(\theta_1 k_1 + \theta_2 k_2) + ixk_3}, \tag{93.36}$$

where x is given by equation (93.32) and θ_1, θ_2 are Cartesian coordinates for the angular position in the small patch in the sky. The Fourier transform of ϵ in the patch is

$$\begin{aligned}\epsilon_{\mathbf{q}} &= \int \epsilon e^{-i\mathbf{q}\cdot\boldsymbol{\theta}} d^2\theta/\Theta^2 \\ &= -\frac{2}{9}\left(\frac{a}{\Theta t}\right)^2 \Sigma\delta_{\mathbf{k}}/k^2[F(xk_1 - q_1)F(xk_2 - q_2)]e^{ik_3x}, \qquad (93.37) \\ & F(x) = (e^{ix\Theta} - 1)/ix.\end{aligned}$$

The power spectrum of ϵ is the mean of the square of this. The means of the cross terms vanish (because the relative phase of $\delta_{\mathbf{k}}$ and $\delta_{\mathbf{k}'}$ varies with the position x_0 of the observer as $\exp i\,(\mathbf{k} - \mathbf{k}') \cdot \mathbf{x}_0)$, and the $|F|^2$ factors approximate delta functions when $q\Theta \gg 1$. This leaves

$$\begin{aligned}\langle |\epsilon_{\mathbf{q}}|^2\rangle &= \frac{2}{81\pi}\left(\frac{a}{t}\right)^4 \frac{V_u}{(\Theta x)^2}\int dk_3|\delta_k|^2/k^4, \\ k^2 &= q^2/x^2 + k_3^2.\end{aligned} \qquad (93.38)$$

This is the general relation between the power spectra of the space density fluctuations and of the angular fluctuations in the background temperature. If $|\delta_{\mathbf{k}}|^2$ does not vary more rapidly than k^3, the main contribution to the integral is at $k \sim q/x$. If $|\delta_{\mathbf{k}}|^2$ is modeled as a power law, the integral can be simplified:

$$\begin{aligned}|\delta_{\mathbf{k}}|^2 &= Ak^n, \qquad n < 3; \\ \int dk_3|\delta_{\mathbf{k}}|^2/k^4 &= A(x/q)^{3-n}H(4-n),\end{aligned} \qquad (93.39)$$

where $H(\gamma)$ is given by equation (52.9).

The mean square value of ϵ is

$$\langle\epsilon^2\rangle = (\Theta/2\pi)^2\int_0^\infty 2\pi q dq\,\langle\epsilon_q^2\rangle. \qquad (93.40)$$

The part of the integral from 0 to q_0 roughly agrees with what would be detected using an antenna with angular resolution q_0^{-1}, so a convenient measure of the background temperature fluctuations on the scale q^{-1} is the contribution to the integral per logarithmic interval of q:

$$\langle(\delta T/T)^2\rangle_q = (\Theta/2\pi)^2 2\pi q^2|\epsilon_{\mathbf{q}}|^2, \qquad (93.41)$$

which is a variant of equation (28.1). The corresponding measure of the space density fluctuations on the scale k^{-1} is

$$\langle(\delta M/M)^2\rangle_k = [V_u/(2\pi)^3]\,4\pi k^3\,|\delta_{\mathbf{k}}|^2. \qquad (93.42)$$

Since $|\epsilon_{\mathbf{q}}|^2$ comes mainly from $|\delta_{\mathbf{k}}|^2$ at $k = q/x$, we can rewrite equations (93.38) and (93.39) with equations (93.41) and (93.42) as

$$\langle(\delta T/T)^2\rangle_q = \frac{2}{81}\left(\frac{xa(t)}{qct}\right)^4 H(4-n)\,\langle(\delta M/M)^2\rangle_{q/x}. \qquad (93.43)$$

This the relation between the growing mass density fluctuations and the angular fluctuations in the radiation temperature. We see again that if the epoch at which the right hand side is evaluated is chosen so the angle $\sim q^{-1}$ subtends ct, then $\delta T/T \sim \delta M/M$.

A fairly standard assumption is that $n = 1$ in equation (93.39) (§ 95 below). This gives $(\delta M/M)^2 \propto k^4$, so it makes $\delta T/T$ independent of q: the background temperature fluctuations are about independent of the angular resolution of the antenna. If

$$\lambda = 2\pi a/k = 2\pi ax/q = \lambda_x, \qquad (93.44)$$

where λ_x is the Jeans length for the matter and radiation (eq. 92.47) and $Z = Z_{dec} = 1300$, equation (93.43) with $n = 1$ indicates

$$\langle(\delta T/T)^2\rangle_q^{1/2} \sim 5\times 10^{-3}\,(\Omega_0 h^2)^{-1}(\delta M/M)_{x,dec}. \qquad (93.45)$$

The last factor is about equal to the net rms fluctuation in density integrated over the entire spectrum because the shorter wavelength part has been stored as acoustic waves (eqs. 92.58 and 92.59). Fluctuations in the background temperature $\delta T/T \gtrsim 10^{-3}$ on angular scales greater than a few degrees probably would have been detected, so we conclude that

$$(\delta M/M)_{x,dec} \lesssim 0.2\,\Omega_0 h^2. \qquad (93.46)$$

As for equations (93.7) and (93.8), this is close to a useful constraint on the possible amplitude of large-scale primeval irregularities. There have not yet been experiments specifically designed to measure $\delta T/T$ on these intermediate angular scales; an interesting improvement in this limit may be possible.

94. Isothermal perturbations

A. *Origin*

As was described in Section 4C, the irregularity in the mass distribution before decoupling can be written as a linear combination of adiabatic and isothermal fluctuations, the former having fixed entropy per baryon and the latter, a homogeneous distribution of radiation. If the scale of the fluctuations is $\lambda \ll ct$, the perturbation to the radiation caused by clumps of matter is down by $\sim(\lambda/ct)^2$ and so can be neglected, so isothermal perturbations remain isothermal. If radiation diffusion can be ignored, entropy is conserved and adiabatic perturbations remain adiabatic. Thus in these approximations one can think of these as linearly independent modes of perturbation.

Analysis of the behavior of isothermal perturbations at $\lambda \gg ct$ requires some change from the calculation in Section 86, for now the perturbation is in part in the pressure. To simplify the equations, let us suppose that $v^\alpha = 0 = \theta$ and that $Z \gg Z_{eq}$, so the expansion of the universe is dominated by the radiation. With $\theta = 0$ the first of equations (86.1) says the fractional perturbations to the densities of matter and radiation are

$$\begin{aligned} \delta_m(t) &= \delta_m(i) + \tfrac{1}{2}\,(h(t) - h(i)), \\ \delta_r(t) &= \delta_r(i) + \tfrac{2}{3}\,(h(t) - h(i)). \end{aligned} \tag{94.1}$$

The evolution of h is given by the first of equations (82.9). On using a radiation-dominated model to approximate the coefficients, we have

$$\frac{\partial^2 h}{\partial t^2} + \frac{1}{t}\frac{\partial h}{\partial t} = \frac{3}{2t^2}\left[\delta_r + \frac{\delta_m}{2}\left(\frac{t}{t_{eq}}\right)^{1/2}\right], \tag{94.2}$$

where the ratio of mean densities of matter and radiation is

$$\mu/\mathscr{E} = (t/t_{eq})^{1/2}. \tag{94.3}$$

The solution at $t \ll t_{eq}$ is

$$\begin{aligned} \delta_m &\simeq \tfrac{1}{2}\,(At + B/t) + \delta_m(i)(1 - (t/t_{eq})^{1/2}/2) - \tfrac{3}{4}\,\delta_r(i), \\ \delta_r &\simeq \tfrac{2}{3}\,(At + B/t) - \tfrac{2}{3}\,\delta_m(i)(t/t_{eq})^{1/2}, \\ At_i + B/t_i &= \tfrac{3}{2}\,\delta_r(i) + \delta_m(i)(t_i/t_{eq})^{1/2}. \end{aligned} \tag{94.4}$$

The most rapidly growing part of the solution, $\delta \propto t$, gives $\delta_r = 4\delta_m/3$. In general, therefore, any small initial fluctuation develops into an adiabatic perturbation. However, this is avoided if some constraint forces $A = 0$. Then equations (94.4) become

$$\delta_m = \delta_m(i)\left[1 - \frac{1}{2}\left(\frac{t}{t_{eq}}\right)^{1/2} + \frac{1}{2}\frac{t_i^{3/2}}{t_{eq}^{1/2}t}\right] + \frac{3}{4}\delta_r(i)\left(\frac{t_i}{t} - 1\right),$$
$$\delta_r = \frac{2}{3}\delta_m(i)\left[\frac{t_i^{3/2}}{t_{eq}^{1/2}t} - \left(\frac{t}{t_{eq}}\right)^{1/2}\right] + \delta_r(i)t_i/t. \tag{94.5}$$

We can isolate from this the part

$$\delta_m \cong \delta_m(i)\left[1 - \frac{1}{2}\left(\frac{t}{t_{eq}}\right)^{1/2}\right], \qquad \delta_r \cong -\frac{2}{3}\left(\frac{t}{t_{eq}}\right)^{1/2}\delta_m(i),$$
$$\frac{\delta\rho}{\rho} \cong \frac{1}{3}\left(\frac{t}{t_{eq}}\right)^{1/2}\delta_m(i). \tag{94.6}$$

The last expression is the fractional perturbation to the total mass density. In this solution the pertubation to the matter is more or less fixed prior to Z_{eq} while the radiation density is smoother, $\delta_r \ll \delta_m$ at $Z \gg Z_{eq}$. Thus what appears on the horizon is nearly pure isothermal at high redshift and a comparable mixture of isothermal and adiabatic at $Z = Z_{eq}$. This effect was pointed out by Mészáros (1975).

One can understand the result by following the method of Section 10: compare the evolution of two homogeneous cosmological models, one to describe the behavior of the background, one to describe the situation in a perturbed patch. The part of the solution $h \propto t$ corresponds to a perturbation in the space curvature R^{-2}, the part $h \propto t^{-1}$ to a perturbation in starting time, and the part $h \propto t^{1/2}$ to a perturbation in the entropy per particle.

It might be noted that an isothermal component also could be produced by nonlinear dissipation. Prior to decoupling, the amplitude of the short wavelength acoustic noise grows as new fluctuations appear on the horizon. In spots where the density fluctuates up to $\delta \sim 1$, there is rapid nonlinear transfer of energy to short wavelengths where it is dumped as entropy (§ 89). The resulting hot spots are an admixture of isothermal and adiabatic fluctuations.

B. Behavior

Let us suppose that at $Z \sim 10^4$ an isothermal component has developed one way or another, with scale $\lambda \ll ct$. If λ is greater than the Jeans length for matter, gravity in this component is balanced by the radiation pressure, which means there is a slight radiation density excess wherever the matter density is high. The radiation tends to diffuse outward, and as it does the isothermal perturbation grows.

We can compute the diffusion rate by using the expansion of the brightness ι in powers of the mean free time t_c (eqs. 92.20 to 92.22). The time derivative $\partial\iota/\partial t$ can be dropped because the distribution is quasi-static. On carrying the expansion to second order in t_c and then integrating ι over angles, one finds the diffusion equation

$$\frac{\partial \delta_m}{\partial t} = -\frac{t_c}{4}\frac{\nabla^2 \delta_r}{a^2}. \tag{94.7}$$

As usual, the space derivatives of δ_r are taken at fixed time t in the time-orthogonal coordinates. The right-hand side of this equation is fixed by the condition that the radiation pressure gradient balance the peculiar gravitational field due to δ_m. To simplify the equations, we will suppose here that $\mu_b \gtrsim \mathscr{E}_b$. Then we have from the ideal fluid equations (92.24) and (92.26) with $\partial\delta_r/\partial t \simeq 0$ and the gravitational field equation (85.10) the balance condition

$$4\pi G\mu_b\delta_m = -\frac{\mathscr{E}_b}{3\mu_b}\frac{\nabla^2\delta_r}{a^2}. \tag{94.8}$$

This gives the growth rate

$$\frac{\partial \delta_m}{\partial t} = 4\pi G\mu_b t_d \delta_m, \tag{94.9}$$

where the dynamic damping time is

$$t_d = \frac{3t_c\mu_b}{4\mathscr{E}_b}. \tag{94.10}$$

Since $G\mu_b \sim t^{-2}$, this diffusion effect is important only when $t_d \gtrsim$ the expansion time t.

The ratio of dynamic to expansion times is

$$t_d/t \sim 7 \times 10^5\,\Omega_0^{1/2}\,h(1 + Z)^{-5/2}x_e^{-1}, \tag{94.11}$$

where x_e is the fractional ionization. At $Z_{dec} = 1300$ this amounts to

$$t_d/t \sim 0.01\,\Omega_0^{1/2}\,hx_e^{-1}. \tag{94.12}$$

In the conventional cosmology the fractional ionization rather quickly drops from $x_e \sim 1$ to $x_e \sim 10^{-5}$ at Z_{dec}, so there is an abrupt switch from $t_d \ll t$, where δ_m is effectively frozen in, to $t_d \gg t$, where radiation drag is negligible and δ_m grows according to the usual zero pressure solution (§ 11). One might imagine that some process keeps the matter ionized past Z_{dec}. With $x_e = 1$, t_d/t reaches unity at redshift

$$1 + Z \sim 200\, \Omega_0^{1/5} h^{2/5}. \tag{94.13}$$

This is the latest epoch at which the background radiation drag can have had an appreciable effect on the dynamics.

There is some reason to speculate that the matter density fluctuations may be strongly nonlinear on small scales (§ 96). We can arrive at a rough approximation to how such fluctuations would behave as follows. In the rest frame of an electron moving at speed v^α through the radiation the radiation brightness is anisotropic (eq. 92.11):

$$I = \frac{aT^4}{4\pi}(1 - 4\gamma_\alpha v^\alpha). \tag{94.14}$$

Thus the electron suffers a drag force

$$F^\alpha = \int \sigma I \gamma^\alpha d\Omega = -\tfrac{4}{3}\, \sigma a_s T^4 v^\alpha. \tag{94.15}$$

Let us approximate a concentration of matter as a distribution spherically symmetric about one point, the mass within the shell of radius r being

$$M(r) = \tfrac{4}{3}\, \pi r^3 \mu_b \delta_m(r), \qquad GM/r \ll 1. \tag{94.16}$$

Since the radiation pressure keeps its distribution quite smooth, the radiation brightness at r appears isotropic to an observer moving away from the origin at speed $H(t)r$. Gravity is just able to hold the radius of the matter shell fixed if the gravitational force per proton agrees with the radiation drag force per electron with $v = Hr$ in equation (94.15). This gives the critical density contrast (for $Z < Z_{eq}$) (Peebles 1965)

$$\delta_c = 4 \times 10^{-6}(1 + Z)^{5/2}\Omega_0^{-1/2}h^{-1}, \quad \delta_c(Z_{dec}) \sim 300\, \Omega_0^{-1/2}h^{-1}, \tag{94.17}$$

which can be compared to equation (94.11). This result assumes the matter is fully ionized, as at $Z > Z_{dec}$, and that the lump does not have subcondensations that would allow the radiation to stream out more readily.

Another contribution to the growth of δ_m might be mentioned. As the universe expands, entropy goes from radiation to matter because the radiation tends to cool as $a(t)^{-1}$ and the plasma as a^{-2}. Where $\delta_m > 0$ the transfer is larger than average, reducing $\mathcal{E}$. Matter and radiation move to even the pressure, increasing δ_m. This is the mock gravity effect Gamow (1949) discussed. We can analyze it in terms of the entropy per particle,

$$s = \frac{4a_s T^3}{3n} + k \ln T^{3/2}/n, \tag{94.18}$$

where a_s is Stefan's constant. If diffusion may be neglected, s is conserved so

$$\frac{T^3}{n} = \frac{T_i^3}{n_i}\left[1 + \frac{3n_i k}{4a_s T_i^3} \ln \left(\frac{T_i}{T}\right)^{3/2} \frac{n}{n_i}\right]. \tag{94.19}$$

The second term in the parentheses is small so to a good approximation this can be reduced to the usual expression

$$T^3 \propto n \propto (1 + Z)^3. \tag{94.20}$$

Since the radiation distribution must be close to uniform to preserve pressure balance, we have then the time variation of the perturbation to the matter density,

$$\delta_m = \delta_m(i)\left[1 + \frac{9}{8}\frac{nk}{a_s T^3} \ln \left(\frac{1 + Z_i}{1 + Z}\right)\right]. \tag{94.21}$$

This result was found by Field (1971). It is quite small in the standard cosmology, for the coefficient is

$$9nk/8a_s T^3 \sim 1 \times 10^{-8} \Omega_0 h^2, \tag{94.22}$$

very nearly independent of redshift.

The diffusion equation (94.9) assumes the matter pressure may be neglected. When the size of the perturbed patch is smaller than the matter Jeans length, the spots where the matter density is high tend to expand, creating a slight depression in the radiation density. As the radiation diffuses into the spot, the perturbation decays. If λ is not much smaller than λ_J, the diffusion is quite slow until Z_{dec}, when the part $\lambda \lesssim \lambda_J$ is strongly dissipated as it starts to slip through the radiation. Since the matter temperature is very nearly uniform at the temperature T of the radiation, the isothermal Jeans length (16.8) applies:

$$\lambda_J = \left(\frac{\pi k T}{G \mu_b m_p}\right)^{1/2} = 8000\, \Omega_0^{-1/2} h^{-1} (1 + Z)^{-1} \text{ pc}. \tag{94.23}$$

The mass within a sphere of radius λ_J is

$$M_J = \tfrac{4}{3} \pi \lambda_J^3 \mu_b \sim 5 \times 10^5\, \Omega_0^{-1/2} h^{-1} \mathrm{M}_\odot, \tag{94.24}$$

independent of redshift (Gamow 1948b). This cutoff in the isothermal perturbations after decoupling is much less than a galaxy mass, but seems interestingly close to the mass of a globular star cluster (Peebles and Dicke 1968).

VI. SCENARIOS

95. Nature of the universe at high redshift

Opinions on what the universe might have been like at high redshift span the range from primeval chaos (for example, Misner 1968) to a distribution well ordered in the large but more or less chaotic or turbulent on small scales (for example, Gamow 1952, Ozernoy and Chernin 1968, Rees 1972), to a universe quite precisely homogeneous and isotropic (for example, Lemaître 1933b, Peebles 1967a, 1972, Clutton-Brock 1974). It is not surprising that the arguments that have been advanced in favor of each scenario all make good points that to some seem compelling. The principle that has been adopted in this book and that argues against all variants of the primeval chaos scenario is that gravity tries to enhance density irregularities, not disperse them: as long as gravity is the dominant force it is hard to see how the universe could do other than grow more irregular. It might be noted that this has been a point of lively debate. For example, Layzer (1954) showed that tides could inhibit the growth of irregularities, at least in a collapsing gas cloud. However, as Hunter (1962) pointed out, that does not happen in linear perturbation theory or in a universe spherically symmetric about one point. These models also show that the higher the redshift and the mean density, the smaller the fractional perturbation on the scale of a fixed number of baryons needed to give gravity ultimate control in a relativistic collapse (§ 87). These points suggest that the universe must be growing more clumpy: the higher the redshift the closer to homogeneous it must have been.

This view has been codified in Chapter V in a particular choice of coordinates by the boundary condition that all perturbations from a Friedman-Lemaître model remain finite as $t \rightarrow 0$. This prescription has the virtue that the boundary conditions are simple and definite, so all can agree at least on the implications of the theory. It determines the expected large-scale pattern of motions of galaxies associated with clustering (§§ 76 to 78), the zeros of the power spectrum in linear perturbation theory (§ 92d), and the large-scale fluctuations in the microwave background (§ 93). One can change all these answers without changing the laws of physics as we now understand them, but at the price of introducing a scenario whose prior history seems contrived in the above view.

With the adopted boundary conditions there is an upper limit on fluctuations of the density averaged over the scale λ (§§ 86, 87):

$$|\delta\rho|/\rho_b \lesssim (ct/\lambda)^2, \qquad \lambda \gtrsim ct, \tag{95.1}$$

at epoch t. For comoving $\lambda \propto a(t)$ this limit approaches zero at $t \rightarrow 0$. A common speculation is that the primeval density fluctuations are a fixed fraction of this limit (Harrison 1970, Peebles and Yu 1970, Zel'dovich 1972),

$$|\delta\rho|/\rho_b = \epsilon(ct/\lambda)^2, \tag{95.2}$$

with $\epsilon < 1$ independent of t and λ. The power spectrum would then vary as (eqs. 70.8, 93.42)

$$|\delta_{\mathbf{k}}|^2 \propto kD_1(t)^2, \qquad ct \lesssim a/k, \tag{95.3}$$

where D_1 is most rapidly growing solution to the perturbation equations at $\lambda \gg ct$ (§ 86). The virtue of this assumption is that no characteristic lengths are built in: for any other power law k^n the perturbation to the geometry diverges at large or small wavelength, while here the divergence is only logarithmic at both ends.[1] However, it is shown in Section 96 that equation (95.3) could not be valid at large scales because it yields the wrong spectrum of galaxy clustering.

Equation (95.1) is a very stringent limit if applied at high redshift, and it is not surprising that people should have invented more lively scenarios. Homogeneous but anisotropic cosmological models have been widely discussed (Gödel 1949, Heckman and Schücking 1962, Misner 1968, Ryan and Shepley 1975 and references therein). The introduction of anisotropy considerably enlarges the range of behavior while keeping the computation fairly simple, which is a great advantage, and has therefore played an important role in teaching us the options available for scenarios. However, the natural tendency once a mathematical solution is found is to try it as an approximation to the real world. Thus Gödel in his pioneering work derived a model that is homogeneous but rotating relative to local inertial frames. This led him to speculate that the rotation of galaxies might come from the global rotation, so that the spin axes of the galaxies would tend to line up. This is a doubtful approach. It is true that the Friedman-Lemaître models require an uncomfortably precise adjustment

[1]It might be noted that if the universe is open, there is a characteristic length in the background cosmological model $cH_0^{-1}(1 + Z)^{-1}$ if $\Omega_0 \ll 1$ (eq. 80.2). Longward of this, space curvature on the hypersurfaces of fixed time cannot be neglected, and, of course, this would seem to be a natural place for a break in the power spectrum.

of initial conditions, but the anisotropic models cause even more discomfort because they require in addition that the initial shear and rotation at each spot be aligned to agree with the underlying global fabric. They also are questionable as models for primeval chaos because they drop only one shoe: the great problem is what happens when density fluctuations are allowed.

One can make a good case for the thought that since the universe earlier than $Z \sim 1000$ is well beyond our ken, perhaps we should not be too optimistic about attempts to divine its nature. The precise isotropy of the radiation background gives strong empirical evidence of the present large-scale homogeneity and isotropy of the universe, and this implies also rather strong limits on fluctuations at decoupling (depending on how much scattering has smoothed out the radiation). One might argue that it is reasonable to consider any scenario that fits these constraints no matter how it looks at much higher redshift. That is the course Ozernoy and Chernin (1967, 1968) adopted. In their scenario matter and radiation at $Z \sim 10000$ are in turbulent motion with eddy sizes $< ct$ and turbulent velocities less than but comparable to the velocity of light. They noted that, according to Lifshitz's (1946) analysis, this situation traces back to diverging fluctuations in the geometry at $Z \to \infty$ (§ 90), but they considered that to be a remote problem outweighed by the advantages turbulence might offer in building structures like galaxies and clusters of galaxies. If future work does show that this scenario can proceed in a self-consistent way from the assumed state at $Z \sim 10000$ to something that matches in detail the clustering that is observed now, it will of course justify the approach.

Another aspect of the philosophy behind the equation (95.1) must be considered. Although $\delta\rho/\rho \to 0$ in the limit $t \to 0$, at the time of the big bang, the perturbations to the geometry remain as permanent wrinkles (§§ 86, 87). These are seeds invoked *ad hoc* in hope of finding a phenomenological account of what we see (Peebles 1967a, Harrison 1968). Of course, the seeds could assume other forms. For example, Harrison (1967c) remarked that at high redshift the mass density in baryons plus antibaryons $B + \overline{B}$ might be quite accurately uniform while the baryon number $B - \overline{B}$ might vary from place to place. After annihilation the local residuum of matter or antimatter could develop under gravity into galaxies and clusters of galaxies. This neatly eliminates the primeval wrinkles in the geometry but that is replaced with another seed, the fluctuations in $B - \overline{B}$ that must be larger than $B^{1/2}$ and must be present at the time of the big bang.

In a similar scheme proposed by Mészáros (1975) some process in the early universe caused the formation of fairly small black holes. This converted some of the mass density originally in radiation into a "zero pressure gas" of black holes. Since energy is conserved, the mass density is not affected but the pressure is reduced so the subsequent evolution is perturbed. The relativistic treatment of this is described in Section 94. The initial values satisfy

$$\mu(i)\delta_m(i) + \mathscr{E}(i)\delta_r(i) = 0, \tag{95.4}$$

because mass is conserved. With this plus the conditions $h = 0 = \partial h/\partial t$ at conversion, one finds $A = 0 = B$ in equation (94.4). The solutions in equations (94.6) thus apply, so the irregularities in the distribution of black holes as measured by δ_m stay about constant to Z_{eq}. Mészáros points out that if the conversion happens at high redshift where $\mathscr{E} \gg \mu$, only a very small fractional amount of energy from the radiation is needed to make a substantial δ_m. On the other hand, the wanted δ_m is large, perhaps 0.01 on a scale of $\sim 10^{10}$ $M_\odot$ which is much more than the shot noise from black holes of one solar mass. We must assume either that protogalaxies nucleate around black holes of substantial mass (a process considered by Ryan 1972 and Gribbin 1974) or that δ_m results from nonrandom clumping of the black holes on scales larger than the horizon when they formed. In either case, of course, we need seeds to form the black holes.

Many authors have considered the idea that galaxy formation was triggered by a primeval magnetic field (Hoyle 1958, Zel'dovich 1965b, 1969, Rees and Reinhardt 1972, Wasserman 1978, § 17 above). Here the primeval seed is the field that had to have existed at the time of the big bang. There are some further complications in this scenario. As noted by Thorne (1967), in the limit $t \to 0$ even a tangled magnetic field would appear homogeneous to every observer because the horizon shrinks to zero. Thus it is reasonable to describe each small patch as a homogeneous anisotropic model. The magnetic field energy makes a fixed fractional perturbation to T_i^j, and this causes a diverging perturbation to the geometry at $t \to 0$: each patch becomes strongly anisotropic in a different orientation (Zel'dovich 1965b, Thorne 1967). As in the case of the primeval turbulence scenarios, a universe with a tangled magnetic field traces back to chaotic space structure at the time of the big bang.

To find an ab initio theory of the origin of galaxies, we would have to establish that the universe is no more or less irregular than is demanded by some new physical principle. As we have seen, there is no limit in general relativity theory to how chaotic the universe might be. A common assumption is that the mass distribution is as close to homogeneous as is allowed by quantum mechanics and the discrete nature of matter and radiation. This is a simple and natural idea, though it does leave us with the puzzle of why the universe is no more chaotic than that. In early discussions it was assumed that departures from homogeneity are forced by thermal or $N^{-1/2}$ fluctuations (Lifshitz 1946, Bonnor 1957). One can imagine that at some epoch t_i each bit of the universe was allowed to come to thermal equilibrium with a separate particle and heat reservoir, each reservoir having the same thermodynamic parameters. That would generate thermal fluctuations $\delta\rho/\rho \sim N_\gamma^{-1/2}$ in regions where N_γ is the expected number of photons. As was discussed in Section 4A, if t_i were small enough, $\delta\rho/\rho$ would have grown to an interesting value now. However, it is important to note that in known laws of physics there is no reason why the universe should have been constructed on this plan. The universe is allowed to be much smoother than this at epochs when ordinary statistical mechanics is a good approximation because mass and energy can be exchanged only over distances comparable to the horizon no matter how rapid the thermal relaxation. Of course, it may happen that the density fluctuations issuing from the moment of the big bang are comparable to the naive estimates from statistical mechanics.

Usually it is easy to deal with the effect of the Hubble expansion in quantum mechanics by the adiabatic approximation because interesting de Broglie oscillation times are much less than the expansion time. The adiabatic approximation must fail when a typical de Broglie wavelength λ is comparable to the horizon ct (Harrison 1967d, Peebles 1968). If at these early times there is a limited number of particle species, then the particle number density is $n \sim \lambda^{-3}$, the typical particle energy is $\sim hc/\lambda$, so the mass density is

$$\rho \sim \lambda^{-4} hc^{-1} \sim (Gt^2)^{-1}, \tag{95.5}$$

where t is the expansion time (eq. 11.5). When $\lambda \sim ct$, the expansion time is on the order of the Planck time,

$$t_p \sim (Gh/c^5)^{1/2} \sim 10^{-43}\ \mathrm{s}. \tag{95.6}$$

At this Planck epoch the usual separation of quantum mechanics and general relativity theory must fail, and we have no firm theoretical guide as to what the universe was like. We are allowed to speculate that the universe that issues from this quantum state was exactly homogeneous (an eigenstate of the translation operators). In this case galaxy formation must have been caused by spontaneous fluctuations in the matter and radiation fields, and, if the divergence of T_i^j vanishes (consistent with general relativity theory), the spontaneous fluctuations must be in the stress part of T_i^j. We saw in Section 86 that pressure fluctuations in the early universe can produce space curvature fluctuations of the wanted amplitude, though it must somehow be contrived that the fluctuations are coherent over scales very much larger than the horizon (Press and Vishniac 1980a, Bardeen 1980). Another possibility is that local interactions cause particles to cluster and that the clustering grows as the universe expands, so as to keep the clustering length a fixed fraction of the distance to the horizon (Press and Schechter 1974). It is unlikely that gravity could do this (§ 27). In a scenario devised by Omnès (1969) and elaborated by Stecker and Puget (1972) irregularities develop through the coalescence and annihilation of pools of matter and antimatter. The idea has considerable appeal because it commences from a simple state—a homogeneous universe with local balance of matter and antimatter—but it proves difficult to see how the coalescence of matter pools and antimatter pools could proceed fast enough to make the growth of the scale of clustering keep pace with the growth of the horizon so as to produce an interesting degree of inhomogeneity now.

The lesson from this review of scenarios is that we are not likely to have a derivation of the existence of galaxies from first principles until we have a much better understanding of physics at the time of the big bang. Meanwhile, the more fruitful approach likely will proceed from an assessment of what the universe is like now and at modest redshift to a deduction of what it must have been like at high redshift. That should be an important datum for the search for the physics of the big bang. Work in this direction is summarized in the next section.

96. Nature of protogalaxies and protoclusters

A. The Situation near decoupling

Speculations on how the large-scale structures we see—galaxies and clusters of galaxies—came into being can be characterized in large part by what the universe is supposed to have been like at the epoch of decoupling of matter and radiation, redshift $Z_{\rm dec} \sim 1300$.

Prior to decoupling the departures from homogeneity may have been well described by linear perturbation theory. If no special constraint prevents it, the density fluctuations are adiabatic (§ 86) and so the spectrum of fluctuations after decoupling is truncated at the Silk mass M_s (eq. 92.50). Since M_s is much greater than the Jeans mass for the matter (eq. 94.24), the initial nonlinear evolution after decoupling is well described by the caustic theory (§ 21, Zel'dovich 1970, 1978).

There are some problems with this picture. The linearity assumption implies the limit in equation (95.1) on the size of primeval inhomogeneities. The power law spectrum in equations (95.2) and (95.3) is a reasonable way to fit the density fluctuations to this condition. The part of the spectrum shorter than the Jeans length λ_x (eq. 92.47) for matter plus radiation tilts to the power law form

$$|\delta_{\mathbf{k}}|^2 \propto k^{-3}, \tag{96.1}$$

as it comes within the horizon (eq. 92.57). The resulting spectrum of mass density fluctuations after decoupling is (eq. 93.42)

$$(\delta M/M)^2 \sim k^3 |\delta_{\mathbf{k}}|^2 \sim \text{constant}, \qquad M_s \lesssim M \lesssim M_x. \tag{96.2}$$

This means there is about equal variance per octave of r on all scales from the Silk mass to the matter-radiation Jeans mass. The analog for a function of one variable is $|\delta_k|^2 \propto k^{-1}$. Figure 96.1 shows an example of a random Gaussian process with this spectrum longward of a sharp cutoff. The function is periodic with period equal to the length plotted. The effect of the short wavelength cutoff is very apparent here as it is for white noise (fig. 26.1): in both cases the cutoff fixes the width of the narrowest spikes. However, there is a considerable difference in the character of the

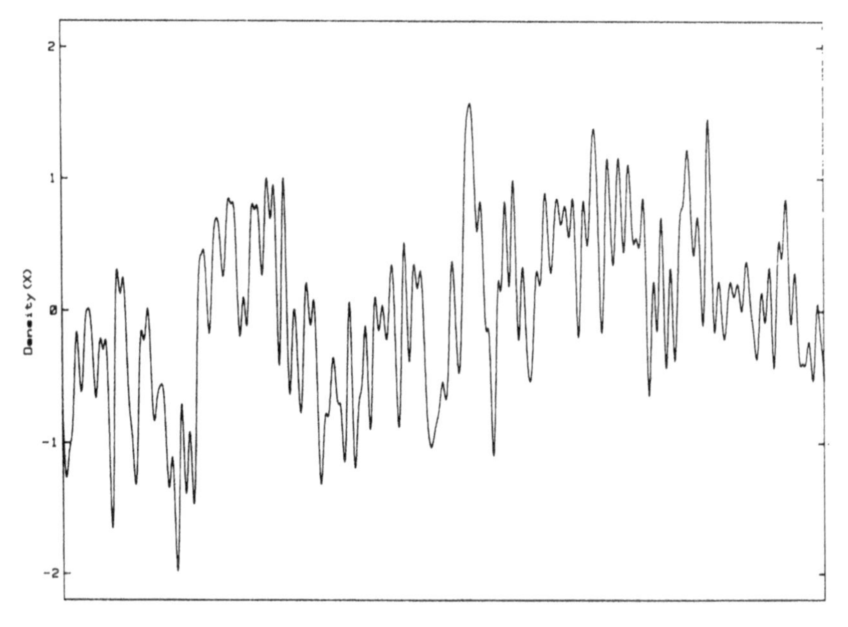

FIG. 96.1. Example of a random Gaussian process with power spectrum $\propto k^{-1}$.

functions. Here an appreciable upward density fluctuation can appear in a single spike with mass comparable to the cutoff (M_s), or in a train of spikes, or in a large clump with size comparable to the long wavelength cutoff of the spectrum. The present value of the long wavelength cutoff is (eq. 92.47)

$$\lambda_x \sim 50\,(\Omega_0 h^2)^{-1}\,\mathrm{Mpc}, \tag{96.3}$$

and, if the parameter $\Omega_0 h^2$ is greater than about 0.1, there is in addition a spike at λ_x (eq. 92.61). With or without the spike the situation is unreasonable, for if the amplitude at decoupling is large enough to make bound systems of size M_s form, then we should also see density fluctuations of large amplitude on the scale λ_x. This strongly contradicts the observation that the galaxy two-point correlation function is less than unity at $r \gtrsim r_0 \sim 4h^{-1}$ Mpc (§ 57). Thus it appears that although the primeval spectrum $|\delta_{\mathbf{k}}|^2 \propto k$ has some attractive theoretical features (§ 95), it could not be right.

If the primeval spectrum increases with the wave number k considerably more rapidly than the first power, then the unwanted fluctuations on scales $\sim \lambda_x$ at decoupling are eliminated and the first generation of mass clouds forms at the Silk mass M_s. However, M_s also is uncomfortably large. We find from equation (92.50) that the value of the cutoff r_s extrapolated to the present epoch is

$$r_s \sim 2/(0.036 + \Omega_0\, h^2)\ \mathrm{Mpc}. \tag{96.4}$$

In a low density universe, $\Omega_0 \lesssim 0.1$, r_s is considerably larger than the characteristic clustering length r_0. Second generation mass concentrations much smaller than r_s can form by the development of caustic surfaces, but it is difficult to see how the formation of the first generation of clouds at M_s could avoid violating the limit on the galaxy two-point correlation function $\xi(r)$. In a dense universe r_s is somewhat less than r_0. However, we want the first generation of clouds to form at redshifts greater than about 3, because quasars are observed at that redshift. Since clustering on larger scales would continue to grow after that epoch because the density parameter is close to unity, it would require careful arrangement of initial conditions to avoid violating the limit on $\xi(r)$ at $r \gtrsim r_0$.

One way out of the problem with the cutoff r_s is to use the hypothesis that much of the mass of the universe is in the form of low pressure material we might call massonium that only weakly interacts with ordinary matter and radiation. One possibility is a gas of neutrinos with nonzero mass (Cowsik and McClelland 1973, Lee and Weinberg 1977,

Steigman *et al.* 1978, Gunn *et al.* 1978, Tremaine and Gunn 1979, Peebles 1980b). Massonium would follow the growth of fluctuations in the distribution of radiation on scales greater than ct (§ 86) but not the oscillation and dissipation of small-scale irregularities (§ 92). The massonium thus could be deposited in a clumpy fashion, gravity would cause the clustering to grow once the mean density exceeds that of the radiation (§ 12), and the hydrogen would fall into the dense spots after the plasma combines and decouples from the radiation. Since the distribution of massonium at decoupling need not have a coherence length, the caustic theory of Section 21 need not give a useful description of how the first clumps of hydrogen form.

Another way out of the above problem is to adopt the hypothesis that the radiation distribution at the epoch of decoupling is smooth while the matter distribution is clumpy (§ 94). This could be the result of the nonlinear dissipation of primeval adiabatic density perturbations discussed in Section 89. The dissipation of the acoustic energy would create entropy in localized spots. The variant of this idea discussed by Eichler (1977) is that the radiation in the early universe was turbulent on small scales and that this turbulence dissipated before decoupling, again creating entropy in patches and so producing isothermal perturbations. As was discussed in Section 94, one can equally well assume that the isothermal perturbations are primeval, imposed by the physics of the big bang.

The constraint in equation (95.1) does not apply to the amplitude of primeval isothermal perturbations because the geometry of the early universe is not perturbed. It was shown in Section 73 that if the matter distribution at decoupling approximates a random Gaussian process with nearly flat power spectrum, then the similarity solution to the BBGKY hierarchy yields correlation functions that are in excellent agreement with the observed galaxy correlation functions. As discussed in Section 73, the main question here is whether the density parameter is large enough that the similarity solution can be a valid approximation.

We can compute the wanted amplitude of mass density fluctuations at decoupling from the observed large-scale clustering of galaxies (Peebles 1969a, 1970a, Press and Schechter 1974, Peebles and Groth 1976). The mean square fluctuation in the mass found within a sphere of radius R, volume V,[2] is

[2]If the power spectrum were increasing with k faster than the first power of k, the value of Δ^2 would be dominated by the spectrum at short wavelengths that enters through the sidelobes of the square window in the integral (§ 28). To prevent this, we could multiply the integrand by a Gaussian window function. That is not needed in the present application because, as it appears, $\xi \geqslant 0$.

$$\Delta^2 = \langle (M - \langle M \rangle)^2 \rangle / \langle M \rangle^2 = \int \xi(r_{12})\, d^3r_1\, d^3r_2 / V^2. \quad (96.5)$$

The integral is over V, and the equation assumes the mass autocorrelation function is usefully approximated by the galaxy two-point correlation function (§§ 36, 63). We shall suppose the sphere expands with the general expansion so $R \propto a(t)$ and $\langle M \rangle$ is constant. Then it was shown in Section 70 that if $\Delta \lesssim 1$, $\Delta(t)$ is a solution to equation (10.3). We have from equations (57.3), (57.5) and (60.3) the present value

$$\Delta^2(Z = 0, R = 10\, h^{-1}\,\text{Mpc}) \simeq 0.4. \quad (96.6)$$

The extrapolation back to decoupling using equations (10.3) and (70.7) gives (Peebles and Groth 1976)

$$\begin{aligned} \Delta^2(\text{dec}) &\sim 1 \times 10^{-6}, \quad \Omega_0 = 1, \quad h = 0.5; \\ &\sim 6 \times 10^{-5}, \quad \Omega_0 = 0.1, \quad h = 0.5. \end{aligned} \quad (96.7)$$

These numbers apply to a comoving sphere with present radius $10\, h^{-1}$ Mpc that contains mean mass

$$\langle M \rangle = 1.2 \times 10^{15}\, \Omega_0 h^{-1}\, \text{M}_\odot. \quad (96.8)$$

In the similarity solution the power spectrum of the mass distribution is supposed to be flat. If so, $\Delta^2 \propto \langle M \rangle^{-1}$, and we find from equations (96.7) and (96.8) that $\Delta^2 = 1$ at decoupling when the radius of the sphere is chosen so the mean value of the mass contained is

$$\begin{aligned} M_n &\sim 2 \times 10^9\, \text{M}_\odot, \quad \Omega_0 = 1; \\ M_n &\sim 1 \times 10^{10}\, \text{M}_\odot, \quad \Omega_0 \sim 0.1. \end{aligned} \quad (96.9)$$

Equations (96.7) assume that the galaxy distribution is a fair measure of the fluctuations in mass on the present scale of $10\, h^{-1}$ Mpc and that these mass fluctuations have grown since decoupling under gravity alone. The boundary condition at decoupling is $d\Delta/dt = 0$. Equations (96.9) depend on the additional assumption that at decoupling the spectrum of the matter density fluctuations is nearly flat. The consequence is that at $Z \sim 1000$ the universe already was strongly inhomogeneous on the mass scale of a dwarf galaxy. If the spectrum is flat well into the nonlinear regime, then at $\langle M \rangle \sim 3 \times 10^5\, \text{M}_\odot$ the rms fluctuation is $\Delta \sim 100$, large enough that the dense spots can resist the radiation drag and stop expanding before decoupling (eq. 94.17). Matter pressure tends to smooth out fluctuations on scales smaller than this. (The Jeans length in equation (94.23) scales as $(1 + \Delta)^{-1/2}$.) A lump that stops expanding before decoupling would do so in a smooth way as the critical density in equation (94.17) falls below the density of the lump and the radiation slowly relaxes its grip. The lump thus might be expected to collapse smoothly and by a large factor before fragmenting so that it might end up as a very dense star cluster or even a coherent superstar of the sort introduced by Hoyle and Fowler (1963). These might be the objects (little bangs) considered by Wagoner, Fowler, and Hoyle (1967) as a way to produce the heavy elements and perhaps also the helium seen in the oldest halo population II stars. As discussed by Doroshkevich, Zel'dovich, and Novikov (1967), they might serve to reheat the matter after decoupling. And, of course, their remnants could contain considerable mass in a form otherwise not easy to detect.

One can argue that it is reasonable to expect that the universe at high redshift should be similar to the universe now, smooth on large scales and clumpy on small scales. The consequence is that star clusters, or objects we might call pregalaxies, existed before there were galaxies. Speculations along this line are discussed by Peebles and Dicke (1968), Hogan (1978, 1979), and White and Rees (1978).

Another line of thought is that galaxies and clusters of galaxies are the remnants of primeval turbulence. In Eichler's (1977) scenario turbulence has decayed prior to decoupling, leaving isothermal density perturbations. In Ozernoy's (1978) scenario some residuum of the turbulence motion survives decoupling but it is weak, $vt \ll l$ for motions v on the scale l, so the motion adds to the growing mode of the linear density perturbation (§ 15). Either scheme may be capable of producing a definite spectrum of residual matter density fluctuations from the physics of the fully developed turbulence at high redshift, which would be a great advantage, though it has not yet been shown in detail how this can be done (Jones 1976, 1977). In earlier versions of the primeval turbulence scenario it was assumed, following the pioneering work of von Weizsäcker (1951), that the matter in a galaxy was driven together by the turbulent motion. However, that would make galaxies form too soon, at the epoch of decoupling of matter and radiation (Peebles 1971a). Limits on the epoch of galaxy formation are discussed next.

B. Protogalaxies and protoclusters

A galaxy might develop out of a coherent and isolated gas cloud collapsing under its own weight (Lemaître 1933b, Gamow and Teller 1939, Hoyle 1953, Arp 1961, Eggen, Lynden-Bell, and Sandage 1962) or driven together by matter currents (Gamow 1952, Oort 1958, 1970, Zel'dovich 1970). The same process could operate on a larger scale, the first generation of gas clouds being protoclusters that fragmented to form galaxies, some clusters dissolving to produce field galaxies. A sequence of this general sort has appealed to many authors (Hubble 1936, Silk 1968b, Zel'dovich 1970, 1978, Icke 1973). In another variant structure develops as a clustering hierarchy, so that before there were galaxies there were pregalaxies of gas and stars and dust (Layzer 1954, Peebles 1965). This gravitational clustering hierarchy scenario is suggested by the galaxy correlation functions (§ 73) and by the continuity between galaxies and clusters of galaxies (§§ 4B, 64).

The epoch at which galaxies might have formed is limited by the fact that at high redshifts galaxies would overlap. We might take the nominal size of a large galaxy to be the Holmberg radius (where the surface brightness is about 3 percent of the night sky),

$$R_H \sim 10\, h^{-1}\ \text{kpc}. \tag{96.10}$$

There are $n \sim 0.03\, h^2$ large galaxies per cubic megaparsec, so the fraction of space within one Holmberg radius of a galaxy is $\sim 10^{-7}$. The fraction would have been unity at redshift

$$1 + Z_1 \sim 200. \tag{96.11}$$

Prior to this epoch the bright central parts of galaxies would have overlapped.

In large spiral galaxies the rotation velocity at $r \sim R_H$ is very nearly constant at

$$\langle v^2 \rangle = (210\ \text{km s}^{-1})^2. \tag{96.12}$$

This is the average over the velocities listed by Faber and Gallagher (1979). The rms deviation of v^2 among the galaxies in this sample is about 50 percent. The fact that the rotation velocity in the outer part of a spiral galaxy is nearly constant implies that the mean mass density varies with distance from the galaxy center as r^{-2}, and we find from equation (96.12)

$$\langle \rho(r) \rangle \sim 6 \times 10^{-23}\, r_{\text{kpc}}^{-2}\ \text{g cm}^{-3}. \tag{96.13}$$

This density evaluated at $r = R_H$ is equal to the mean mass density of the universe (eq. 97.13) at redshift

$$1 + Z_2 \sim 30/\Omega_0^{1/3}. \tag{96.14}$$

If galaxies form by gravitational attraction, then the part of a spiral outside the Holmberg radius could not have been attached to the galaxy at redshifts greater than Z_2. The limit Z_2 for elliptical galaxies is thought to be comparable to the number for spirals because the velocity dispersion in the central part of a large elliptical is similar to equation (96.12); however, the density run in the outer part of an elliptical galaxy is not known.

If a protogalaxy forms and collapses by gravity, then we can reckon when this happened from the minimum density of the cloud and that with a cosmological model to fix redshift as a function of time yields the redshift at maximum expansion of the cloud (Partridge and Peebles 1967, Silk 1968b, Peebles 1968). The relation between minimum density and the time of maximum expansion in the spherical model is given by equation (19.52). It is doubtful that the numerical factor in this relation is significant because it is doubtful that a protogalaxy would be even close to spherically symmetric at the nominal epoch of maximum expansion (§§ 20, 21). We can arrive at a rough estimate by observing that equations (96.10) and (96.12) yield a rotation time of $3 \times 10^8\ h^{-1}$ years. In the gravitational instability picture the discs would have formed when the universe was that old, or older if the material collapsed by an appreciable factor. Setting the rotation time equal to the cosmic time in the Einstein-de Sitter model (eq. 11.5), we find

$$1 + Z_3 \sim 10, \qquad \Omega_0 = 1. \tag{96.15}$$

In a very low density cosmological model where $t \propto (1 + Z)^{-1}$ we find

$$1 + Z_3 \sim 30, \qquad \Omega_0 \sim 0. \tag{96.16}$$

This refers to the formation of the disc of a large spiral. In the gravitational hierarchical clustering scenario there may have been star clusters at higher redshifts. If galaxies formed out of coherent gas clouds, star formation may have commenced much later than Z_3, depending how far the cloud collapsed before stars formed. The flat rotation curves of large spirals suggest these galaxies have extended halos, the mass within radius r varying roughly as r to $r \sim 50\ h^{-1}$ kpc (Ostriker, Peebles, and Yahil 1974, Einasto, Kaasik, and Saar 1974, Rubin, Ford, and Thonnard 1980). Increasing the radius from R_H to $50\ h^{-1}$ Mpc would increase the dynamic time by a factor of 5, reducing the redshift to $Z_3 \sim 3$ to 6, which is interestingly close to the epoch of maximum abundance of quasars. It is worth noting also that we can reduce Z_3 still more by further increasing the scale. The Local Group of galaxies is thought to have just passed the epoch of maximum expansion because the Andromeda nebula is moving toward us at a speed of 100 km s^{-1}, making the crossing time in the group comparable to the Hubble time. Here, of course, we have the sure knowledge that strong subcondensations formed well before the collapse of the protogroup. One can imagine that the central parts of the Milky Way galaxy formed as a coherent collapsing gas cloud while the halo and companions in the group were added later by accretion. Models for spherical accretion were discussed in Section 25. This produces a halo of reasonable shape but does require very special initial conditions if the model is to be realistic. In another model the accretion is part of the process of development of the clustering hierarchy (§ 26). Here the mean density is high in the neighborhood of a galaxy but there is a considerable fluctuation from galaxy to galaxy. That fits with the fluctuating abundances of neighbors of galaxies; it is not clear whether a fluctuating halo mass can be reconciled with the observed small scatter among rotation velocities of large spirals (eq. 96.12) and the similarly small scatter among velocity dispersions in large ellipticals.

Hoyle (1953) was the first to consider in detail the temperature of a collapsing gas cloud the size and mass of a galaxy. He showed that the gas could not be hot enough to support the cloud so there would be almost free collapse interrupted by episodes of shock formation and strong compaction as different parts of the (nonspherical) cloud ran into each other. The orders of magnitude go as follows. At temperatures $T \gtrsim 10^4$ K the main radiative energy loss in a gas of hydrogen and helium is the bremsstrahlung emission by the electrons accelerated in the fields of the ions with a substantial contribution by radiative recombination at $T \sim 10^4$. A reasonable approximation to the cooling time at temperature T (degrees K) and density n (protons cm^{-3}) is

$$t_c \sim 10^4 T^{1/2} n^{-1} \text{ y}. \tag{96.17}$$

Let us consider as an example 10^{11} solar masses of gas, the nominal mass of the Milky Way galaxy, uniformly distributed in a sphere of radius 10 kpc. The density is $n \sim 1$ proton cm^{-3}, the free gravitational collapse time is $t_f \sim 10^8$ y, and the plasma temperature needed to prevent collapse and support the cloud is $T \sim 10^6$ K. If the cloud is this hot, the cooling time is $t_c \sim 10^7$ years, appreciably shorter than the collapse time. Hoyle concluded therefore that the cloud could not stay hot: it must cool to $T \sim 10^4$ K, the threshold for collisional ionization of atomic hydrogen by free electrons, and the residual ionization must adjust itself to make the energy loss rate balance the rate at which energy is being supplied by compression. This means the conversion from gas to stars must be rapid: within a few collapse times the gas either must be converted to stars or else dumped into the disc, and since there are no known disc stars with low heavy element abundance, nucleosynthesis of the elements must have been essentially completed during the collapse.

A new element was introduced by Ostriker (1974), who pointed out that if the gas cloud started out from a radius only moderately larger than was assumed above, the cooling time would be much longer. If in the above example the radius is increased to 100 kpc, the cooling time of a cloud hot enough to support itself increases to $\sim 3 \times 10^9$ y, somewhat longer than the collapse time and almost as long as the age of the universe. Thus Ostriker points out that a cloud somewhat more dilute than this would be "hung up," unable to form a galaxy. If the cloud can contract, it does so at first slower than free fall, so it is nearly in pressure equilibrium and subcondensations cannot form. When the density reaches the point that the cooling time falls below the free collapse time, the nature of the contraction changes: the pressure drops and nonspherical fluctuations grow, develop into shocks, and produce the first generations of stars. Since the star orbits suffer negligible drag, these first generations would mark the radius at which the cooling time fell below the collapse time.

Ostriker's scheme has been elaborated by Gott and Thuan (1976), Silk (1977b), and Rees and Ostriker (1977). It yields some contraints on scenarios that are interesting and may prove important. However, it starts from the assumption that there are coherent gas clouds in the wanted range of mass and radius. Substructure in these primeval clouds is not allowed because that would increase the mean square density and hence decrease the cooling time. A considerable elaboration of the scenario would be needed to explain why any appreciable fraction of the mass of the universe should have been in that particular form.

Our understanding of the nature of a protogalaxy eventually must be strongly affected by two lines of research that have not been analyzed here. First is the attempt to discover galaxies in the process of forming either at the present epoch (Burbidge, Burbidge, and Hoyle 1963, Sargent and Searle 1971) or at high redshift (Davis and Wilkinson 1974, Partridge 1974, Davis 1980). The search depends on an opinion of what to look for, and there is a considerable range of ideas. Thus Burbidge, Burbidge, and Hoyle (1963) considered the features that might distinguish galaxies forming now. Meier (1976) argued from the results of the spherical collapse models of Larson (1974) that a young galaxy might look like a quasar. Earlier Field (1964) and Weyman (1966) had discussed models for compact young galaxies. If the protogalaxy is not spherical, one might instead look for a highly irregular object bright in patches where the matter happens to be piling up at the time (Partridge and Peebles 1967). If the object is highly irregular, it could be difficult to draw the line between one protogalaxy and its neighbors: that might be so in particular if stars formed before galaxies. Here the sky at $Z \sim 10$ could simply appear mottled.

The second important subject is the attempt to deduce the origin of galaxies from the details of their present structures. Because the phenomena one can bring to bear are so much more numerous, this is a much richer problem than the approach from cosmology that has been the subject of this book, but of course the richness considerably muddies the waters.

There is a broad range of ideas on the origins of galaxies and clusters of galaxies because it proves so easy to invent detailed scenarios and so difficult to put them to the test. This book lists a number of elements of the theory and observations that do seem to be on firm ground. We may hope that the list will continue to grow so that we will see a progressive narrowing of the range of scenarios. On the other hand, we must still bear in mind Bondi's caution that "there are probably few features of theoretical cosmology that could not be completely upset and rendered useless by new observational discoveries," (Bondi, p. 169). For the present subject we might add, "or by a good new idea."

APPENDIX

97. Models and notation

This book deals with departures from an ideal homogeneous and isotropic Friedmann-Lemaître cosmological model. The homogeneity and isotropy of the background model imply that all physical variables can be expressed as functions of the proper cosmic time t kept by an observer at rest in a patch of fluid, the standard convention being that $t = 0$ at the singular epoch of the big bang. The proper distance between two chosen particles in the model, reckoned along a hypersurface of fixed t, must scale with time as

$$r(t) = a(t)x, \tag{97.1}$$

where x is a constant for the pair and $a(t)$ is the universal expansion parameter. The wavelength of a free photon stretches with other lengths as $a(t)$, so the wavelength λ_0 observed now, epoch t_0, of radiation emitted at epoch t at wavelength λ by an object comoving with the fluid is

$$\lambda_0 = \lambda a(t_0)/a(t) = \lambda(1 + Z), \tag{97.2}$$

where Z is the cosmological redshift. The redshift often is used as a label of an epoch.

The rate of increase of proper separation of a pair of particles in the model is

$$\frac{dr}{dt} = \frac{\dot{a}}{a} r = H(t)r. \tag{97.3}$$

The dot means the derivative of a with respect to cosmic time. $H(t)$ is Hubble's constant at epoch t: the present value is written as

$$H_0 = 100\, h \text{ km s}^{-1} \text{ Mpc}^{-1},$$

$$H_0^{-1} \sim 3 \times 10^{17}\, h^{-1}\text{s} \sim 1 \times 10^{10}\, h^{-1} \text{ y}. \tag{97.4}$$

The dimensionless parameter h reflects the uncertainty in this important quantity. It is generally thought to be in the range

$$0.5 \lesssim h \lesssim 1. \tag{97.5}$$

Equation (97.3) gives the redshift of an object at proper distance $r \ll cH_0^{-1}$:

$$Z \simeq H_0 r/c. \tag{97.6}$$

This is a good approximation at $Z \ll 1$. A convenient measure of the distance to the horizon is the Hubble length

$$r_H = cH_0^{-1} = 3000\, h^{-1}\,\mathrm{Mpc}. \tag{97.7}$$

This is about the distance free radiation has travelled since the big bang.

The evolution of the model is determined by the dynamic equation (eq. 7.13)

$$\frac{\ddot{a}}{a} = -\frac{4}{3}\pi G(\rho_b + 3p_b/c^2) + \Lambda/3, \tag{97.8}$$

with the energy equation

$$d\rho_b/dt = -3(\rho_b + p_b/c^2)\,\dot{a}/a. \tag{97.9}$$

The mass density is ρ, the pressure is p, and the subscripts refer to the background model.

Equations (97.8) and (97.9) can be integrated once:

$$\frac{\dot{a}^2}{a^2} = \frac{8}{3}\pi G\rho_b + \frac{\Lambda}{3} - \frac{R^{-2}}{a^2}. \tag{97.10}$$

The constant of integration R^{-2} appears in the expression for the line element,

$$ds^2 = g_{ij}\,dx^i dx^j = c^2dt^2 - \frac{a^2dx^2}{1 - x^2/R^2c^2} - a^2x^2(d\theta^2 + \sin^2\theta\, d\phi^2). \tag{97.11}$$

The coordinates x, θ, ϕ are comoving, fixed to fluid elements. To simplify equations, I have given R units of time rather than length. The proper radius of curvature of the hypersurface t = constant is $a|R|c$. In a closed model $R^{-2} > 0$, in an open model, $R^{-2} < 0$. The Einstein-de Sitter model has $R^{-2} = \Lambda = 0$ with $p = 0$.

The density parameter is the ratio of the mean mass density $\rho_b(t)$ to the density in an Einstein-de Sitter model with the same Hubble constant,

$$\Omega(t) = \frac{8}{3}\pi G\rho_b/H^2. \tag{97.12}$$

The present mean density can then be written as

$$\rho_0 = 1.88 \times 10^{-29}\, \Omega_0 h^2 \text{ g cm}^{-3}, \tag{97.13}$$

where the present value of the density parameter is thought to be in the range

$$0.03 \lesssim \Omega_0 \lesssim 1. \tag{97.14}$$

There is fairly strong evidence that the universe contains a homogeneous sea of blackbody radiation, temperature

$$T(t) = T_0(1 + Z), \qquad T_0 \simeq 2.7 \text{ K}. \tag{97.15}$$

The indicated variation with redshift assumes negligible change in the entropy of the radiation, a good approximation at $Z \lesssim 10^9$ (eqs. 94.19 and 94.22). (At $Z \sim 3 \times 10^9$ roughly half the entropy is taken up in electron-positron pairs.) The mean mass density in the radiation is

$$\mathcal{E}_b(t) = \mathcal{E}_0 (1 + Z)^4, \quad \mathcal{E}_0 = a_s T_0^4/c^2 = 4.5 \times 10^{-34} \text{ g cm}^{-3}. \tag{97.16}$$

Since this is well below the density from galaxies (eqs. 97.13 and 97.14), we can write the ratio of mass densities in nonrelativistic matter and radiation as

$$\frac{\mu}{\mathcal{E}} = \frac{4.21 \times 10^4\, \Omega_0 h^2}{1 + Z}. \tag{97.17}$$

It is thought that there is in addition to this electromagnetic radiation a uniform background of neutrinos, mass density comparable to $\mathcal{E}$ (or higher if the lepton number exceeds the baryon number of the universe by some eight orders of magnitude; *PC*, chapter VIII). Since the net density of such a relativistic mass of neutrinos and perhaps also gravitational radiation is so very indirectly reckoned, the numerical estimates in Sections 92–96 take the net density to be the electromagnetic part only: this simple convention likely is about right or else much too small. Following the same principle, I have taken the primeval matter to be pure hydrogen. If helium is present, it slightly reduces the mean number of electrons per baryon, but the effect is negligible compared to the main uncertainties. The net mass density and pressure are written as

$$\rho = \mu + \mathscr{E}, \qquad p = \mathscr{E} c^2/3, \tag{97.18}$$

with μ given by equation (97.13).

The peculiar velocity $\mathbf{v}$ discussed in Chapter II is the proper velocity, km s^{-1}, relative to an observer to whom the redshifts near the horizon appear isotropic. Of course, this is not meaningful for irregularities on scales larger than the horizon: in Chapter V, $\mathbf{v}$ is the proper velocity relative to a time-orthogonal coordinate system. The dimensionless density contrast is

$$\delta(\mathbf{x}, t) = \rho(\mathbf{x}, t)/\rho_b(t) - 1. \tag{97.19}$$

The value of δ depends on how the hypersurfaces of fixed t are assigned. In Chapter II, where regions of size $r \ll cH^{-1}$ are considered, the standard Newtonian synchronization of time can be used. The relativistic treatment in Chapter V is based on the prescription that $\mathbf{x}$, t are time-orthogonal coordinates (§ 81).

Some aspects of the evolution of the models might be noted. The simplest case assumes the pressure and cosmological constant both are negligibly small. Then equation (97.9) says $\rho_b \propto a^{-3}$, and we have from equation (97.10) the convenient relations

$$\begin{aligned} \Omega(t)^{-1} - 1 &= (\Omega_0^{-1} - 1)/(1 + Z), \\ H^2(t) &= H_0^2\,(1 + Z)^2(1 + \Omega_0 Z), \\ a(t)\,|R| &= H^{-1}\,|1 - \Omega(t)|^{-1/2}. \end{aligned} \tag{97.20}$$

This last equation indicates that in a low density cosmological model, $\Omega \ll 1$, the radius of curvature is comparable to the Hubble length (eq. 97.7). The parametric solution to equation (97.10) is

$$\begin{aligned} a &= A(1 - \cos\eta), & t &= B(\eta - \sin\eta); & \Omega &> 1 \\ a &= A(\cosh\eta - 1), & t &= B(\sinh\eta - \eta); & \Omega &< 1. \end{aligned} \tag{97.21}$$

The constants A and B satisfy the relation

$$A = \tfrac{4}{3}\pi G\rho_b a^3 |R|^2, \qquad B = A|R|. \tag{97.22}$$

One finds from equation (97.10) the relations

$$\begin{aligned} 1 + \cos\eta &= 2\Omega^{-1}, \qquad \Omega > 1, \\ \cosh\eta + 1 &= 2\Omega^{-1}, \qquad \Omega < 1, \end{aligned} \tag{97.23}$$

and

$$(BH)^2 = \Omega^{-1}/(4|\Omega^{-1} - 1|^3). \tag{97.24}$$

The limiting case $\Omega = 1$ is the Einstein-de Sitter model, where

$$a \propto t^{2/3}. \tag{97.25}$$

Analytic expressions for coordinate distance in the models as a function of redshift are given in equations (56.7) and (56.8).

If $\Lambda = 0$ but the pressure is not negligible, the general character of the evolution is the same. Whatever the equation of state, if $R^{-2} < 0$ so the geometry is open, then $0 < \Omega(t) < 1$ (eqs. 97.10 and 97.12), and $\dot{a}^2 > 0$: an expanding model starts at $a = 0$ and never stops expanding. If $R^{-2} > 0$, the geometry is closed, $\Omega(t) > 1$, and there is a maximum value of $a(t)$ at which $\dot{a} = 0$. The model expands from $a = 0$, reaches a point of maximum expansion, and then collapses back to a singularity at $a = 0$.

There are several historically interesting models with $\Lambda \neq 0$. In the Lemaître model, $\Lambda > 0$, $R^{-2} > 0$, and $a(t)$ has an inflection point at $a = a_e$, say. By equation (97.8) the density and pressure at the inflection point satisfy the relation

$$\Lambda = 4\pi G(\rho_b(e) + 3p_b(e)/c^2). \tag{97.26}$$

The expansion rate at the inflection point is (eq. 97.10)

$$\dot{a}_e^2 = R_e^{-2} - R^{-2}, \qquad R_e^{-2} = 4\pi G(\rho_b(e) + p_b(e)/c^2)a_e^2. \tag{97.27}$$

In the Lemaître model $R > R_e$, so $\dot{a}_e^2 > 0$: the model expands from $a = 0$ and never stops expanding. The dwell time at $a \sim a_e$ is increased by making R very close to R_e. If $R = R_e$, we have the unstable solution $a = a_e$, which is the original static Einstein model. The Eddington model assumes $R = R_e$, so $a(t)$ approaches a_e in the distant past.

LIST OF ABBREVIATIONS

AA: Astronomy and Astrophysics.
AJ: The Astronomical Journal.
AN: Astronomische Nachrichten.
Ap J: The Astrophysical Journal.
Ap Space Sci: Astrophysics and Space Science.
Astron Zh: the English translation is published in Soviet Astronomy.
BAN: Bulletin of the Astronomical Institutes of the Netherlands.
CR: Comptes Rendus de l'Académie des Sciences, Paris.
IAU Symposium No. 9: *Paris Symposium on Radio Astronomy,* ed. R. N. Bracewell. Stanford, Calif.: Stanford University Press, 1959.
IAU Symposium No. 63: *Confrontation of Cosmological Theories with Observational Data,* ed. M. S. Longair. Dordrecht, Holland: D. Reidel. 1974.
IAU Symposium No. 79: *The Large-Scale Structure of the Universe,* eds. M. Longair and J. Einasto. Dordrecht, Holland: D. Reidel. 1978.
IAU Symposium No. 92: *Objects of High Redshift,* eds. G. O. Abell and P. J. E. Peebles. Dordrecht, Holland: D. Reidel. 1980.
IAU and International Union of Theoretical and Applied Mechanics Symposium: *Problems of Cosmical Aerodynamics.* Dayton, Ohio: Central Air Documents Office. 1949.
MN: Monthly Notices of the Royal Astronomical Society.
PASP: Publications of the Astronomical Society of the Pacific.
Proc NAS: Proceedings of the National Academy of Sciences, Washington, D.C.
QJRAS: Quarterly Journal of the Royal Astronomical Society.
RMP: Reviews of Modern Physics.
Solvay Conference No. 11: *La Structure et l'Evolution de l'Univers,* ed. R. Stoops. Brussels: 1958.
Solvay Conference No. 13: *The Structure and Evolution of Galaxies* London: Interscience. 1965.
Z Ap: Zeitschrift für Astrophysik.
Zh ETF: the English translation is published in Soviet Physics JETP.

REFERENCES

Aarseth, S. J., and Hills, J. G. 1972. *AA*. **21,** 255 (§73).

Abell, G. O. 1958. *Ap J Suppl*. **3,** 211 (§§2, 29, 77).

Adams, J. B., Mjolsness, R., and Wheeler, J. A. 1958. In the Proceedings of the 11th Solvay Conference, p. 113 (§4).

Adams, P. J., and Canuto, V. 1975. *Phys Rev. D*. **12,** 3793 (§86).

Alpher, R. A., Bethe, H. A., and Gamow, G. 1948. *Phys Rev*. **73,** 803 (§4).

Ambartsumian, V. A. 1944. *Doklady Acad Nauk USSR*. **44,** 223 (§§29, 58).

______. 1958. In the Proceedings of the 11th Solvay Conference, p. 241 (§4).

______. 1965. In the Proceedings of the 13th Solvay Conference, p. 1 (§4).

Anile, A. M., and Motta, S. 1976. *Ap J*. **207,** 685 (§93).

Arp, H. 1961. *Science*. **134,** 810 (§96).

______. 1970. *Nature*. **225,** 1033 (§4).

Bardeen, J. 1980. preprint (§§81, 86, 95).

Barrow, J. D. 1977. *MN*. **179,** 47p (§90).

Bhavsar, S. P. 1978. *Ap J*. **222,** 412 (§29).

Binney, J. 1974. *MN*. **168,** 73 (§23).

Birkhoff, G. D. 1923. *Relativity and Modern Physics*. Cambridge, Mass.: Harvard University Press. (§3).

Bisnovatyi-Kogan, G. S., and Zel'dovich, Ya. B. 1970. *Astron Zh*. **47,** 942; 1971. *Soviet Astron*. **14,** 758 (§68).

Bogart, R. S., and Wagoner, R. V. 1973. *Ap J*. **181,** 609 (§29).

Bok, B. J. 1934. *Bull Harvard Obs*. **895,** 1 (§29).

Bondi, H. 1947. *MN*. **107,** 410 (§3).

______. 1952. *Cosmology*. Cambridge: Cambridge University Press. (§§ 2, 96).

Bonnor, W. B. 1956. *Z Ap*. **39,** 143 (§4).

______. 1957. *MN*. **117,** 111 (§§4, 95).

______. 1967. In *Relativity Theory and Astrophysics*, ed. J. Ehlers. Lectures in Applied Math. No. 8. Providence: Am. Math. Soc. (§4).

______. 1974. *MN*. **167,** 55 (§2).

Bonometto, S. A., and Lucchin, F. 1976. *Ap J.* **206,** 391 (§92).
———. 1978. *AA.* **67,** L7 (§42).
Boynton, P. 1978. In IAU Symposium No. 79. (§93).
———. 1980. In IAU Symposium No. 92. (§93).
Brecher, K., and Silk, J. 1969. *Ap J.* **158,** 91 (§13).
Brown, G. S. 1974. Doctoral Dissertation. University of Texas, Austin. (§2).
Burbidge, E. M., Burbidge, G. R., and Hoyle, F. 1963. *Ap J.* **138,** 873 (§96).
Byalko, A. V. 1969. *ZhETF.* **55,** 317; *Soviet Phys JETP.* **28,** 168 (§13).
Callan, C., Dicke, R. H., and Peebles, P. J. E. 1965. *Am J Phys.* **33,** 105 (§3).
Carlitz, R., Frautschi, S., and Nahm, W. 1973. *AA.* **26,** 171 (§§4, 28).
Carpenter, E. F. 1938. *Ap J.* **88,** 344 (§4).
Carr, B. J. 1978. *Comments on Astrophys.* **7,** 161 (§87).
Chandrasekhar, S., and Münch, G. 1952. *Ap J.* **115,** 103 (§§29, 58).
Charlier, C.V.L. 1908. *Arkiv för Mat Astron och Fys.* **4,** 1 (§3).
———. 1922. *Arkiv för Mat Astron och Fys.* **16,** 1; *Medd Lund Obs.* No. 98 (§§3, 62).
Chernin, A. D. 1970. *Pisma ZhETP.* **11,** 317; *Soviet Phys JETP Letters.* **11,** 210 (§23).
Chibisov, G. V. 1972. *Astron Zh.* **49,** 74; *Soviet Astron.* **16,** 56 (§92).
Chibisov, G. V., and Ozernoy, L. M. 1969. *Astrophys Letters.* **3,** 189 (§93).
Chincarini, G., and Rood, H. J. 1975. *Nature.* **257,** 294 (§§76, 77).
Clutton-Brock, M. 1974. *Ap Space Sci.* **30,** 395 (§95).
Clutton-Brock, M. and Peebles, P.J.E. 1980. Unpublished. (§8).
Courant, R. and Hilbert, D. 1953. *Methods of Mathematical Physics.* New York: Interscience (§53).
Cowsik, R. and McClelland, J. 1973. *Ap J.* **180,** 7 (§96).
Curtis, H. D. 1918. *Publ Lick Obs.* **13,** 11 (§2).
Dautcourt, G. 1969. *MN.* **144,** 255 (§92).
———. 1970. *AN.* **292,** 113 (§92).
———. 1977. *AN.* **298,** 253 (§56).
Davis, M. 1980. In IAU Symposium No. 92. (§96).
Davis, M., and Geller, M. J. 1976. *Ap J.* **208,** 13 (§§47, 56, 63).
Davis, M., Geller, M. J., and Huchra, J. 1978. *Ap J.* **221,** 1 (§§57, 76).
Davis, M., Groth, E. J., and Peebles, P.J.E. 1977. *Ap J.* **212,** L107 (§§57, 71, 79).

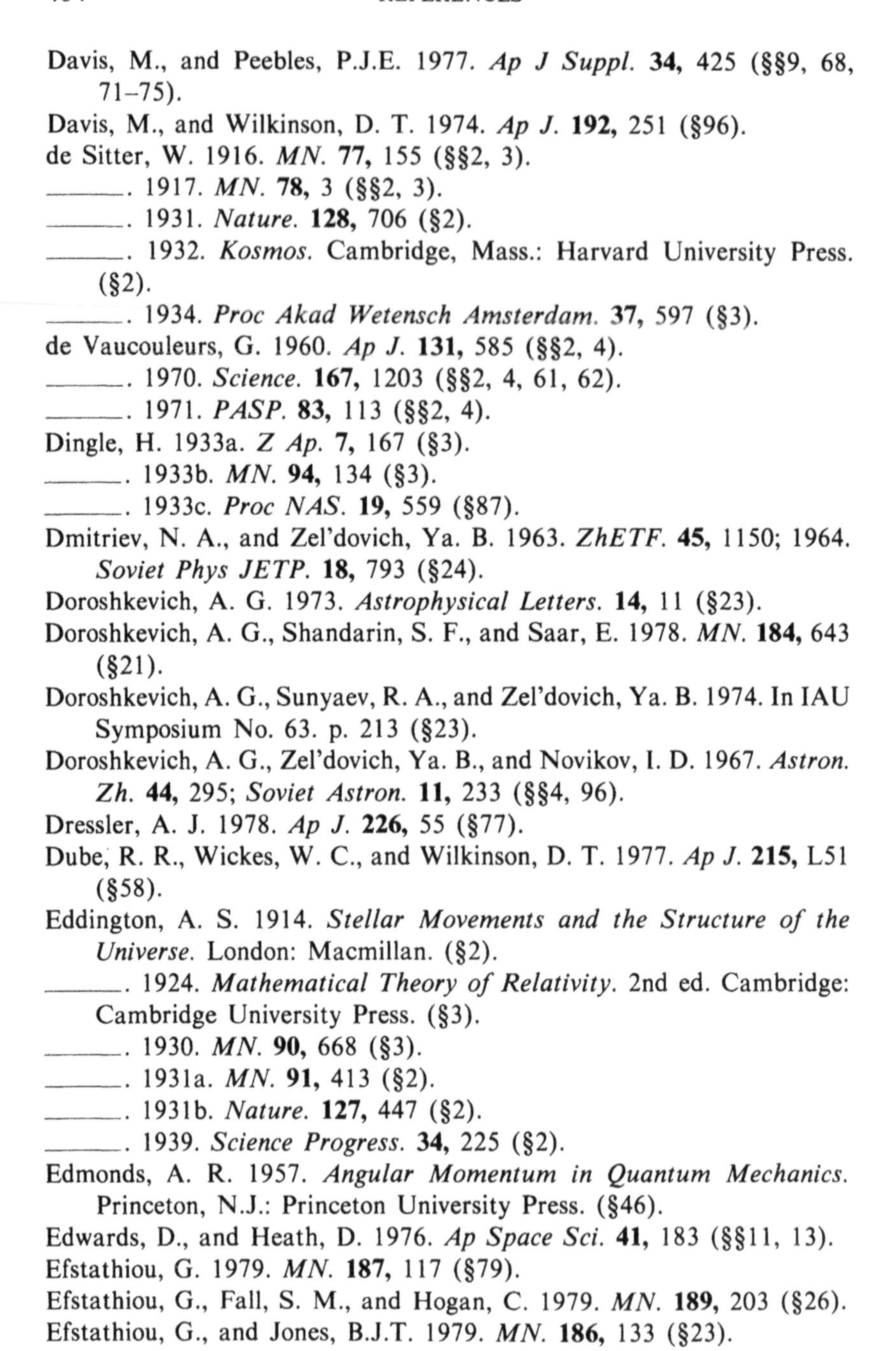

Davis, M., and Peebles, P.J.E. 1977. *Ap J Suppl.* **34,** 425 (§§9, 68, 71–75).

Davis, M., and Wilkinson, D. T. 1974. *Ap J.* **192,** 251 (§96).

de Sitter, W. 1916. *MN.* **77,** 155 (§§2, 3).

______. 1917. *MN.* **78,** 3 (§§2, 3).

______. 1931. *Nature.* **128,** 706 (§2).

______. 1932. *Kosmos.* Cambridge, Mass.: Harvard University Press. (§2).

______. 1934. *Proc Akad Wetensch Amsterdam.* **37,** 597 (§3).

de Vaucouleurs, G. 1960. *Ap J.* **131,** 585 (§§2, 4).

______. 1970. *Science.* **167,** 1203 (§§2, 4, 61, 62).

______. 1971. *PASP.* **83,** 113 (§§2, 4).

Dingle, H. 1933a. *Z Ap.* **7,** 167 (§3).

______. 1933b. *MN.* **94,** 134 (§3).

______. 1933c. *Proc NAS.* **19,** 559 (§87).

Dmitriev, N. A., and Zel'dovich, Ya. B. 1963. *ZhETF.* **45,** 1150; 1964. *Soviet Phys JETP.* **18,** 793 (§24).

Doroshkevich, A. G. 1973. *Astrophysical Letters.* **14,** 11 (§23).

Doroshkevich, A. G., Shandarin, S. F., and Saar, E. 1978. *MN.* **184,** 643 (§21).

Doroshkevich, A. G., Sunyaev, R. A., and Zel'dovich, Ya. B. 1974. In IAU Symposium No. 63. p. 213 (§23).

Doroshkevich, A. G., Zel'dovich, Ya. B., and Novikov, I. D. 1967. *Astron. Zh.* **44,** 295; *Soviet Astron.* **11,** 233 (§§4, 96).

Dressler, A. J. 1978. *Ap J.* **226,** 55 (§77).

Dube, R. R., Wickes, W. C., and Wilkinson, D. T. 1977. *Ap J.* **215,** L51 (§58).

Eddington, A. S. 1914. *Stellar Movements and the Structure of the Universe.* London: Macmillan. (§2).

______. 1924. *Mathematical Theory of Relativity.* 2nd ed. Cambridge: Cambridge University Press. (§3).

______. 1930. *MN.* **90,** 668 (§3).

______. 1931a. *MN.* **91,** 413 (§2).

______. 1931b. *Nature.* **127,** 447 (§2).

______. 1939. *Science Progress.* **34,** 225 (§2).

Edmonds, A. R. 1957. *Angular Momentum in Quantum Mechanics.* Princeton, N.J.: Princeton University Press. (§46).

Edwards, D., and Heath, D. 1976. *Ap Space Sci.* **41,** 183 (§§11, 13).

Efstathiou, G. 1979. *MN.* **187,** 117 (§79).

Efstathiou, G., Fall, S. M., and Hogan, C. 1979. *MN.* **189,** 203 (§26).

Efstathiou, G., and Jones, B.J.T. 1979. *MN.* **186,** 133 (§23).

Eggen, O. J., Lynden-Bell, D., and Sandage, A. R. 1962. *Ap J.* **136,** 748 (§96).
Eichler, D. 1977. *Ap J.* **218,** 579 (§96).
Einasto, J. 1978. In IAU Symposium No. 79. (§21).
Einasto, J., Kaasik, A., and Saar, E. 1974. *Nature.* **250,** 309 (§96).
Einstein, A. 1917. *S-B Preuss Akad Wiss.* p. 142 (§§1, 3, 4).
______. 1922. *Ann Phys.* **69,** 436 (§3).
______. 1933. *Structure Cosmologique de l'Espace.* Paris: Hermann et Cie. (§2).
Einstein, A., and Straus, E. G. 1945. *RMP.* **17,** 120; **18,** 148 (§3).
Ellis, R. S. 1980. In IAU Symposium No. 92. (§§2, 57).
Faber, S. M. and Dressler, A. J. 1977. *AJ.* **82,** 187 (§77).
Faber, S. M. and Gallagher, J. S. 1979. *Ann Rev Astron Astrophys.* **17,** 135 (§§64, 96).
Fabian, A. C. 1972. *Nature Phys Sci.* **237,** 19 (§2).
Fall, S. M. 1975. *MN.* **172,** 23 (§74).
______. 1976a. Doctoral Dissertation. Oxford University. (§56).
______. 1976b. *MN.* **176,** 181 (§74).
______. 1978. *MN.* **185,** 165 (§63).
______. 1979. *RMP.* **51,** 21 (§§52, 79).
Fall, S. M., Geller, M. J., Jones, B.J.T., and White, S.D.M. 1976. *Ap J.* **205,** L121 (§39).
Fall, S. M. and Severne, G. 1976. *MN.* **174,** 241 (§68, 74).
Fall, S. M. and Tremaine, S. 1977. *Ap J.* **216,** 682 (§§52, 53, 57).
Fath, E. A. 1914. *AJ.* **28,** 75 (§2).
Field, G. B. 1964. *Ap J.* **140,** 1434 (§96).
______. 1965. *Ap J.* **142,** 531 (§4).
______. 1971. *Ap J.* **165,** 29 (§92, 94).
______. 1975. In *Galaxies and the Universe.* ed. A. Sandage, M. Sandage, and J. Kristian. p. 359. Chicago: University of Chicago Press. (§23).
Field, G. B. and Shepley, L. C. 1968. *Ap Space Sci.* **1,** 309 (§92).
Fournier d'Albe, E. E. 1907. *Two New Worlds.* London: Longmans Green. (§§3, 62).
Fry, J. N. 1979. Doctoral Dissertation. Princeton University. (§77).
Fry, J. N., and Peebles, P.J.E. 1978. *Ap J.* **221,** 19 (§§29, 36, 55–57).
______. 1980a. *Ap J.* **238,** (§§44, 77).
______. 1980b. *Ap J.* **236,** 343 (§79).
Gamow, G. 1948a. *Phys Rev.* **74,** 505 (§4).
______. 1948b. *Nature.* **162,** 680 (§§4, 94).
______. 1949. *RMP.* **21,** 367 (§94).

———. 1952. *Phys Rev.* **86,** 251 (§§4, 95, 96).
———. 1953. *Kong Dan Vid Selsk.* **27,** No. 10 (§4).
———. 1954. *Proc NAS.* **40,** 480 (§4).
Gamow, G., and Teller, E. 1939. *Nature.* **143,** 116; *Phys Rev.* **55,** 654 (§§4, 96).
Geller, M. J. 1975. Unpublished. (§58).
Geller, M. J., and Peebles, P.J.E. 1973. *Ap J.* **184,** 329 (§§75, 76).
Gilbert, I. H. 1965. Doctoral Dissertation. Harvard University. (§§68, 74).
———. 1966. *Ap J.* **144,** 233 (§68).
Gödel, K. 1949. *RMP.* **21,** 447 (§95).
Gold, T., and Hoyle, F. 1959. In IAU Symposium No. 9. p. 583 (§4).
Gott, J. R. 1975. *Ap J.* **201,** 296 (§25).
Gott, J. R., and Rees, M. J. 1975. *AA.* **45,** 365 (§§19, 20, 71).
Gott, J. R., and Thuan, T. X. 1976. *Ap J.* **204,** 649 (§96).
Gott, J. R., and Turner, E. L. 1977. *Ap J.* **216,** 357 (§29).
———. 1979. *Ap J.* **232,** L79 (§57).
Gott, J. R., Turner, E. L., and Aarseth, S. J. 1979. *Ap J.* **234,** 13 (§§79, 96).
Gott, J. R., Wrixon, G. T., and Wannier, P. 1973. *Ap J.* **186,** 777 (§19).
Gregory, S. A., and Thompson, L. A. 1978. *Ap J.* **222,** 784 (§77).
Gribbin, J. 1974. *Nature.* **252,** 445 (§§25, 95).
Groth, E. J. 1977. Unpublished. (§59).
Groth, E. J., and Peebles, P.J.E. 1975. *AA.* **41,** 143 (§§11, 12).
———. 1977. *Ap J.* **217,** 385 (§§2, 47, 56, 57, 61).
Groth, E. J., Peebles, P.J.E., Seldner, M., and Soneira, R. M. 1977. *Scientific American.* **237,** Nov. p. 76 (§79).
Gunn, J. E. 1965. Doctoral Dissertation. California Institute of Technology. (§58).
———. 1977. *Ap J.* **218,** 592 (§25).
Gunn, J. E. and Gott, J. R. 1972. *Ap J.* **176,** 1 (§§19, 25).
Gunn, J. E., Lee, B. W., Lerche, I., Schramm, D. N., and Steigman, G. 1978. *Ap J.* **233,** 105 (§96).
Guyot, M. and Zel'dovich, Ya. B. 1970. *AA.* **9,** 227 (§§10–12).
Haggerty, M. J. 1970. *Physica.* **50,** 391 (§24).
———. 1971. *Ap J.* **166,** 257 (§2).
Haggerty, M. J. and Janin, G. 1974. *AA.* **36,** 415 (§23).
Harrison, E. R. 1967a. *RMP.* **39,** 862 (§4).
———. 1967b. *Mem Soc R Sci Liège.* **14,** 15 (§4).
———. 1967c. *Phys Rev Letters.* **18,** 1011 (§95).
———. 1967d. *Nature.* **215,** 151 (§95).

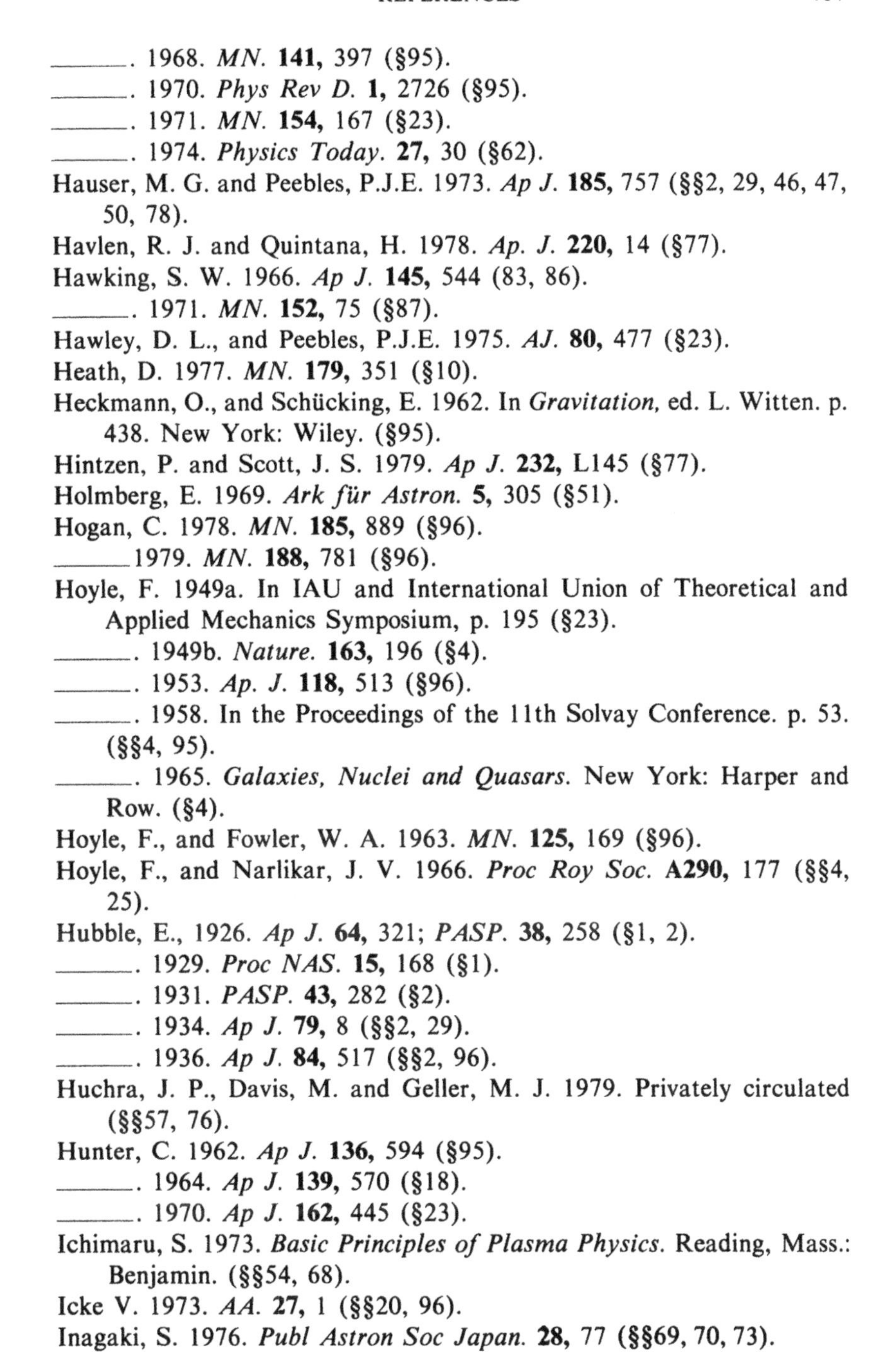

———. 1968. *MN.* **141,** 397 (§95).
———. 1970. *Phys Rev D.* **1,** 2726 (§95).
———. 1971. *MN.* **154,** 167 (§23).
———. 1974. *Physics Today.* **27,** 30 (§62).
Hauser, M. G. and Peebles, P.J.E. 1973. *Ap J.* **185,** 757 (§§2, 29, 46, 47, 50, 78).
Havlen, R. J. and Quintana, H. 1978. *Ap. J.* **220,** 14 (§77).
Hawking, S. W. 1966. *Ap J.* **145,** 544 (83, 86).
———. 1971. *MN.* **152,** 75 (§87).
Hawley, D. L., and Peebles, P.J.E. 1975. *AJ.* **80,** 477 (§23).
Heath, D. 1977. *MN.* **179,** 351 (§10).
Heckmann, O., and Schücking, E. 1962. In *Gravitation,* ed. L. Witten. p. 438. New York: Wiley. (§95).
Hintzen, P. and Scott, J. S. 1979. *Ap J.* **232,** L145 (§77).
Holmberg, E. 1969. *Ark für Astron.* **5,** 305 (§51).
Hogan, C. 1978. *MN.* **185,** 889 (§96).
———1979. *MN.* **188,** 781 (§96).
Hoyle, F. 1949a. In IAU and International Union of Theoretical and Applied Mechanics Symposium, p. 195 (§23).
———. 1949b. *Nature.* **163,** 196 (§4).
———. 1953. *Ap. J.* **118,** 513 (§96).
———. 1958. In the Proceedings of the 11th Solvay Conference. p. 53. (§§4, 95).
———. 1965. *Galaxies, Nuclei and Quasars.* New York: Harper and Row. (§4).
Hoyle, F., and Fowler, W. A. 1963. *MN.* **125,** 169 (§96).
Hoyle, F., and Narlikar, J. V. 1966. *Proc Roy Soc.* **A290,** 177 (§§4, 25).
Hubble, E., 1926. *Ap J.* **64,** 321; *PASP.* **38,** 258 (§1, 2).
———. 1929. *Proc NAS.* **15,** 168 (§1).
———. 1931. *PASP.* **43,** 282 (§2).
———. 1934. *Ap J.* **79,** 8 (§§2, 29).
———. 1936. *Ap J.* **84,** 517 (§§2, 96).
Huchra, J. P., Davis, M. and Geller, M. J. 1979. Privately circulated (§§57, 76).
Hunter, C. 1962. *Ap J.* **136,** 594 (§95).
———. 1964. *Ap J.* **139,** 570 (§18).
———. 1970. *Ap J.* **162,** 445 (§23).
Ichimaru, S. 1973. *Basic Principles of Plasma Physics.* Reading, Mass.: Benjamin. (§§54, 68).
Icke V. 1973. *AA.* **27,** 1 (§§20, 96).
Inagaki, S. 1976. *Publ Astron Soc Japan.* **28,** 77 (§§69, 70, 73).

Irvine, W. M. 1961. Doctoral Dissertation. Harvard University. (§24).
———. 1965. *Ann Phys* (N.Y.). **32,** 322 (§24).
Jackson, J. C. 1972. *MN.* **156,** 1P (§76).
Jahnke, E., and Emde, E. 1945. *Tables of Functions.* New York: Dover (§46).
Jeans, J. 1902. *Phil Trans.* **199A,** 49 (§§4, 16).
———. 1928. *Astronomy and Cosmogony.* Cambridge: Cambridge University Press. (§4).
Jones, B.J.T. 1976. *RMP.* **48,** 107 (§§23, 96).
———. 1977. *MN.* **180,** 151 (§96).
Jones, B.J.T., and Peebles, P.J.E. 1972. *Comments on Astrophys and Space Phys.* **4,** 121 (§4).
Kantowski, R. 1969. *Ap J.* **155,** 1023 (§14).
Kellogg, O. 1953. *Foundations of Potential Theory.* New York: Dover. (§20).
Kiang, T. 1967. *MN.* **135,** 1 (§4).
Kiang, T., and Saslaw, W. C. 1969. *MN.* **143,** 129 (§§2, 4).
Kihara, T. 1968. *Publ Astron Soc Japan.* **20,** 220 (§19).
Kihara, T., and Saki, K. 1970. *Publ Astron Soc Japan.* **22,** 1 (§22).
Kirshner, R. P., Oemler, A., and Schechter, P. L. 1978. *AJ.* **83,** 1549 (§§57, 76).
Kristian, J., and Sachs, R. K. 1966. *Ap J.* **143,** 379 (§2).
Kron, R. 1978. Doctoral Dissertation. University of California, Berkeley. (§2).
Lake, G. and Tremaine, S. 1980. preprint (§§51, 57).
Landau, L. D. and Lifshitz, E. M. 1959. *Fluid Mechanics.* London: Pergamon (§89).
———. 1979. *Classical Theory of Fields.* 4th. ed. London: Pergamon. (§§6, 21).
Larson, R. B. 1974. *MN.* **166,** 585 (§96).
Layzer, D. 1954. *AJ.* **59,** 170 (§§4, 26, 95, 96).
———. 1956. *AJ.* **61,** 383 (§§29, 33, 38).
———. 1963. *Ap J.* **138,** 174 (§24).
———. 1964. *Ann Rev Astron Astrophys.* **2,** 341 (§24).
Lee, B. W., and Weinberg, S. 1977. *Phys Rev Letters.* **39,** 165 (§96).
Lemaître, G. 1927. *Ann Soc Sci Bruxelles.* **47A,** 49; 1931. *MN,* **91,** 483 (§1, 3).
———. 1931. *MN.* **91,** 490 (§3, 6).
———. 1933a. *CR.* **196,** 903 and 1085 (§§3, 4, 87).
———. 1933b. *Ann Soc Sci Bruxelles.* **A53,** 51 (§§3, 4, 21, 26, 87, 95, 96).
———. 1934. *Proc NAS.* **20,** 12 (§§4, 26).

______. 1958. In the Proceedings of the 11th Solvay Conference. p. 1. (§4).
Liang, E.P.T. 1977a. *Ap J.* **211,** 361 (§89).
______. 1977b. *Ap J.* **216,** 206 (§89).
Liang, E.P.T., and Baker, K. 1977. *Phys Rev Letters.* **39,** 191 (§89).
Lifshitz, E. M. 1946. *ZhETF.* **16,** 587; *J Phys.* **10,** 116 (§§4, 86, 90, 95).
Lifshitz, E. M., and Khalatnikov, I. M. 1963. *Usp Fiz Nauk.* **80,** 391; *Adv Phys.* **12,** 185 (§21).
Limber, D. N. 1953. *Ap J.* **117,** 134 (§29).
______. 1954. *Ap J.* **119,** 655 (§§4, 29, 47, 51, 57).
______. 1957. *Ap J.* **125,** 9 (§§29, 33).
Lin, C. C., Mestel, L., and Shu, F. H. 1965. *Ap J.* **142,** 1431 (§20).
Longair, M. S., and Sunyaev, R. A. 1969. *Nature.* **223,** 719 (§93).
Lynden-Bell, D. 1964. *Ap J.* **139,** 1195 (§20).
McClelland, J., and Silk, J. 1977. *Ap J.* **217,** 331 (§§40, 61).
McCrea, W. H. 1939. *Z Ap.* **18,** 98 (§2).
______. 1964. *MN.* **128,** 335 (§4).
______. 1968. *Science.* **160,** 1295 (§3).
McCrea, W. H., and McVittie, G. C. 1931. *MN.* **92,** 7 (§3).
McCrea, W. H., and Milne, E. A. 1934. *Q J Math (Oxford).* **5,** 73 (§3).
McVittie, G. C. 1932. *MN.* **92,** 500 (§3).
______. 1961. *Fact and Theory in Cosmology.* London: Eyre and Spottiswoode. (§2).
Mandelbrot, B. B. 1975. *CR.* **280A,** 1551 (§62).
______. 1977. *Fractals: Form, Chance and Dimension.* San Francisco: W. H. Freeman. (§§3, 62).
Mattig, W. 1958. *AN.* **284,** 109 (§56).
Meier, D. L. 1976. *Ap J.* **203,** L103 (§96).
Mészáros, P. 1974. *AA.* **37,** 225 (§§11, 12).
______. 1975. *AA.* **38,** 5 (§§94–96).
Michie, R. W. 1967. Kitt Peak National Observatory Contr. No. 440. (§§4, 92).
Milne, E. A. 1933a. *Z Ap.* **6,** 1 (§3).
______. 1933b. *Z Ap.* **7,** 180 (§3).
______. 1934. *Q J Math (Oxford).* **5,** 64 (§3).
______. 1935. *Relativity, Gravitation and World Structure.* Oxford: Clarendon Press. (§§3, 81).
Misner, C. W. 1967. *Nature.* **214,** 40 (§3).
______. 1968. *Ap J.* **151,** 431 (§§3, 95).
Miyoshi, K., and Kihara, T. 1975. *Publ Astron Soc Japan.* **27,** 333 (§79).

Montgomery, D. C., and Tidman, D. A. 1964. *Plasma Kinetic Theory.* New York: McGraw-Hill. (§§32, 68).
Mowbray, A. G. 1938. *PASP.* **50,** 275 (§29).
Nariai, H. 1969. *Progr Theor Phys.* **41,** 1379 (§13).
Nariai, H., and Fujimoto, M. 1972. *Progr Theor Phys.* **47,** 105 (§20).
Nariai, H., and Tomita, K. 1971. *Suppl Progr Theor Phys.* **49,** 83 (§4).
Ne'eman, Y. 1965. *Ap J.* **141,** 1303 (§4).
Neyman, J. 1962. In *Problems of Extragalactic Research,* ed. G. C. McVittie. p. 294. New York: Macmillan. (§§29, 30).
Neyman, J., and Scott, E. L. 1952. *Ap J.* **116,** 144 (§§40, 61).
———. 1955. *AJ.* **60,** 33 (§29).
Neyman, J., Scott, E. L., and Shane, C. D. 1954. *Ap J Suppl.* **1,** 269 (§29).
———. 1956. *Proc Third Berkeley Symposium Math Stat and Probability.* **3,** 75 (§§4, 29, 47, 57).
Noonan, T. W. 1977. *AA.* **54,** 57 (§78).
Novikov, I. D. 1964a. *ZhETF.* **46,** 686; *Soviet Phys JETP.* **19,** 467 (§§4, 86).
———. 1964b. *Astron Zh.* **41,** 1075; 1965. *Soviet Astron.* **8,** 857 (§4).
Oke, J. B., and Sandage, A. 1968. *Ap J.* **154,** 21 (§§56, 58).
Olson, D. W. 1976. *Phys Rev D.* **14,** 327 (§86).
Olson, D. W., and Sachs, R. K. 1973. *Ap J.* **185,** 91 (§22).
Omer, G. C. 1949. *Ap J.* **109,** 164 (§2).
———. 1965. *Proc NAS.* **53,** 1 (§2).
Omnès, R. 1969. *Phys Rev Letters.* **23,** 38 (§95).
Oort, J. H. 1958. In the Proceedings of the 11th Solvay Conference. p. 163. (§§4, 96).
———. 1970. *AA.* **7,** 381 (§§23, 96).
Ostriker, J. P. 1974. Paper presented at the seventh Texas conference. (§96).
Ostriker, J. P., Peebles, P.J.E., and Yahil, A. 1974. *Ap J.* **193,** L1 (§96).
Ostriker, J. P. and Turner, E. L. 1979. *Ap J.* **234,** 785 (§57).
Ozernoy, L. M. 1974. In IAU Symposium, No. 63. p. 227 (§90).
———. 1978. In IAU Symposium, No. 79. p. 427 (§96).
Ozernoy, L. M., and Chernin, A. D. 1967. *Astron Zh.* **44,** 1131; 1968. *Soviet Astron.* **11,** 907, (§§4, 95).
———. 1968. *Astron Zh.* **45,** 1137; *Soviet Astron,* **12,** 901 (§§4, 90, 95).
Parijski, Yu. N. 1978. In IAU Symposium, No. 79. (§93).

Parker, E. N. 1975. *Ap J.* **202,** 523 (§§4, 17).
Parry, W. E. 1977. *Physics Letters.* **60A,** 265 (§53).
Partridge, R. B. 1974. *Ap J.* **192,** 241 (§96).
______. 1980. *Physica Scripta.* **21,** 624 (§93).
Partridge, R. B., and Peebles, P.J.E. 1967. *Ap J.* **147,** 868; **148,** 377 (§§19, 96).
Peebles, P.J.E. 1965. *Ap J.* **142,** 1317 (§§4, 26, 94, 96).
______. 1967a. *Ap J.* **147,** 859 (§§3, 4, 95).
______. 1967b. Paper presented at the fourth Texas conference. (§4).
______. 1968. *Nature.* **220,** 237 (§§4, 95).
______. 1969a. *J Roy Astron Soc Canada.* **63,** 4 (§§19, 96).
______. 1969b. *Ap J.* **157,** 1075 (§4).
______. 1969c. *Ap J.* **155,** 393 (§§18, 23).
______. 1970a. *Phys Rev D.* **1,** 397 (§§89, 96).
______. 1970b. *AJ.* **75,** 13 (§73).
______. 1971a. *Ap Space Sci.* **11,** 443 (§§4, 21, 96).
______. 1971b. *AA.* **11,** 377 (§23).
______. 1971c. *Physical Cosmology*. Princeton, N.J.: Princeton University Press. (§§4, 58, 79, 90).
______. 1972. *Comments on Astrophys and Space Phys.* **4,** 53 (§§3, 87, 95).
______. 1973a. *Publ Astron Soc Japan.* **25,** 291 (§23).
______. 1973b. *Ap J.* **185,** 413 (§§41, 46, 48).
______. 1973c. *Ap J.* **180,** 1 (§91).
______. 1973d. In *Fundamental Interactions in Physics and Astrophysics.* ed. G. Iverson, A. Perlmutter, and S. Mintz. p. 318. New York: Plenum. (§79).
______. 1974a. *Ap J.* **189,** L51 (§§4, 26, 29).
______. 1974b. *AA.* **32,** 197 (§§4, 61).
______. 1974c. *AA.* **32,** 391 (§§27, 28).
______. 1974d. *Ap Space Sci.* **31,** 403 (§60).
______. 1974e. *Ap J Suppl.* **28,** 37 (§§44, 77).
______. 1975. *Ap J.* **196,** 647 (§§36, 54, 57).
______. 1976a. *Ap J.* **205,** 318 (§14).
______. 1976b. *Ap Space Sci.* **45,** 3 (§§71, 75, 76).
______. 1976c. *Ap J.* **205,** L109 (§§75, 76).
______. 1978. *AA.* **68,** 345 (§73).
______. 1979a. *MN.* **189,** 89 (§§61, 73).
______. 1979b. *AJ.* **84,** 730 (§§57, 76).
______. 1980a. To be published. (§§57, 76).
______. 1980b. *Physical Cosmology,* ed. J. Audouze. Amsterdam: North-Holland. (§96).

Peebles, P.J.E., and Dicke, R. H. 1968. *Ap. J.* **154,** 891 (§§4, 26, 94, 96).
Peebles, P.J.E., and Groth, E. J. 1975. *Ap J.* **196,** 1 (§§29, 54, 57, 61).
———. 1976. *AA.* **53,** 131 (§§28, 42, 59, 70, 79, 96).
Peebles, P.J.E., and Hauser, M. G. 1974. *Ap J Suppl.* **28,** 19 (§§2, 29, 46, 47, 50, 52, 57).
Peebles, P.J.E., and Wilkinson, D. T. 1968. *Phys Rev.* **174,** 2168 (§93).
Peebles, P.J.E., and Yu, J. T. 1970. *Ap J.* **162,** 815 (§§88, 92, 93, 95).
Pence, W. 1976. *Ap J.* **203,** 39 (§56).
Petrosian, V., Salpeter, E. E., and Szekeres, P. 1967. *Ap J.* **147,** 1222 (§13).
Phillipps, S., Fong, R., Ellis, R. S., Fall, S. M., and MacGillivray, H. T. 1978. *MN.* **182,** 673 (§56).
Press, W. H., and Lightman, A. P. 1978. *Ap J.* **219,** L73 (§20).
Press, W. H., and Schechter, P. 1974. *Ap J.* **187,** 425 (§§4, 26, 28, 95, 96).
Press, W. H. and Vishniac, E. T. 1980a. *Ap J.* **239,** (§§81, 86, 95).
———. 1980b. *Ap J.* **236,** 323 (§92).
Rawson-Harris, D. 1969. *MN.* **143,** 49 (§13).
Raychaudhuri, A. 1952. *Phys Rev.* **86,** 90 (§4).
———. 1955. *Phys Rev.* **98,** 1123 (§22).
Rees, M. J. 1972. *Phys Rev Letters.* **28,** 1669 (§95).
Rees, M. J., and Ostriker, J. P. 1977. *MN.* **179,** 541 (§96).
Rees, M. J., and Reinhardt, M. 1972. *AA.* **19,** 189 (§95).
Rees, M. J., and Sciama, D. W. 1968. *Nature.* **217,** 511 (§§2, 93).
Rindler, W. 1956. *MN.* **116,** 662 (§3).
Robertson, H. P. 1933. *RMP.* **5,** 62 (§2).
———. 1935. *Ap J.* **82,** 284 (§3).
Rubin, V. C. 1954. *Proc NAS.* **40,** 541 (§§4, 29, 47, 51, 57).
———. 1979. *Comments on Astrophysics.* **8,** 79 (§64, 79).
Rubin, V. C., Ford, W. K., and Thonnard, N. 1980. *Ap J.* **238,** (§96).
Rubin, V. C., Thonnard, N., Ford, W. K., and Roberts, M. S. 1976. *AJ.* **81,** 719 (§2).
Rudnicki, K., Dworak, T. Z., Flin, P., Baranowski, B., and Sendrakowski, A. 1973. *Acta Cosmologica.* **1,** 7 (56, 57).
Ryan, M. P. 1972. *Ap J.* **177,** L79 (§§25, 95).
Ryan, M. P., and Shepley, L. C. 1975. *Homogeneous Relativistic Cosmologies.* Princeton, N.J.: Princeton University Press. (§95).
Sachs, R. K., and Wolfe, A. M. 1967. *Ap J.* **147,** 73 (§93).

Sakai, K. 1969. *Progr Theor Phys.* **41,** 1461 (§81).
Sandage, A., Tammann, G. A., and Hardy, E. 1972. *Ap J.* **172,** 253 (§62).
Sanford, R. F. 1917. *Bull Lick Obs.* **9,** 80 (§2).
Sargent, W.L.W., and Searle, L. 1971. *Comments on Astrophys and Space Phys.* **3,** 111 (§96).
Sargent, W.L.W., and Turner, E. L. 1977. *Ap J.* **212,** L3 (§§20, 76).
Saslaw, W. C. 1967. *MN.* **136,** 39 (§4).
______. 1968. *MN.* **141,** 77 (§4).
______. 1972. *Ap J.* **177,** 17 (§68).
Schwartz, D. A. 1970. *Ap J.* **162,** 439 (§2).
Schwartz, D. A., Murray, S. S., and Gursky, H. 1976. *Ap J.* **204,** 315 (§2).
Sciama, D. W. 1955. *MN.* **115,** 3 (§4).
______. 1964. *QJRAS.* **5,** 196 (§4).
Scott, E. L. 1962. In *Problems of Extrgalactic Research.* ed. G. C. McVittie. p. 269. New York: Macmillan. (§29).
Seares, F. H. 1925. *Ap J.* **62,** 168 (§2).
Seldner, M. 1977. Doctoral Dissertation. Princeton University. (§47).
Seldner, M., and Peebles, P.J.E. 1977a. *Ap J.* **215,** 703 (§§2, 44, 47, 77).
______. 1977b. *Ap J.* **214,** L1 (§77).
______. 1978. *Ap J.* **225,** 7 (§§2, 44).
______. 1979. *Ap J.* **227,** 30 (§44).
Seldner, M., Siebers, B., Groth, E. J., and Peebles, P.J.E. 1977. *AJ.* **82,** 249 (§57).
Sen, N. R. 1934. *Z Ap.* **9,** 215 (§3).
Shane, C. D., and Wirtanen, C. A. 1950. *Proc Amer Phil Soc.* **94,** 13 (§2).
______. 1967. *Publ Lick Obs.* **22,** part 1 (§§2, 57).
Shanks, T., Fong, R., Ellis, R. S., and MacGillivray, H. T. 1980. *MN.* In press (§57).
Shapley, H. 1933. *Harvard Bull.* **890,** 1 (§2).
______. 1934. *MN.* **94,** 791 (§2).
______. 1935. *Proc NAS.* **21,** 587 (§29).
______. 1938a. *Proc NAS.* **24,** 282 (§2).
______. 1938b. *Proc NAS.* **24,** 527 (§2).
Shapley, H., and Ames, A. 1932. *Harvard Annals.* **88,** Part II (§2).
Shectman, S. A. 1973. *Ap J.* **179,** 681 (§58).
______. 1974. *Ap J.* **188,** 233 (§29, 58).
Shklovsky, I. 1967. *Ap J.* **150,** L1 (§13).
Silk, J. 1967. *Nature.* **215,** 1155 (§4).

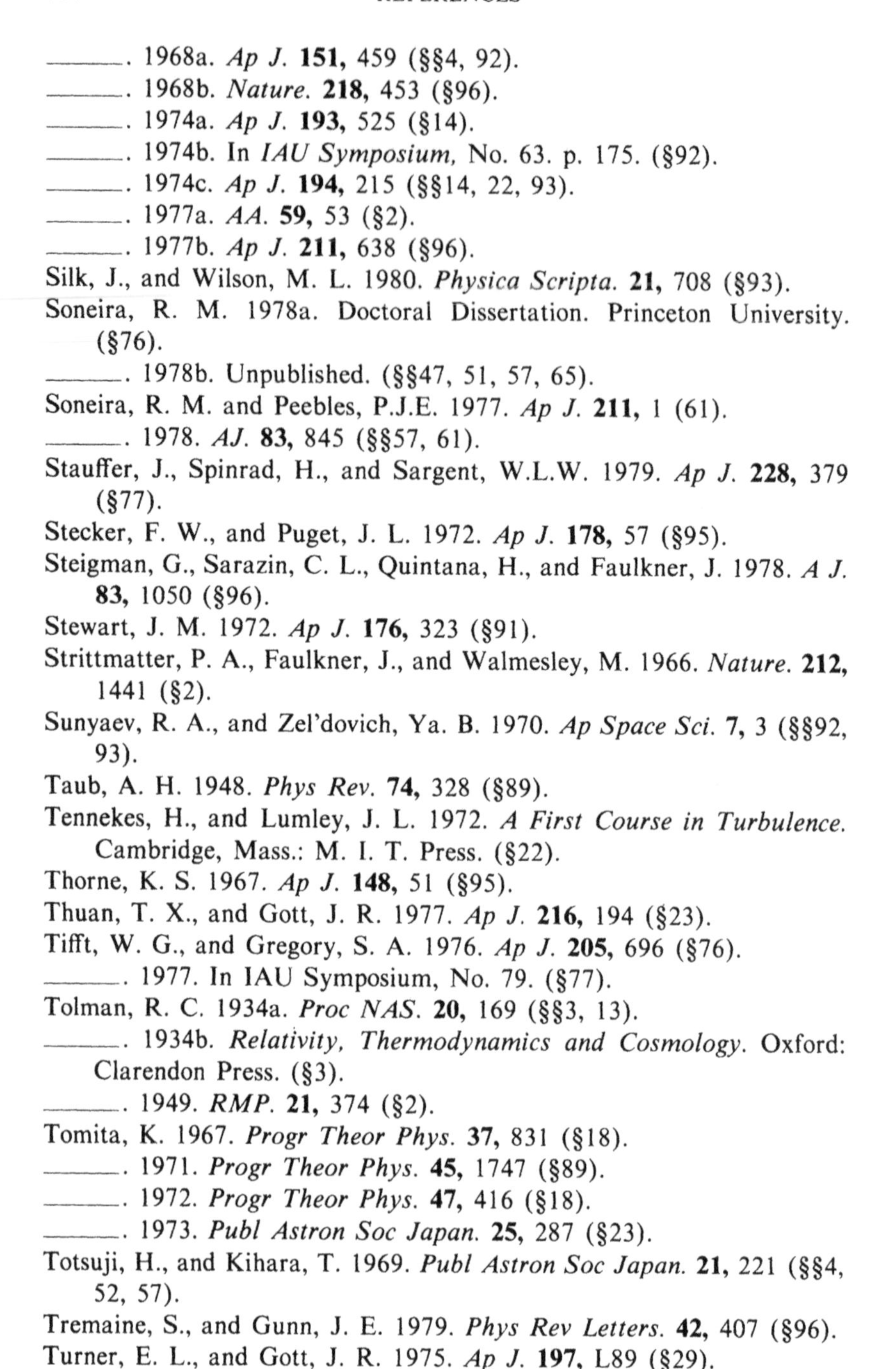

———. 1968a. *Ap J.* **151,** 459 (§§4, 92).
———. 1968b. *Nature.* **218,** 453 (§96).
———. 1974a. *Ap J.* **193,** 525 (§14).
———. 1974b. In *IAU Symposium,* No. 63. p. 175. (§92).
———. 1974c. *Ap J.* **194,** 215 (§§14, 22, 93).
———. 1977a. *AA.* **59,** 53 (§2).
———. 1977b. *Ap J.* **211,** 638 (§96).
Silk, J., and Wilson, M. L. 1980. *Physica Scripta.* **21,** 708 (§93).
Soneira, R. M. 1978a. Doctoral Dissertation. Princeton University. (§76).
———. 1978b. Unpublished. (§§47, 51, 57, 65).
Soneira, R. M. and Peebles, P.J.E. 1977. *Ap J.* **211,** 1 (61).
———. 1978. *AJ.* **83,** 845 (§§57, 61).
Stauffer, J., Spinrad, H., and Sargent, W.L.W. 1979. *Ap J.* **228,** 379 (§77).
Stecker, F. W., and Puget, J. L. 1972. *Ap J.* **178,** 57 (§95).
Steigman, G., Sarazin, C. L., Quintana, H., and Faulkner, J. 1978. *A J.* **83,** 1050 (§96).
Stewart, J. M. 1972. *Ap J.* **176,** 323 (§91).
Strittmatter, P. A., Faulkner, J., and Walmesley, M. 1966. *Nature.* **212,** 1441 (§2).
Sunyaev, R. A., and Zel'dovich, Ya. B. 1970. *Ap Space Sci.* **7,** 3 (§§92, 93).
Taub, A. H. 1948. *Phys Rev.* **74,** 328 (§89).
Tennekes, H., and Lumley, J. L. 1972. *A First Course in Turbulence.* Cambridge, Mass.: M. I. T. Press. (§22).
Thorne, K. S. 1967. *Ap J.* **148,** 51 (§95).
Thuan, T. X., and Gott, J. R. 1977. *Ap J.* **216,** 194 (§23).
Tifft, W. G., and Gregory, S. A. 1976. *Ap J.* **205,** 696 (§76).
———. 1977. In IAU Symposium, No. 79. (§77).
Tolman, R. C. 1934a. *Proc NAS.* **20,** 169 (§§3, 13).
———. 1934b. *Relativity, Thermodynamics and Cosmology.* Oxford: Clarendon Press. (§3).
———. 1949. *RMP.* **21,** 374 (§2).
Tomita, K. 1967. *Progr Theor Phys.* **37,** 831 (§18).
———. 1971. *Progr Theor Phys.* **45,** 1747 (§89).
———. 1972. *Progr Theor Phys.* **47,** 416 (§18).
———. 1973. *Publ Astron Soc Japan.* **25,** 287 (§23).
Totsuji, H., and Kihara, T. 1969. *Publ Astron Soc Japan.* **21,** 221 (§§4, 52, 57).
Tremaine, S., and Gunn, J. E. 1979. *Phys Rev Letters.* **42,** 407 (§96).
Turner, E. L., and Gott, J. R. 1975. *Ap J.* **197,** L89 (§29).

Tyson, J. A., and Jarvis, J. F. 1979. *Ap J.* **230,** L153 (§2).
van Albada, G. B. 1960. *BAN.* **15,** 165 (§4, 9, 14, 68).
______. 1962. In *Problems of Extragalactic Research,* ed. G. C. McVittie. p. 411. New York: Macmillan. (§4).
van den Bergh, S. 1961. *PASP.* **73,** 46 (§2).
Vishniac, E. T., and Press, W. H. 1978. Unpublished. (§27).
von Weizsäcker, C. F. 1949. In IAU and International Union of Theoretical and Applied Mechanics Symposium. pp. 158 and 200 (§4).
______. 1951. *Ap J.* **114,** 165 (§§4, 96).
Wagoner, R. V., Fowler, W. A., and Hoyle, F. 1967. *Ap J.* **148,** 3 (§96).
Walker, A. G. 1936. *Proc London Math Soc.* **42,** 90 (§3).
Wasserman, I. 1978. *Ap J.* **224,** 337 (§§17, 95).
Webster, A. 1976a. *MN.* **175,** 61 (§46).
______. 1976b. *MN.* **175,** 71 (§§2, 29).
Weinberg, S. 1971. *Ap J.* **168,** 175 (§92).
______. 1972. *Gravitation and Cosmology.* New York: Wiley. (§11, 85).
Wertz, J. R. 1971. *Ap J.* **164,** 227 (§2).
Wesson, P. S. 1976. *Ap Space Sci.* **42,** 477 (§2).
Weyl, H. 1922. *Space Time Matter,* tr. H. L. Brose. 4th ed. Reprint. New York: Dover Publications. (§3).
Weyman, R. 1966. Privately circulated paper. University of Arizona. (§96).
White, S.D.M. 1976. *MN.* **177,** 717 (§73).
______. 1979. *MN.* **186,** 145 (§§29, 38, 39).
White, S.D.M., and Rees, M. J. 1978. *MN.* **183,** 341 (§96).
Wickes, W. C., and Peebles, P.J.E. 1976. *Bull Amer Astron Soc.* **8,** 543 (§27).
Wolfe, A. M. 1970. *Ap J.* **159,** L61 (§2).
Wolfe, A. M., and Burbidge, G. R. 1970. *Nature.* **228,** 1170 (§2).
Yu, J. T. 1968. Doctoral Dissertation. Princeton University. (§86).
Yu, J. T., and Peebles, P.J.E. 1969. *Ap J.* **158,** 103 (§§29, 41, 46).
Zel'dovich, Ya. B. 1961. *ZhETF.* **41,** 1609; 1962. *Soviet Phys JETP.* **14,** 1143. (§86).
______. 1964. *Astron Zh.* **41,** 873; 1965. *Soviet Astron.* **8,** 700. (§20).
______. 1965a. *Adv Astron Astrophys.* **3,** 241 (§§4, 10, 24, 28).
______. 1965b. *ZhETF.* **48,** 986; *Soviet Phys JETP.* **21,** 656 (§95).
______. 1967. *Usp Fiz Nauk.* **89,** 647; *Soviet Phys Usp.* **9,** 602 (§4).
______. 1969. *Astron Zh.* **46,** 775; 1970. *Soviet Astron.* **13,** 608. (§95).

———. 1970. *Astrofizika.* **6,** 319; *AA.* **5,** 84 (§§21, 96).
———. 1972. *MN.* **160,** 1p (§95).
———. 1974. In IAU Symposium No. 63. p. 250 (§28).
———. 1978. In IAU Symposium No. 79. p. 409 (§§4, 21, 96).
Zel'dovich, Ya. B., and Novikov, I. D. 1970. *Astrofizika.* **6,** 379 (§90).
Zwicky, F. 1953. *Helv Phys Acta.* **26,** 241 (§29).
Zwicky, F., Herzog, E., Wild, P., Karpowicz, M., and Kowal, C. T. 1961–1968. *Catalogue of Galaxies and Clusters of Galaxies.* 6 vol. Pasadena, Calif.: California Institute of Technology. (§§4, 57).

INDEX

Milton Keynes UK
Ingram Content Group UK Ltd.
UKHW032141170324
439604UK00009B/1326